Nanocomposites of Polymers and Inorganic Nanoparticles

Nanocomposites of Polymers and Inorganic Nanoparticles

Editor

Walter Remo Caseri

MDPI • Basel • Beijing • Wuhan • Barcelona • Belgrade • Manchester • Tokyo • Cluj • Tianjin

Editor
Walter Remo Caseri
Vladimir-Prelog-Weg 5
Switzerland

Editorial Office
MDPI
St. Alban-Anlage 66
4052 Basel, Switzerland

This is a reprint of articles from the Special Issue published online in the open access journal *Nanomaterials* (ISSN 2079-4991) (available at: https://www.mdpi.com/journal/nanomaterials/special_issues/polymers_inorganic_nano).

For citation purposes, cite each article independently as indicated on the article page online and as indicated below:

LastName, A.A.; LastName, B.B.; LastName, C.C. Article Title. *Journal Name* **Year**, *Volume Number*, Page Range.

ISBN 978-3-0365-0352-3 (Hbk)
ISBN 978-3-0365-0353-0 (PDF)

© 2021 by the authors. Articles in this book are Open Access and distributed under the Creative Commons Attribution (CC BY) license, which allows users to download, copy and build upon published articles, as long as the author and publisher are properly credited, which ensures maximum dissemination and a wider impact of our publications.

The book as a whole is distributed by MDPI under the terms and conditions of the Creative Commons license CC BY-NC-ND.

Contents

About the Editor vii

Preface to "Nanocomposites of Polymers and Inorganic Nanoparticles" ix

Weronika Górka, Tomasz Kuciel, Paula Nalepa, Dorota Lachowicz, Szczepan Zapotoczny and Michał Szuwarzyński
Homogeneous Embedding of Magnetic Nanoparticles into Polymer Brushes during Simultaneous Surface-Initiated Polymerization
Reprinted from: *Nanomaterials* **2019**, *9*, 456, doi:10.3390/nano9030456 1

Pallavi Pandit, Matthias Schwartzkopf, André Rothkirch, Stephan V. Roth, Sigrid Bernstorff and Ajay Gupta
Structure–Function Correlations in Sputter Deposited Gold/Fluorocarbon Multilayers for Tuning Optical Response
Reprinted from: *Nanomaterials* **2019**, *9*, 1249, doi:10.3390/nano9091249 13

Yong You, Ling Tu, Yajie Wang, Lifen Tong, Renbo Wei and Xiaobo Liu
Achieving Secondary Dispersion of Modified Nanoparticles by Hot-Stretching to Enhance Dielectric and Mechanical Properties of Polyarylene Ether Nitrile Composites
Reprinted from: *Nanomaterials* **2019**, *9*, 1006, doi:10.3390/nano9071006 31

Andrew Harrison, Tien T. Vuong, Michael P. Zeevi, Benjamin J. Hittel, Sungsool Wi and Christina Tang
Rapid Self-Assembly of Metal/Polymer Nanocomposite Particles as Nanoreactors and Their Kinetic Characterization
Reprinted from: *Nanomaterials* **2019**, *9*, 318, doi:10.3390/nano9030318 47

Jin-Hae Chang
Comparative Analysis of Properties of PVA Composites with Various Nanofillers: Pristine Clay, Organoclay, and Functionalized Graphene
Reprinted from: *Nanomaterials* **2019**, *9*, 323, doi:10.3390/nano9030323 61

Jaroslav Hornak, Pavel Trnka, Petr Kadlec, Ondřej Michal, Václav Mentlík, Pavol Šutta, Gergely Márk Csányi and Zoltán Ádám Tamus
Magnesium Oxide Nanoparticles: Dielectric Properties, Surface Functionalization and Improvement of Epoxy-Based Composites Insulating Properties
Reprinted from: *Nanomaterials* **2018**, *6*, 381, doi:10.3390/nano8060381 77

Makoto Takafuji, Maino Kajiwara, Nanami Hano, Yutaka Kuwahara and Hirotaka Ihara
Preparation of High Refractive Index Composite Films Based on Titanium Oxide Nanoparticles Hybridized Hydrophilic Polymers
Reprinted from: *Nanomaterials* **2019**, *9*, 514, doi:10.3390/nano9040514 95

Zuopeng Qu, Lei Wang, Hongyu Tang, Huaiyu Ye and Meicheng Li
Effect of Nano-SnS and Nano-MoS$_2$ on the Corrosion Protection Performance of the Polyvinylbutyral and Zinc-Rich Polyvinylbutyral Coatings
Reprinted from: *Nanomaterials* **2019**, *9*, 956, doi:10.3390/nano9070956 105

Kanthasamy Raagulan, Ramanaskanda Braveenth, Lee Ro Lee, Joonsik Lee, Bo Mi Kim, Jai Jung Moon, Sang Bok Lee and Kyu Yun Chai
Fabrication of Flexible, Lightweight, Magnetic Mushroom Gills and Coral-Like MXene–Carbon Nanotube Nanocomposites for EMI Shielding Application
Reprinted from: *Nanomaterials* **2019**, *9*, 519, doi:10.3390/nano9040519 . . . . **119**

Yuji Ohkubo, Tomonori Aoki, Satoshi Seino, Osamu Mori, Issaku Ito, Katsuyoshi Endo and Kazuya Yamamura
Improved Catalytic Durability of Pt-Particle/ABS for H_2O_2 Decomposition in Contact Lens Cleaning
Reprinted from: *Nanomaterials* **2019**, *9*, 342, doi:10.3390/nano9030342 . . . . **141**

Stefano Caimi, Antoine Klaue, Hua Wu and Massimo Morbidelli
Effect of SiO_2 Nanoparticles on the Performance of PVdF-HFP/Ionic Liquid Separator for Lithium-Ion Batteries
Reprinted from: *Nanomaterials* **2018**, *8*, 926, doi:10.3390/nano8110926 . . . . **153**

Sylvie Ribeiro, Tânia Ribeiro, Clarisse Ribeiro, Daniela M. Correia, José P. Sequeira Farinha, Andreia Castro Gomes, Carlos Baleizão and Senentxu Lanceros-Méndez
Multifunctional Platform Based on Electroactive Polymers and Silica Nanoparticles for Tissue Engineering Applications
Reprinted from: *Nanomaterials* **2018**, *8*, 933, doi:10.3390/nano8110933 . . . . **165**

Lili Lin, Jingqi Ma, Quanjing Mei, Bin Cai, Jie Chen, Yi Zuo, Qin Zou, Jidong Li and Yubao Li
Elastomeric Polyurethane Foams Incorporated with Nanosized Hydroxyapatite Fillers for Plastic Reconstruction
Reprinted from: *Nanomaterials* **2018**, *8*, 972, doi:10.3390/nano8120972 . . . . **185**

About the Editor

Walter Remo Caseri studied chemistry at ETH Zürich, Switzerland, in 1983 and was subsequently educated as a Swiss radiation protection expert. Thereafter, he finished his Ph.D. in 1987 in the area of organometallic chemistry and catalysis. After a year at the Max Planck Institute for Polymer Research in Mainz, Germany, working in the area of the synthesis of inorganic polymers and their processing to ultrathin layers, he habilitated at the Institute of Polymers at ETH Zurich in the field of the synthesis and characterization of nanocomposites and chemistry at surfaces. Subsequently, he acted during 1993–1996 as leader of the Surface Protection Group at EMPA Dübendorf and thereafter joined ETH Zürich again, where he was awarded with the title of professor in 2006. He is active in the synthesis and investigation of materials of nanocomposites and polymers, surface chemistry and adhesion, and more recently also in the investigation of Roman concrete.

Preface to "Nanocomposites of Polymers and Inorganic Nanoparticles"

This special issue is devoted to nanocomposites consisting of inorganic nanoparticles which are embedded in a polymer matrix. The collection of articles in this volume reflects the versatility of this class of nanocomposites. In fact, the diversity of both organic polymers and inorganic compounds suited for nanoparticles gives rise to the creation of a virtually indefinite number of nanocomposites with a broad spectrum of characteristics and applications. With regard to distinguished materials properties, often the inorganic component is in the focus, while the polymer acts as the matrix in which the nanoparticles are dispersed. The different nature of inorganic compounds applied for nanocomposites is clearly evident from the articles in this Special Issue, which include metals, metal oxides, metal sulfides, clays, hydroxyapatite, carbon nanotubes and graphene. The polymer matrix, on the other hand, can provide not only mechanical stability and processability but also electrical conductivity and biocompatibility.

Nanocomposites have a history dating back more than 100 years, but they began to attract broader attention only in the early 1990s, which led to an ongoing boom. While initially the preparation of nanocomposites was in the foreground, the focus in the scientific literature has shifted more and more to material properties and potential applications, although innovative synthetic approaches are still being developed. All these aspects emerge in the articles of this Special Issue.

A key feature in nanocomposite preparation is the prevention of nanoparticle agglomeration or control of the arrangement of nanoparticles in the composite structure. This is of particular importance in the area of optics (cf. article by Pandit et al.) and superparamagnetism (cf. article by Górka et al.), as those properties are influenced by interparticle distances or clustering. While particle agglomeration in superparamagnetic nanocomposites is usually not desired, as this diminishes the superparamagnetic effect, on the other hand, particle agglomeration or controlled particle assembly can sometimes be used for specific changes in the optical behavior of a material. The prevention of particle agglomeration is especially difficult in nanocomposites with high particle contents since the probability that particles come into contact with each other increases with an increasing particle fraction. In such cases, agglomeration can be prevented by coating the surfaces of the particles, which has also been applied with a view to nanocomposites with enhanced dielectric and mechanical properties (cf. article by You et al. and Hornak et al.).

As already indicated above, the variety of properties of nanocomposites offers possibilities for applications in a variety of areas. With regard to the usage of nanocomposites for chemical reactions, catalysis attracts attention since a number of nanoparticles can act as catalytically active centers, as demonstrated in this Special Issue by a model system for the reduction of 4-nitrophenol (cf. article by Harrison et al.) and for the decomposition of hydrogen peroxide in the context of cleaning contact lenses (cf. article by Ohkubo et al.). Moreover, particular physical properties promote the deployment of nanocomposites in technical fields, such as gas barriers (cf. article by Chang), optics (e.g., materials with high refractive index, cf. article by Takafuji et al.), corrosion protection (cf. article by Qu et al.), electromagnetic inference (EMI) shielding (cf. article by Raagular et al.) and lithium ion batteries (cf. article by Caimi et al.), as well as in the area of biomedicine, such as tissue engineering (cf. article by Ribeiro et al.) and plastic surgery (cf. article by Lin et al.). Accordingly, it is evident that the area of composites of polymers and inorganic nanoparticles is of vital interest in various disciplines, particularly chemistry, physics, biomedicine and materials science, and that such nanocomposites are a focal point for innovative science and a source of inspiration for currently relevant economic topics as well as for envisaged technologies of the future.

Walter Remo Caseri
Editor

Article

Homogeneous Embedding of Magnetic Nanoparticles into Polymer Brushes during Simultaneous Surface-Initiated Polymerization

Weronika Górka [1,2], Tomasz Kuciel [2], Paula Nalepa [2], Dorota Lachowicz [3], Szczepan Zapotoczny [2,*] and Michał Szuwarzyński [3,*]

1. Faculty of Physics, Jagiellonian University, Astronomy and Applied Computer Science, S. Łojasiewicza 11, 30-348 Krakow, Poland; weronika.gorka@doctoral.uj.edu.pl
2. Faculty of Chemistry, Jagiellonian University, Gronostajowa 2, 30-387 Krakow, Poland; kuciel@chemia.uj.edu.pl (T.K.); paula.nalepa@student.uj.edu.pl (P.N.)
3. Academic Centre for Materials and Nanotechnology, AGH University of Science and Technology, A. Mickiewicza 30, 30-059 Krakow, Poland; dbielska@agh.edu.pl
* Correspondence: zapotocz@chemia.uj.edu.pl (S.Z.); szuwarzy@agh.edu.pl (M.S.); Tel.: +48-12-617-35-28 (S.Z. & M.S.)

Received: 26 February 2019; Accepted: 15 March 2019; Published: 19 March 2019

Abstract: Here we present a facile and efficient method of controlled embedding of inorganic nanoparticles into an ultra-thin (<15 nm) and flat (~1.0 nm) polymeric coating that prevents unwanted aggregation. Hybrid polymer brushes-based films were obtained by simultaneous incorporation of superparamagnetic iron oxide nanoparticles (SPIONs) with diameters of 8–10 nm into a polycationic macromolecular matrix during the surface initiated atom transfer radical polymerization (SI-ATRP) reaction in an ultrasonic reactor. The proposed structures characterized with homogeneous distribution of separated nanoparticles that maintain nanometric thickness and strong magnetic properties are a good alternative for commonly used layers of crosslinked nanoparticles aggregates or bulk structures. Obtained coatings were characterized using atomic force microscopy (AFM) working in the magnetic mode, secondary ion mass spectrometry (SIMS), and X-ray photoelectron spectroscopy (XPS).

Keywords: polymer brushes; nanoparticles; SPION; thin magnetic films; ATRP; hybrid polymer/ inorganic composites

1. Introduction

During the last decade great attention has been devoted to hybrid inorganic–polymer systems grafted or adsorbed to/from the surface [1,2] and their potential applications in cell culturing [3], fabrication of antibacterial coatings [4], sensors [5], photovoltaic devices [6], field-effect transistors [7], light-emitting diodes [8] and magnetic inks [9]. The incorporation of nanoparticles (NPs) offers advantages of combining properties of both entities: unique magnetic, electronic, thermal or optical properties of nanosized inorganic objects with physical and chemical properties of polymeric matrices [10–12]. A number of techniques could be used for controlled fabrication of well-defined hybrid nanocomposites such as: coating of inorganic particles by polymeric chains using "grafting to" and "grafting from" methods [13,14], layer-by-layer assemblies [4], in situ synthesis of NPs in between the surface-grafted polymer chains [15,16], the self-organization process of copolymer thin films blended with nanoparticles [17], and the polymerization of monomers in the presence of inorganic particles [18]. Polymer brushes [19], as coatings composed of chains tethered by one end to a surface, may be particularly useful matrices that work as a macromolecular capping agent for preventing aggregation due to the inherently extended chains' conformation in the brushes. Because of that, polymer brushes, by linking a specific molecules or functional groups in demanded order, can be used for an advanced applications: conjugated side chains for conducive electricity [20,21], crown ethers for selective ion sensing [22], growth factors for cell culturing [23], or chromophore molecules for light absorption [24,25]. For better control on the size and composition of NPs it is often desired to form them ex situ and the polymer matrix volume should be limited for the formation of functional nanocomposites. There have been numerous attempts at incorporation of NPs into polymeric matrices but obtaining a thin layer with a thickness comparable to the diameter of the nanoparticles while maintaining their homogenous distribution in the layer is still a challenge [26–28] and typically much thicker polymer matrices have been used for that purpose [29,30].

Here we present an approach based on surface-initiated polymerization of a cationic monomer in the presence of superparamagnetic iron oxide nanoparticles (SPIONs) that leads to the formation of a magnetic nanocomposites layer with NPs homogenously distributed within the thin polyelectrolyte brushes. Such smooth magnetic ultrathin layers have high application potential but can hardly be obtained by embedding nanoparticles in already prepared brushes.

2. Materials and Methods

2.1. Materials

Superparamagnetic iron oxide nanoparticles (SPIONs): iron(III) chloride hexahydrate (p.a.), iron(II) chloride tetrahydrate (p.a.) were purchased from Sigma Aldrich (St. Louis, MO, USA). Ammonium hydroxide (25%, p.a.) was purchased from Chempur (Piekary Slaskie, Poland).

Polymer brushes: (3-Aminopropyl)triethoxysilane (APTES, 99%), α-bromoisoubtyryl bromide (BIB, 98%), triethylamine (Et$_3$N, ≥99.5%), N,N,N′,N″,N″-Pentamethyldiethylenetriamine (PMDETA, 99%), (3-acrylamidoproplyl)trimethylamonnium chloride (APTAC, 75% solution in water), copper (I) bromide (99.999%) were all purchased from Sigma Aldrich (St. Louis, MO, USA). Toluene (p.a.), methanol (p.a.), tetrahydrofuran (p.a,), isopropanol (p.a.), dichloromethane (high-performance liquid chromatography, HPLC grade, 99.8%) were purchased from Chempur (Piekary Slaskie, Poland). Polished Prime Silicon Wafers were obtained from Cemat Silicon SA (Warszawa, Poland) and ITO glass with the ITO layer thickness of 150 nm was purchased from VisionTek System LTD (Cheshire, United Kingdom). Ammonia solution (25% p.a.) and sulfuric acid (≥95% p.a.) were obtained from POCH S. A. (Gliwice, Poland). Hydrogen peroxide (30% p.a.) was purchased from Stanlab (Lublin, Poland).

2.2. Methods

A transmission electron microscope (TEM) Tecnai TF 20 X-TWIN (FEI, Hillsboro, OR, USA) was used for determination of the shape and the size of the nanoparticles. The aqueous dispersions of SPIONs were sonicated for 5 min before deposition on ultrathin carbon coated copper grid and air-dried at room temperature. The TEM images were analyzed by fitting round circles around the edges of the nanoparticles and measuring their diameters that lead to determination of the mean particle size.

Sizes and zeta potentials of the nanoparticles were determined using dynamic light scattering (DLS, Malvern Nano ZS light-scattering apparatus, Malvern Instrument Ltd., Worcestershire, UK; measurements at 173° scattering angle) at 25 °C. The time-dependent autocorrelation function of the photocurrent was acquired every 10 s, with 15 acquisitions for each run. The sample was illuminated by a 633 nm laser. The z-averaged mean diameters (d_z) and distribution profiles of the samples were collected using the software provided by Malvern. The zeta potential was measured using the technique of laser Doppler velocimetry (LDV).

Atomic force microscope (AFM) images were obtained with a Dimension Icon atomic force microscope (Bruker, Santa Barbara, CA, USA) working in the air or water in the PeakForce Tapping (PFT) mode using standard silicon cantilevers of nominal spring constant of 0.4 N/m for air measurements and 0.7 N/m for liquids. Magnetic force microscope (MFM) images were acquired using the same microscope and magnetic Co/Cr coated standard silicon cantilevers of nominal spring constant of 2 N/m. All MFM images were captured in the lift mode at 50 nm lift height. A potential (5 V) between the tip and the sample was applied for the measurements of the brushes in order to compensate for their positive charge that could otherwise contribute to the magnetic phase signal. The cantilevers were magnetized with a small magnet before the measurements. Quantitative nanomechanical mapping (QNM) measurements were done using the previously calibrated silicon AFM tip. The results obtained were averaged for at least 10 locations on each sample. The thickness measurements were performed at the edges of the scratched layers.

Secondary ion mass spectrometry (SIMS) experiments were performed on an ION TOF TOFSIMS V (Munster, Germany) instrument, equipped with bismuth manganium liquid metal ion source and C_{60} ion source. The depth profiles of samples were obtained in interlaced dual beam mode; 20 keV C_{60}^+ ion beam was used to sputter a 500 × 500 µm^2 area and Bi_3^+ 30 keV ion beam was used to analyse a 300 × 300 µm^2 area concentric to the sputtered surface. For the surface characterization Bi_3^+ 30 keV ion beam was used with the ion dose density lower than 10^{12} ions/cm^2 to ensure static SIMS conditions. TOF-SIMS spectra were acquired from 5 non-overlapping 500 × 500 µm^2 area.

X-ray photoelectron spectroscopy (XPS) analysis were carried out in a PHI VersaProbeII Scanning XPS system (ULVAC-PHI, Chigasaki, Japan) using monochromatic Al Kα (1486.6 eV) X-rays focused on a 100 µm spot and scanned over the sample area of 400 × 400 µm. The photoelectron take-off angle was 30° and the pass energy in the analyser was set to 23.50 eV to obtain high energy resolution spectra for the C 1s, N 1s, O 1s, Cl 2p, Si 2p and Fe 2p regions. All XPS spectra were charged to the unfunctionalized, saturated carbon (C–C) C 1s peak at 284.8 eV. The pressure in the analytical chamber was less than 4×10^{-8} mbar. The deconvolution of spectra was performed using PHI MultiPak software (v.9.7.0.1). The spectrum background was subtracted using the Shirley method.

2.3. Synthesis of Superparamagnetic Iron Oxide Nanoparticles (SPIONs)

Briefly, the precursor solutions of the salts: 0.1622 g $FeCl_3·6H_2O$ and 0.0596 g $FeCl_2·4H_2O$ (the molar ratios of ions Fe(III): Fe(II) = 2:1, pH = 2.37) were dissolved in 50 mL of deionized water (Scheme 1A1). The solution was deoxygenated by purging with argon and sonicated (Sonic-6, Polsonic, Warszawa, Poland, 480 W, 1 s pulse per every 5 s) for 10 min in a thermostatic bath at 20°C. Afterward, 5 mL of 5 M $NH_{3(aq)}$ were added dropwise, and the suspension of the formed polydisperse nanoparticles was further sonicated for 30 min (Scheme 1A2). Purification of the formed SPIONs was performed using magnetic chromatography (Scheme 1A3) until a stable dispersion of the nanoparticles without reaction remains was obtained (Scheme 1A4).

2.4. Synthesis of Poly(APTAC) Brushes

Silicon wafers were firstly purified by immersing in a "piranha" solution (a mixture of H_2SO_4 and H_2O_2 at a 3:1 ratio) (Caution! "Piranha solution" should be handled with extreme care) for 1 h, subsequently rinsed by water and dried out in a stream of argon. Such prepared substrates were immediately immersed into the solution of amide-silane initiator (APTES, one drop in 5 mL of toluene) and left for 24 h in a sealed flask under argon atmosphere at room temperature. Afterwards the substrates were rinsed with copious amounts of toluene and dichloromethane. Then 2-isobromobutyryl bromide (BIB; 0.06 mL), triethylamine (Et$_3$N; 0.07 mL) and dichloromethane (CH$_2$Cl$_2$; 10 mL) were added over the substrates under an argon atmosphere, at room temperature, and left for 1 h in a sealed flask (Scheme 1B1.). Poly(APTAC) brushes were obtained using the ATRP method. (3-acrylamidoproplyl)trimethylamonnium chloride monomer (APTAC; 3 mL, 75% solution in water) was added into the mixture of deionized water (0.375 mL) and isopropanol (1.125 mL) with dissolved CuBr (20 mg) under an argon atmosphere at room temperature. Then N,N,N',N'',N''-pentamethyldiethylenetriamine (PMDETA; 0.1 mL) was injected into the deoxygenated solution, mixed and left for 72 h (Scheme 1B2.). The samples were then carefully cleaned in a mixture of water:propan-2-ol (1:1, v/v), toluene and methanol by bubbling for circa 5 min in each solution and finally dried in a stream of argon.

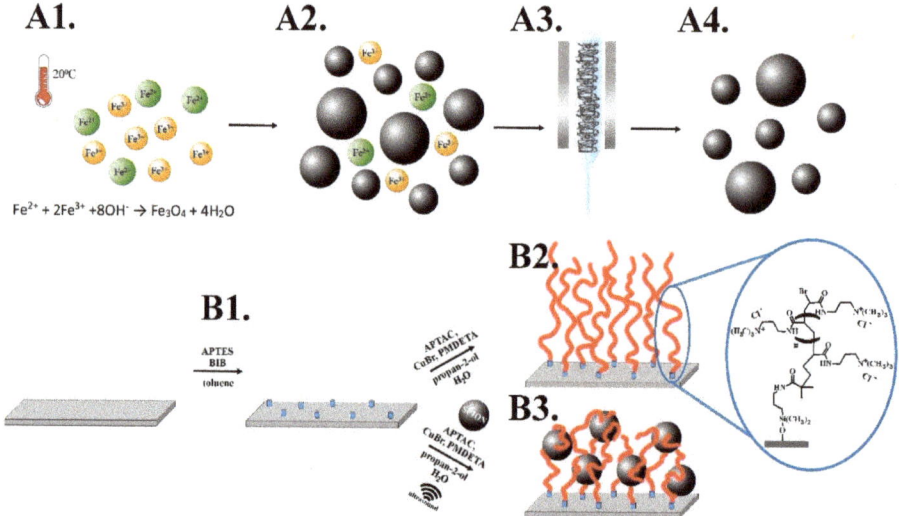

Scheme 1. A. Synthesis of magnetic nanoparticles: 1. Fe^{2+}, Fe^{3+} salts mixture at room temperature (RT) in H_2O, 2. Mixture of nanoparticles and post-reaction remains, 3. Magnetic chromatography, 4. Purified nanoparticles. B. Synthesis of poly(APTAC) and poly(APTAC)+SPIONs brushes: 1. Amide-silane initiator adsorption on a silicon substrate, 2. Formation of poly(APTAC) brushes via surface initiated atom transfer radical polymerization (SI-ATRP) of APTAC, 3. Formation of poly(APTAC)+SPIONs via SI-ATRP of APTAC in the presence of the magnetic nanoparticles.

2.5. Synthesis of Poly(APTAC) Brushes with Embedded SPIONs (Poly(APTAC)+SPIONs)

Brushes with incorporated magnetic nanoparticles were obtained similarly to the method described above but 0.2 mL of a stable dispersion of SPIONs in water was injected into a sealed reaction flask at the beginning of the ATPR reaction (Scheme 1B3.) and the first 1 h of the polymerization was conducted in the ultrasonic reactor (pulsed sonication as described in Section 2.3). Before injection, a solution of SPIONs was sonicated for 10 min (continuous sonication).

3. Results and Discussion

In the presented approach we first synthesized superparamagnetic iron oxide nanoparticles (SPIONs) with well-defined diameters of 8–10 nm (Figure S1, Supplementary Materials) and strong magnetic properties [31]. As shown in Scheme 1 the synthesis of SPIONs was carried out by coprecipitation of respective iron salts in an aqueous medium according to the method described previously [32,33]. The surface of the obtained SPIONs was shown to be strongly negatively charged (zeta potential, $\xi = -47.6 \pm 0.4$ mV) so their suspension can be considered as stable (Figure S2, Supplementary Materials) that is crucial for the successful running of the polymerization. Moreover, magnetic properties of the formed NPs were studied previously and confirmed here using magnetic force microscopy (MFM, Figure S3, Supplementary Materials). The concentration of iron in the suspension of SPIONs was calculated to be 0.87 mg/mL (Figure S4, Supplementary Materials).

Polymer brushes were grafted from a silicone surface via surface initiated atom transfer radical polymerization (SI-ATRP) using (3-acrylamidoproplyl)trimethylamonium chloride (APTAC) as a monomer (see Scheme 1 and Materials and Methods for details). The brushes with incorporated SPIONs were obtained in a similar polymerization procedure but a stable dispersion of SPIONs was injected into a sealed reaction flask at the beginning of ATRP and for the first 1 h the process was conducted in an ultrasonic reactor (pulsed sonication). The obtained polycationic brushes (poly(APTAC)) and the same brushes with incorporated magnetic nanoparticles (poly(APTAC)+SPIONs), were characterized using AFM. Thicknesses of the dry coatings were determined to be 27.9 ± 0.8 nm and 12.0 ± 1.1 nm, respectively (Figure 1).

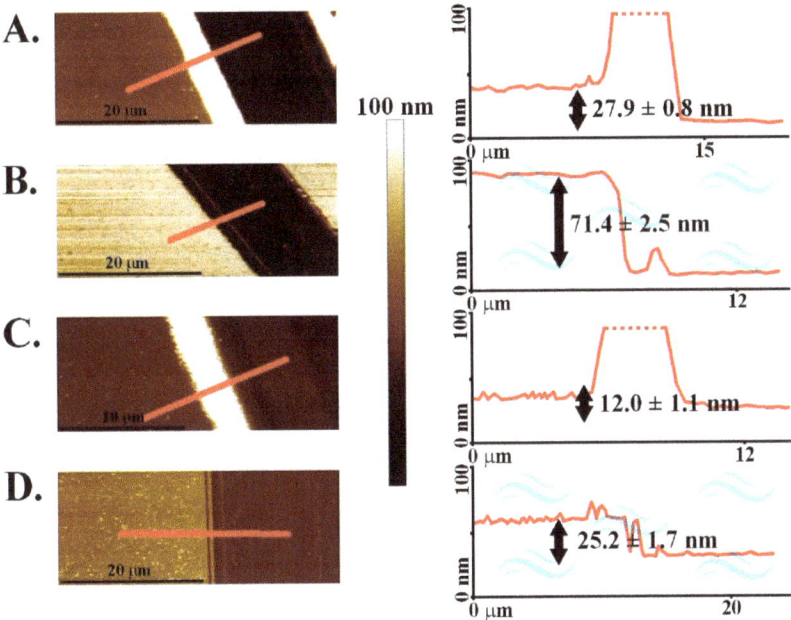

Figure 1. Atomic force microscopy (AFM) topography images of the brushes with the corresponding cross-sections: (**A**) (poly(APTAC) in the air, (**B**) poly(APTAC) in the water, (**C**) poly(APTAC)+SPIONs in the air, (**D**) poly(APTAC)+SPIONs in the water.

The topography images (Figure 2) showed distinct differences between both layers. However, the calculated RMS roughness for poly(APTAC) brushes (0.8 ± 0.1 nm) was only slightly smaller than the value for the layer with incorporated SPIONs (1.2 ± 0.2 nm). There are a few objects sticking out from the smooth underlying layer in the poly(APTAC)+SPIONs sample. The AFM imaging indicates that while some nanoparticles sit on the top of the layer (features up to 10 nm above the layer), some other are partially or completely immersed in the brushes matrix. Importantly, practically no aggregates of SPIONs could be observed on the surface of poly(APTAC)+SPIONs thanks to the applied procedure. Moreover, the incorporation of SPIONs in the already prepared poly(APTAC) brushes, by depositing them under sonication, failed as indicated by the AFM measurements (Figure S5, Supplementary Materials). Stability over time of the obtained structures has been confirmed (see Figure S6, Supplementary Materials). After one month of storage at room temperature, no peeling process of brushes layer and no removal of SPIONs from the polymeric matrix was observed.

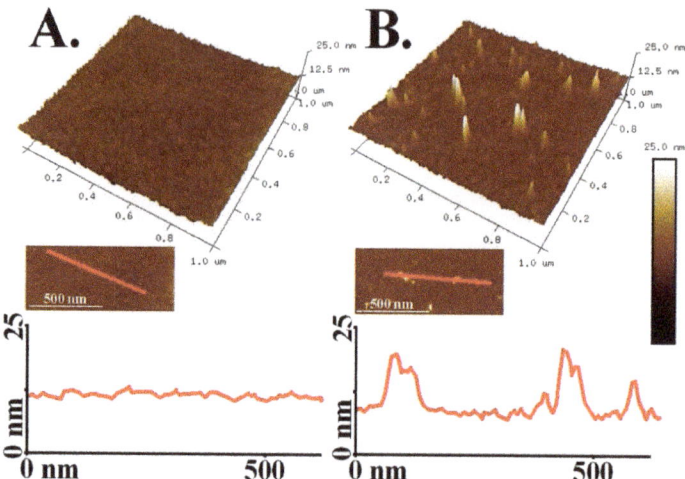

Figure 2. AFM images of poly(APTAC) brushes (**A**) and poly(APTAC)+SPIONs brushes (**B**) with the corresponding cross-sections.

The results of the quantitative nanomechanical mapping (QNM) measurements may indicate efficient incorporation of the nanoparticles into the brush structure. The average DMT modulus (calculated using Derjaguin-Muller-Toporov model) [34] of poly(APTAC) brushes grafted from silicon was determined to be 8.8 ± 1.1 MPa, while for poly(APTAC)+SPIONs it was found to be twice as large (17.6 ± 1.3 MPa) (Figure 3). The reported values can be reliably compared as the average indentations applied during the QNM measurements were kept very small (ca. 1 nm for poly(APTAC)+SPIONs, Figure 4) limiting the influence of the underlying substrate. It seems that the incorporated nanoparticles that are wrapped by the grafted polymer chains significantly increase the modulus of the whole layer. There are clearly some spots on the surface of poly(APTAC)+SPIONs with much higher DMT modulus (116–173 MPa) that can be correlated with SPIONs located at the top of the brushes (see Figure 2B). The measured values are three orders of magnitude smaller than the values reported for SPION in literature (151–192 GPa) [35] indicating that they are supported here by soft brushes rather than adsorbed directly on the silicon surface. However, the AFM cantilever selected for the measurements of the soft polymer layer cannot be reliably used for the determination of elasticity of much harder inorganic nanoparticles so the values of DMT modules determined for the SPION sticking out from the surface should be treated with caution. It seems that the incorporated nanoparticles mechanically integrate the whole layer by cross-linking of the polymeric chains formed in their proximity to electrostatic interactions.

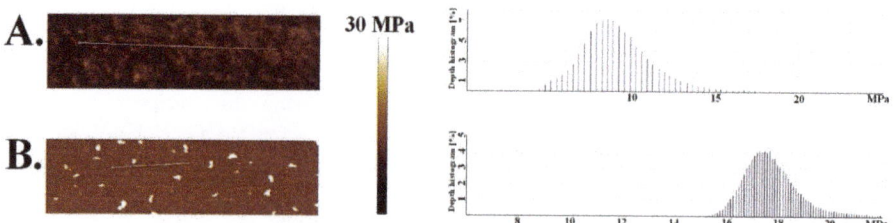

Figure 3. Quantitative nanomechanical mapping (QNM) measurements of poly(APTAC) (**A**) and poly(APTAC)+SPIONs (**B**) with the corresponding bearing analysis histograms.

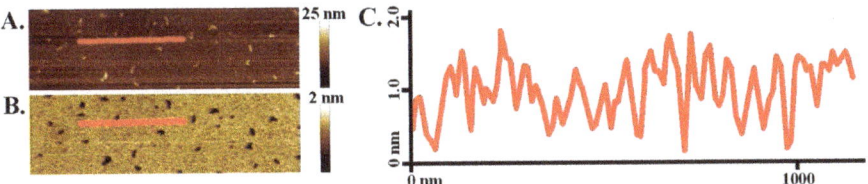

Figure 4. Poly(APTAC)+SPIONs brushes AFM images: topography (**A**) and indentation (**B**) with corresponding indentation image cross-section (**C**).

The coating profiling using the secondary ion mass spectrometry (SIMS) confirmed distribution of SPIONs in the whole volume of the poly(APTAC)+SPIONs layer (Figure 5). The signals of CH_3N^+ and $C_2H_2N^+$ ions, characteristic of poly(APTAC) brushes, appeared for the measurements of both coatings. Their intensities stay at the beginning constant during sputtering until they decrease when the signal of Si^+ from the silicon substrate becomes more intense. For the poly(APTAC)+SPIONs the signal of iron from magnetic nanoparticles is clearly visible for the whole thickness profile while no such signal can be observed for poly(APTAC) brushes. Moreover, the Fe^+ profile indicates homogeneous distribution of SPIONs along the brush thickness and even somehow higher content of SPIONs in the proximity of the silicon substrate. This observation is opposite to those of the systems prepared by incubation of polymer brushes in the suspension of nanoparticles for which the distribution is typically not homogeneous and even decreases gradually toward the substrate [28,36].

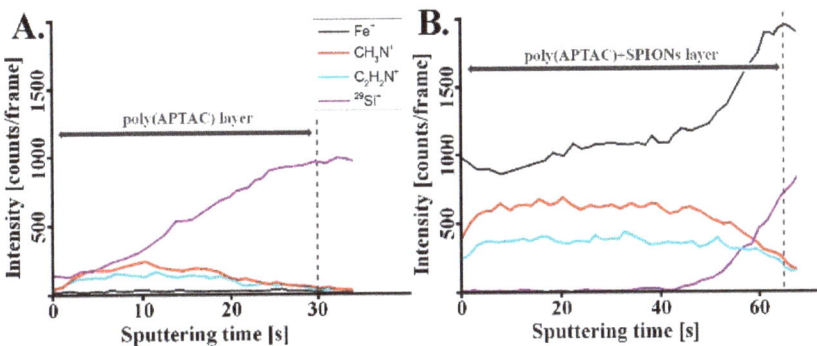

Figure 5. Secondary ion mass spectrometry (SIMS) spectra for polymeric brushes: (**A**) bare poly(APTAC), (**B**) poly(APTAC)+SPIONs.

The amount of SPIONs loaded into the polymeric brushes was indicated using X-ray photoelectron spectroscopy (XPS). As shown in Figure 6 and Figure S7 (Supporting Materials) the bands with characteristic binding energies—C 1s: 284–286 eV, N 1s: 399–402 eV, O 1s: 531–533 eV—appeared for both samples. Moreover two bands—Cl 2s: 270 eV, Cl 2p: 200 eV—of chlorine that are present as counterions for positively charged poly(APTAC) are shown as well. The intensity of the band at 709 eV (Fe $2p^{3/2}$, iron from SPIONs) was used to estimate the total amount of iron in the sample to be ca. 0.7% (ca. 1% SPIONs) that correlates with a thin layer of SPIONs in the brushes.

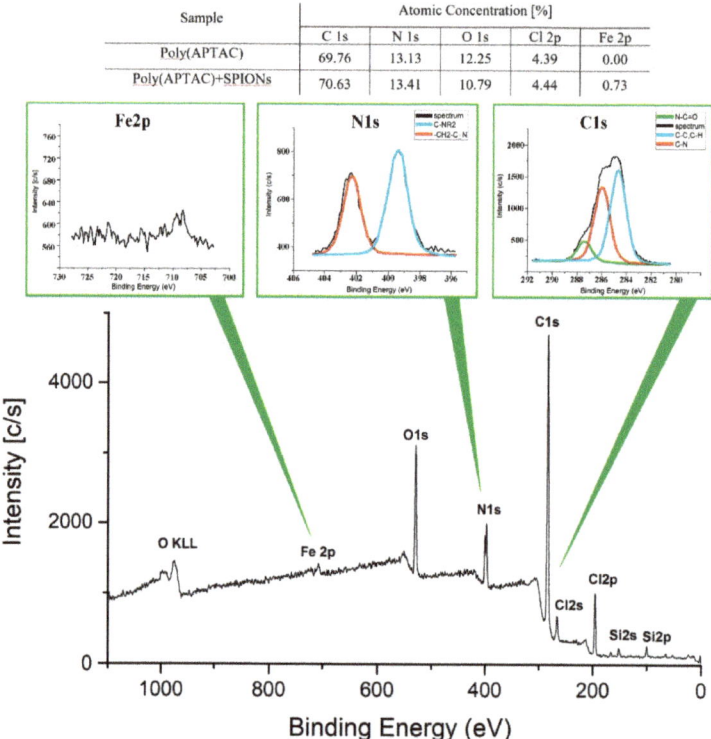

Figure 6. X-ray photoelectron spectroscopy (XPS) spectrum of poly(APTAC)+SPIONs with fragments corresponding to binding energies for Fe2p, N1s and C1s.

The thickness of the layers obtained was measured in both air and water using AFM (Figure 1). The polycationic chains are well-solvated in water that leads to their stretching due to repulsive interaction of the cationic side groups. This implies the appearance of poly(APTAC) brushes as a much thicker layer in water than in a dry state. In fact, the thickness of poly(APTAC) brushes increased more than 2.5 times reaching 71.4 ± 2.5 nm in water while for poly(APTAC)+SPIONs the increase was smaller reaching 2.1 times (thickness in water: 25.2 ± 1.7 nm). It seems that the embedded SPIONs limited the conformational freedom of the neighboring poly(APTAC) chains in the brushes due to mutual interactions and/or the formation of more entanglements of the chains that is consistent with the QNM results.

Finally, the magnetic properties of the poly(APTAC)+SPIONs hybrid brushes obtained were studied. Magnetic force microscopy (MFM) confirmed the presence of magnetic domains that can be assigned to the well-separated SPIONs placed in the polymeric matrix (Figure 7). Heterogeneity of the poly(APTAC)+SPIONs sample is also clearly visible in the adhesion image (Figure 7B2) that can be also related to the presence of variously entangled polymer chains at various distances from the embedded SPIONs. Both magnetic signal and adhesion heterogeneity are not visible in the poly(APTAC) sample (Figure 7A2,A3).

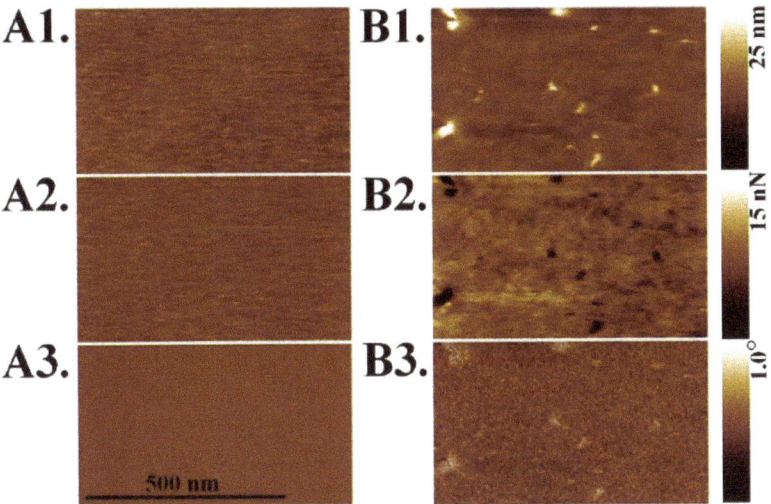

Figure 7. PF-MFM images of poly(APTAC) brushes (**A1–A3**) and poly(APTAC)+SPIONs brushes (**B1–B3**): 1. topography, 2. adhesion, 3. magnetic phase.

4. Conclusions

We reported here on a facile and efficient method for homogenous incorporation of magnetic nanoparticles (SPIONs) in the ultrathin (<15 nm) cationic polymer brush layer poly(APTAC) during their formation via surface-initiated polymerization. We showed magnetic signals of the well-separated SPIONs in the formed nanocomposite layer using magnetic force microscopy. A twice higher elastic modulus of the formed SPIONs-containing brushes as compared to the parent poly(APTAC) brushes was indicated using quantitative nanomechanical mapping (QNM) in spite of the low content of SPIONs (ca. 1%). We showed that application of a pulsed sonication during ATRP does not block formation of the surface-grafted brushes while it limits the unwanted aggregation of the added nanoparticles. The proposed method utilizes the nanoparticles formed ex situ, the properties of which can be determined and optimized prior to incorporation of them into a polymer layer. The proposed method can be treated as a model for formation of other hybrid structures composed of charged nanoparticles interacting electrostatically with oppositely charged polymers grafted from surfaces. As such, the method can be useful in the preparation of nanocomposite layers with non-aggregated small nanoparticles well-distributed in a polymeric matrix that are of high interest in e.g., nanoelectronic, photopholtaic applications.

Supplementary Materials: The following are available online at http://www.mdpi.com/2079-4991/9/3/456/s1: Figure S1: The exemplary high-resolution transmission electron microscope (HR TEM) image of SPIONs; Figure S2: Histograms of zeta potential values (A.) and hydrodynamic diameter (B.) of the suspension of SPIONs; Figure S3: AFM images of bare SPIONs: A. topography, B.-D. magnetic phase with a lift distance of 50 nm (B.), 100 nm (C.) and 250 nm (D.); Figure S4: The ultraviolet-visible (UV-Vis) absorption spectrum of phenanthroline complex with Fe(II) used for the determination of the iron content in the SPIONs. Inset: the calibration line; Figure S5: AFM images with cross-sections of bare poly(APTAC) brushes (A.) and the same brushes after 5 min of ultrasound treatment in SPIONs suspension (B.); Figure S6: AFM topography images of poly(APTAC)+SPIONs samples after synthesis (A.) and one month (B.); Figure S7: XPS spectrum of poly(APTAC).

Author Contributions: S.Z. and M.S. initiated and designed all of the experiments, W.G., P.N. and T.K. performed synthesis, D.L. and M.S. analyzed the data, W.G., S.Z. and M.S. wrote the manuscript.

Funding: This research was funded by TEAM programme (grant number: TEAM/2016-1/9) of the Foundation for Polish Science co-financed by the European Union under the European Regional Development Fund.

Acknowledgments: Marta Gajewska (Academic Centre for Materials and Nanotechnology, AGH) is acknowledged for TEM measurements, Paweł Dąbczyński (Faculty of Physics, Astronomy and Applied Computer Science, JU) is acknowledged for SIMS measurements and Hartmut Stadler (Bruker Nano Surfaces Division) for help in PF-MFM and QNM measurements.

Conflicts of Interest: The authors declare no competing financial interest.

References

1. Zhang, H.; Han, J.; Yang, B. Structural Fabrication and Functional Modulation of Nanoparticle–Polymer Composites. *Adv. Funct. Mater.* **2010**, *20*, 1533–1550. [CrossRef]
2. Al-Hussein, M.; Koenig, M.; Stamm, M.; Uhlmann, P. The Distribution of Immobilized Platinum and Palladium Nanoparticles within Poly (2-vinylpyridine) Brushes. *Macromol. Chem. Phys.* **2014**, *215*, 1679–1685. [CrossRef]
3. Zhang, Q.; Su, K.; Chan-Park, M.B.; Wuc, H.; Wang, D.; Xu, R. Development of high refractive ZnS/PVP/PDMAA hydrogel nanocomposites for artificial cornea implants. *Acta Biomater.* **2014**, *10*, 1167–1176. [CrossRef]
4. Zhu, X.; Loh, X.J. Layer-by-layer assemblies for antibacterial applications. *Biomater. Sci.* **2015**, *3*, 1505–1518. [CrossRef] [PubMed]
5. Ferhan, A.R.; Kim, D.H. Nanoparticle polymer composites on solid substrates for plasmonic sensing applications. *Nano Today* **2016**, *11*, 415–434. [CrossRef]
6. Yan, J.; Ye, Q.; Wang, X.; Yua, B.; Zhou, F. CdS/CdSe quantum dot co-sensitized graphene nanocomposites via polymer brush templated synthesis for potential photovoltaic applications. *Nanoscale* **2012**, *4*, 2109–2116. [CrossRef] [PubMed]
7. Long, D.X.; Choi, E.Y.; Noh, Y.Y. Manganese oxide nanoparticle as a new p-type dopant for high-performance polymer field-effect transistors. *ACS Appl. Mater. Interfaces* **2017**, *9*, 24763–24770. [CrossRef] [PubMed]
8. Tu, M.L.; Su, Y.K.; Chen, R.T. Hybrid light-emitting diodes from anthracene-contained polymer and CdSe/ZnS core/shell quantum dots. *Nanoscale Res. Lett.* **2014**, *9*, 611. [CrossRef]
9. Basly, B.; Alnasser, T.; Aissou, K.; Fleury, G.; Pecastaings, G.; Hadziioannouu, G.; Duguet, E.; Goglio, G.; Mornet, S. Optimization of Magnetic Inks Made of $L1_0$-Ordered FePt Nanoparticles and Polystyrene-*block*-Poly(ethylene oxide) Copolymers. *Langmuir* **2015**, *31*, 6675–6680. [CrossRef] [PubMed]
10. Nie, G.; Li, G.; Wang, L.; Zhang, X. Nanocomposites of polymer brush and inorganic nanoparticles: Preparation, characterization and application. *Polym. Chem.* **2016**, *7*, 753–769. [CrossRef]
11. Sarkar, B.; Alexandridis, P. Block copolymer–nanoparticle composites: Structure, functional properties, and processing. *Prog. Polym. Sci.* **2015**, *40*, 33–62. [CrossRef]
12. Douadi-Masrouki, S.; Frka-Petesic, B.; Save, M.; Charleux, B.; Cabuil, V.; Sandre, V. Incorporation of magnetic nanoparticles into lamellar polystyrene-b-poly-(n-butyl methacrylate) diblock copolymer films: Influence of the chain end-groups on nanostructuration. *Polymer* **2010**, *51*, 4673–4685. [CrossRef]
13. Oren, R.; Liang, Z.; Barnard, J.S.; Warren, S.C.; Wiesner, U.; Huck, W.T.S. Organization of nanoparticles in polymer brushes. *J. Am. Chem. Soc.* **2009**, *131*, 1670–1671. [CrossRef]
14. Choi, W.S.; Koo, H.Y.; Kim, J.Y.; Huck, W.T.S. Collective behavior of magnetic nanoparticles in polyelectrolyte brushes. *Adv. Mater.* **2008**, *20*, 4504–4508. [CrossRef]
15. Cui, T.; Zhang, J.; Wang, J.; Cui, F.; Chen, W.; Xu, F.; Wang, Z.; Zhang, K.; Yang, B. CdS-Nanoparticle/Polymer Composite Shells Grown on Silica Nanospheres by Atom-Transfer Radical Polymerization. *Adv. Funct. Mater.* **2005**, *15*, 481–486. [CrossRef]
16. Benetti, E.M.; Sui, X.; Zapotoczny, S.; Vancso, G.J. Surface-Grafted Gel-Brush/Metal Nanoparticle Hybrids. *Adv. Funct. Mater.* **2010**, *20*, 939–944. [CrossRef]
17. Aissou, K.; Fleury, G.; Pecastaings, G.; Alnasser, T.; Mornet, S.; Goglio, G.; Hadziioannou, G. Hexagonal-to-Cubic Phase Transformation in Composite Thin Films Induced by FePt Nanoparticles Located at PS/PEO Interfaces. *Langmuir* **2011**, *27*, 14481–14488. [CrossRef]
18. Shang, Q.; Liu, H.; Gao, L.; Xiao, G. Synthesis and Characterization of Film-forming Polymer/SiO_2 Nanocomposite via Surfactant-Free Emulsion Polymerization. *Asian J. Chem.* **2013**, *25*, 5347–5350. [CrossRef]

19. Zoppe, J.O.; Ataman, N.C.; Mocny, P.; Wang, J.; Moraes, J.; Klok, H.A. Surface-initiated controlled radical polymerization: State-of-the-art, opportunities, and challenges in surface and interface engineering with polymer brushes. *Chem. Rev.* **2017**, *117*, 1105–1318. [CrossRef]
20. Szuwarzyński, M.; Kowal, J.; Zapotoczny, S. Self-templating surface-initiated polymerization: A route to synthesize conductive brushes. *J. Mater. Chem.* **2012**, *22*, 20179–20181. [CrossRef]
21. Szuwarzyński, M.; Wolski, K.; Zapotoczny, S. Enhanced stability of conductive polyacetylene in ladder-like surface-grafted brushes. *Polym. Chem.* **2016**, *7*, 5664–5670. [CrossRef]
22. Schüwer, N.; Klok, H.A. A potassium-selective quartz crystal microbalance sensor based on crown-ether functionalized polymer brushes. *Adv. Mater.* **2010**, *22*, 3251–3255. [CrossRef]
23. Psarra, E.; Foster, E.; König, U.; You, J.; Ueda, Y.; Eichhorn, K.J.; Müller, M.; Stamm, M.; Revzin, A.; Uhlmann, P. Growth factor-bearing polymer brushes-versatile bioactive substrates influencing cell response. *Biomacromolecules* **2015**, *16*, 3530–3542. [CrossRef]
24. Szuwarzyński, M.; Wolski, K.; Pomorska, A.; Uchacz, T.; Gut, A.; Łapok, Ł.; Zapotoczny, S. Photoactive Surface-Grafted Polymer Brushes with Phthalocyanine Bridging Groups as an Advanced Architecture for Light-Harvesting. *Eur. J. Chem.* **2017**, *23*, 11239–11243. [CrossRef]
25. Wolski, K.; Szuwarzyński, M.; Kopeć, M.; Zapotoczny, S. Ordered photo- and electroactive thin polymer layers. *Eur. Polym. J.* **2015**, *65*, 155–170. [CrossRef]
26. Taccola, S.; Pensabene, V.; Fujie, T.; Takeoka, S.; Pugno, N.M.; Mattoli, V. On the injectability of free-standing magnetic nanofilms. *Biomed. Microdevices* **2017**, *19*, 51. [CrossRef]
27. Le Ouay, B.; Guldin, S.; Luo, Z.; Allegri, S.; Stellacci, F. Freestanding Ultrathin Nanoparticle Membranes Assembled at Transient Liquid–Liquid Interfaces. *Adv. Mater. Interfaces* **2016**, *3*, 1600191. [CrossRef]
28. Boyaciyan, D.; Braun, L.; Löhmann, O.; Silvi, L.; Schneck, E.; von Klitzing, R. Gold nanoparticle distribution in polyelectrolyte brushes loaded at different pH conditions. *J. Chem. Phys.* **2018**, *149*, 163322. [CrossRef]
29. Tamai, T.; Watanabe, M.; Ikeda, S.; Kobayashi, Y.; Fujiwara, Y.; Matsukawa, K.J. A Photocurable Pd Nanoparticle/Silica Nanoparticle/Acrylic Polymer Hybrid Layer for Direct Electroless Copper Deposition on a Polymer Substrate. *J. Photopolym. Sci. Technol.* **2012**, *25*, 141–146. [CrossRef]
30. Porel, S.; Singh, S.; Harsha, S.; Rao, D.N.; Radhakrishnan, T.P. Nanoparticle-embedded polymer: In situ synthesis, free-standing films with highly monodisperse silver nanoparticles and optical limiting. *Chem. Mater.* **2005**, *17*, 9–12. [CrossRef]
31. Lachowicz, D.; Kaczyńska, A.; Wirecka, R.; Kmita, A.; Szczerba, W.; Bodzoń-Kułakowska, A.; Sikora, M.; Karewicz, A.; Zapotoczny, S. A Hybrid System for Magnetic Hyperthermia and Drug Delivery: SPION Functionalized by Curcumin Conjugate. *Materials* **2018**, *11*, 2388. [CrossRef]
32. Szpak, A.; Kania, G.; Skórka, T.; Tokarz, W.; Zapotoczny, S.; Nowakowska, M. Stable aqueous dispersion of superparamagnetic iron oxide nanoparticles protected by charged chitosan derivatives. *J. Nanopart. Res.* **2013**, *15*, 1372. [CrossRef]
33. Kania, G.; Kwolek, U.; Nakai, K.; Yusa, S.I.; Bednar, J.; Wójcik, T.; Chłopicki, S.; Skórka, T.; Szuwarzyński, M.; Szczubiałka, K.; et al. Stable polymersomes based on ionic–zwitterionic block copolymers modified with superparamagnetic iron oxide nanoparticles for biomedical applications. *J. Mater. Chem. B* **2015**, *3*, 5523–5531. [CrossRef]
34. Dokukin, M.E.; Sokolov, I. Quantitative mapping of the elastic modulus of soft materials with HarmoniX and PeakForce QNM AFM modes. *Langmuir* **2012**, *28*, 16060–16071. [CrossRef]
35. Nicholls, J.R.; Hall, D.J.; Tortorelli, P.F. Hardness and modulus measurements on oxide scales. *Mater. High Temp.* **1994**, *12*, 141–150. [CrossRef]
36. Christau, S.; Möller, T.; Yenice, Z.; Genzer, J.; von Klitzing, R. Brush/gold nanoparticle hybrids: Effect of grafting density on the particle uptake and distribution within weak polyelectrolyte brushes. *Langmuir* **2014**, *30*, 13033–13041. [CrossRef]

© 2019 by the authors. Licensee MDPI, Basel, Switzerland. This article is an open access article distributed under the terms and conditions of the Creative Commons Attribution (CC BY) license (http://creativecommons.org/licenses/by/4.0/).

Article

Structure–Function Correlations in Sputter Deposited Gold/Fluorocarbon Multilayers for Tuning Optical Response

Pallavi Pandit [1,*], Matthias Schwartzkopf [1], André Rothkirch [1], Stephan V. Roth [1,2], Sigrid Bernstorff [3] and Ajay Gupta [4,*]

1. Deutsches Elektronen-Synchrotron (DESY), Notkestraße 85, D-22607 Hamburg, Germany
2. KTH Royal Institute of Technology, Department of Fibre and Polymer Technology, Teknikringen 56-58, SE-100 44 Stockholm, Sweden
3. Elettra-Sincrotrone Trieste, SS 14, Km 163.5, I-34149 Basovizza, Trieste, Italy
4. Center for Spintronic Materials, Amity University, UP Noida 201 313, India
* Correspondence: pallavi.pandit@desy.de (P.P.); agupta2@amity.edu (A.G.)

Received: 26 June 2019; Accepted: 26 August 2019; Published: 3 September 2019

Abstract: A new strategy to nanoengineer gold/fluorocarbon multilayer (ML) nanostructures is reported. We have investigated the morphological changes occurring at the metal–polymer interface in ML structures with varying volume fraction of gold (Au) and the kinetic growth aspect of the microscale properties of nano-sized Au in plasma polymer fluorocarbon (PPFC). Investigations were carried out at various temperatures and annealing times by means of grazing incidence small-angle and wide-angle X-ray scattering (GISAXS and GIWAXS). We have fabricated a series of MLs with varying volume fraction (0.12, 0.27, 0.38) of Au and bilayer periodicity in ML structure. They show an interesting granular structure consisting of nearly spherical nanoparticles within the polymer layer. The nanoparticle (NP) morphology changes due to the collective effects of NPs diffusion within ensembles in the in-plane vicinity and interlayer with increasing temperature. The in-plane NPs size distinctly increases with increasing temperature. The NPs become more spherical, thus reducing the surface energy. Linear growth of NPs with temperature and time shows diffusion-controlled growth of NPs in the ML structure. The structural stability of the multilayer is controlled by the volume ratio of the metal in polymer. At room temperature, UV-Vis shows a blue shift of the plasmon peak from 560 nm in ML Au/PTFE_1 to 437 nm in Au/PTFE_3. We have identified the fabrication and postdeposition annealing conditions to limit the local surface plasmon resonance (LSPR) shift from $\Delta\lambda_{LSPR} = 180$ nm (Au/PTFE_1) to $\Delta\lambda_{LSPR} = 67$ nm (Au/PTFE_3 ML)) and their optical response over a wide visible wavelength range. A variation in the dielectric constant of the polymer in presence of varying Au inclusion is found to be a possible factor affecting the LSPR frequency. Our findings may provide insights in nanoengineering of ML structure that can be useful to systematically control the growth of NPs in polymer matrix.

Keywords: nanocomposite; metal–polymer interface; multilayer; structure–function correlation; indirect band gap; GISAXS; GIWAXS; UV-Vis

1. Introduction

Easy processability, high flexibility, and tunable physical properties make nanocomposites very attractive for a broad range of applications. Recently, the combination of metal nanoparticles with dielectric media, such as polymers, has gained great pertinence both in fundamental as well as technological aspects. Their fine control and possible tuning of physical properties can lead to the fabrication of materials with novel functional, electric, and optical properties and engender their accessibility to various applications in the field of optics, electronics, and biomedicine [1–5]. These physical properties of the metal are strongly morphology-dependent [6–9]. In particular, the metal undergoes a significant property change compared to the bulk due to the quantum confinement towards nanoscale and their large surface to volume ratio. Interfacial energy variation may also enhance their functionality [1]. Incorporation of metallic nanoparticles (NP) into a polymer improves the functionality of the polymer. Governed by the metal–polymer interactions, which generally differ from the polymer–polymer interactions, the properties of the composite material are thus dominated by their interfacial interactions [10–13]. The morphologies of the embedded metal NPs can be artificially modified by controlling their processing parameters, such as the preparation method, rate of deposition, thermal annealing, etc. [6,9,10,14]. Furthermore, the volume fraction of the metal in the polymer matrix plays a significant role in deciding the morphological structure. The conductivity of metal–polymer nanocomposites (MPNC) varies from insulating to conducting as a function of metal concentration. The resistivity drops by several orders of magnitude near percolation [15]. The conductivity depends exponentially on the cluster separation near the percolation threshold (insulator metal transition), which is proved by varying the metal–polymer volume fraction $\varphi = \frac{V_m}{V_p}$, with V_m & V_P being the volume of metal and polymer, respectively. Furthermore, in-plane growth of NPs and their effect on the properties is limited by the percolation threshold as the metal layer above the percolation threshold gains a three-dimensional structure [1,6,16]. Varying the metallic volume fraction in the dielectric matrix also influences the refractive index and this can alter the optical properties of nanocomposite [17]. Additionally, thermal annealing of MPNC can artificially modify the morphological structure due to enhancement of atomic mobility and diffusivity at higher temperature [18]. The resulting structure of NPs is then driven by nucleation, thermal mobility, and growth. However, the fabrication and the control of the growth of NPs for desired applications of such materials is still a major challenge. To study the individual particle growth, emphasis has been given to the MPNC multilayer (ML) structure of fixed volume fraction below the percolation threshold of metal in varying polymer layers. Such arrangement allows inter alia control growth of NPs in the polymer matrix. MPNC MLs have been prepared via sputter deposition, having nanofabrication capability of sequential deposition of metal and polymer with the same precision [16,19]. Particularly, gold nanoparticles (Au NPs) modified with plasma polymer fluorocarbon (PPFC) [20] (sputtered Polytetrafluoroethylene (PTFE)) MLs have been prepared by alternating sputtering from an Au and a PTFE target. Au NPs in the PPFC matrix show long-term stability, brilliant optical properties, and lead to a spectral optical shift upon variation in their size, shape, and surrounding dielectric matrix [21]. PTFE has high chemical stability and comparatively high glass transition temperature with heat resistance capacity. It has low surface free energy that supports growing spherical NPs and a high sputter yield with low crosslinking tendency [22], making this polymer especially suitable for sputter deposition and for such studies. The correlation of the Au NPs' morphology with the resulting optical properties of the MLs has been investigated in this study. The structural properties of the nanocomposites and kinetic growth of NPs in polymer matrix have been studied using grazing incidence small and wide-angle X-ray scattering (GISAXS & GIWAXS). GISAXS yields the statistically pertinent horizontal and vertical correlation information of the changing electron density distribution [23,24] during growth, while GIWAXS can provide information about NPs' crystallinity and kinetic growth of NPs in polymer [25,26]. UV-Vis spectroscopy has been used to study the optical properties of nanocomposites [27,28]. The optical behavior of the nanocomposite evolves upon metal volume fraction variation due to the rearrangement of the NPs in the polymeric matrix. The range of the plasmon peak shift depends upon the density of implanted NPs and the thickness of the intermediate polymer layer. To further contextualize the structure–function correlations, we have performed isochronal thermal annealing of the ML structure and then compared the growth tendencies to the optical behavior. Hereby, the first question addressed is the effect of the metal volume fraction on the structural and optical properties of the nanocomposite. The second is the effect of temperature-induced Au NP growth behavior in PPFC matrix. The relation of structural and optical properties during this growth is discussed.

2. Experimental Details

2.1. Sample Preparation

The ML structure was prepared on optically polished Si (100) substrate (Si-Mat, Kaufering, Germany). Prior to the deposition, the Si wafers were ultrasonically cleaned with acetone, isopropanol, and deionized water for 10 min, each followed by piranha cleaning. This strongly oxidizing chemical cleaning removes most organic matter and ion contamination from the substrate [29]. MPNC MLs were prepared by radio frequency (rf) ion beam sputtering, using two alternating independent gold (Mateck, Berlin, Germany, purity 99.99%) and PTFE targets (Disk 2800 g/mol; Science Fellow Industries, Indore, India, purity 98%). Ion bombardment was initiated by a plasma glow discharge using a 3 cm broad-beam Kaufman-type hot-cathode at a low argon pressure (2.5 sccm), which thus generates Ar ions. The base pressure in the chamber was 3×10^{-7} mbar. The chamber was flushed with pure Ar (99.995%) a few times before deposition to minimize oxygen and water vapor contamination. The accelerating voltage and current were 1000 V and 30 mA, respectively. The target was kept at 45° with respect to the argon beam direction. The substrate was kept parallel to the target at a distance of 150 mm. The deposition pressure was 2.6×10^{-3} mbar. A schematic diagram for sputtering setup involved is shown in Figure S1. Three nanocomposites were prepared by varying the Au volume fraction. The deposition rate of each of the materials (Au and PTFE) was determined separately at the same pressure and gas flow rate which was used for alternate sputtering. Deposited PTFE is named hereafter plasma polymer fluorocarbon (PPFC), due to its structural and chemical changes during sputtering. The effective thickness rates monitored were $J_{Au} = 0.92 \pm 0.03$ Å/s and $J_{PTFE} = 3.4 \pm 0.2$ Å/s. The filling factor of gold in three samples was 0.12, 0.27, and 0.38, which is below the gold percolation threshold of 0.45 as reported for gold in amorphous fluoroplastics [15]. For the fixed filling factor, nanocomposite density and volume fraction were calculated in terms of bilayer period in three MLs. The volume fraction was estimated by considering the film thickness ratio of Au and PTFE inside the ML structure. The thickness estimation for the various volume fractions has been detailed in the SI. The in-plane thickness of each gold layer was kept constant at 1.0 ± 0.1 nm in the entire three MLs, whereas the thickness of the polymer layer varied in the three samples (as 19.0, 9.0, and 5.6 nm, respectively) to offer different volume fractions of Au in PPFC. The total thickness of the ML stack in all three cases was 140 ± 5 nm. The structures of the films were as follows: Si/SiO$_2$/[Au(1 nm)/PPFC(19 nm)]$_7$; [Au(1 nm)/PPFC(9 nm)]$_{14}$ and [Au(1 nm)/PPFC(5.6 nm)]$_{21}$. The indices in subscript denote the total number of bilayers in the three MLs. The samples named thereafter are Au/PPFC_1, Au/PPFC_2, and Au/PPFC_3. The schematic arrangement of NPs in the three MLs is shown in (SI) Figure S2a–c, respectively.

2.2. Characterization

A monochromatic X-ray beam with a wavelength of 0.154 nm was used for the simultaneous GISAXS and GIWAXS measurements at the SAXS beamline (BL 5.2), Elettra synchrotron, Trieste [30]. The incident beam had a cross section of 1.2 mm (H) × 100 μm (V). The small angle scattering signal was recorded with a 2D image plate detector (MAR 300; pixel size of 150 × 150 μm^2). The image plate was kept in the forward direction at a distance of 1850 ± 1 mm from the sample, to measure the scattered intensity in forward direction. The intensity of the incident and specular beam near $q_y = 0$ nm^{-1} was attenuated with a partly transparent aluminum filter in the beam path. Measurements were performed at constant grazing incidence angle $\alpha_i \sim 0.45°$. For in-plane GIWAXS measurements, a Pilatus 100 K (Dectris Ltd., Baden, Switzerland; pixel size of 172 × 172 μm^2) detector was kept at a distance of 327.7 ± 0.5 mm in the film plane (tilted 55°; a sketch of the detector position used for experiment is shown in Figure S3.) This detector distance had been fixed to cover the required angular range. The exposure time for each measurement was set to 300 s for GISAXS and GIWAXS. To address the possible effects of X-ray irradiation on the structure, the ML was initially exposed to the photon beam for a longer time: up to 20 min exposure, no significant change in the scattering pattern was observed. Beyond 20 min, a small broadening of the diffraction peaks and a decrease in peak intensities were observed. Accordingly, to avoid any effect of radiation damage during the measurements, after every scan the sample was shifted across the beam by 2 mm (sample size 20 × 20 mm^2) to expose an unexposed area (fresh area) of the film. The angular detector ranges have been converted from pixel to the scattering vector q using as two standard samples calibration reference; rat tail tendon collagen for GISAXS and Cu foil for GIWAXS. The width of the calibrant Cu diffraction lines (W_{ins}) was used as instrumental width for estimating the experimental crystalline size in the GIWAXS data of MLs using the Debye–Scherrer formula [31]. The MLs were isochronally annealed at 373 K, 473 K, and 573 K in situ during the measurement to study the kinetic growth of particles with time and temperature. A miniature boron nitride (BN) furnace (mounted on the sample rotation stage) was used for controlled heating of the sample, which was kept in a protective atmosphere of nitrogen gas. The sample temperature was maintained with an accuracy of ±0.5 K. The optical reflectance spectra of the MPNCs were measured ex situ with a UV-Vis spectrophotometer (Perkin Elmer, Waltham, WA, USA. Model: Lambda 950). The scans (normal incidence; reflectance mode) were recorded in the wavelength range of 200 nm to 800 nm. The main interest was to see the variation in the surface plasmon resonance of Au.

3. Results and Discussion

3.1. Structural Properties at Room Temperature

An overview of the setup used for simultaneous GISAXS and GIWAXS measurements at the SAXS beamline (BL 5.2) Elettra, Trieste and their extracted 1D information is shown in Figure 1.

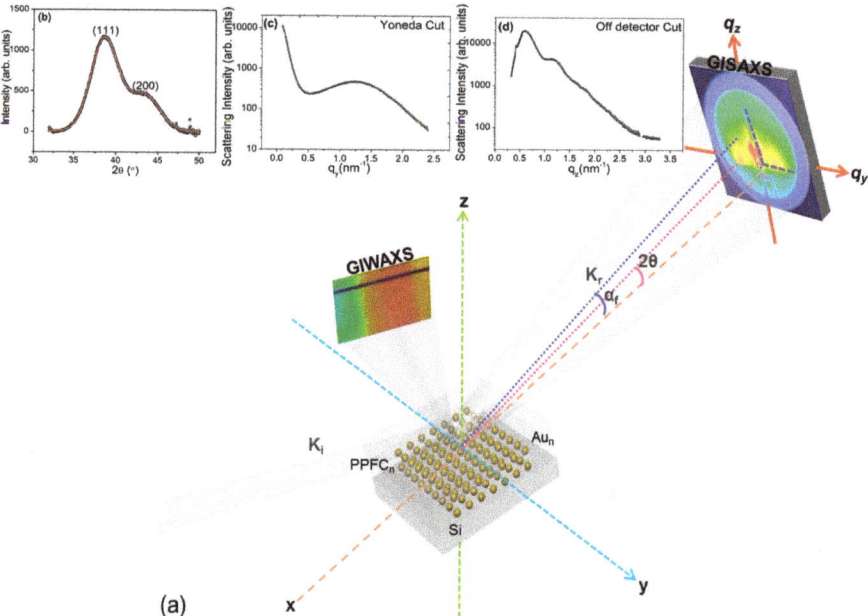

Figure 1. (a) Schematic view of the GISAXS experiment combined with in-plane-GIWAXS; (b) the extracted 1D GIWAXS profile was taken as line cut as indicated in the 2D GIWAXS image, (c) Yoneda cut (q_y = 0.236) and (d) off-detector cut (along q_z) as indicated by blue dashed lines in the 2D GISAXS pattern. In the GISAXS pattern, the origin of the coordinates of q_y and q_z is indicated by the direct beam position.

3.2. GIWAXS

The 1D diffraction pattern (Figure 1b) was extracted from the 2D in-plane diffraction image of the sample in θ–2θ geometry. The GIWAXS pattern confirms the formation of well-established nanocrystalline particles due to the appearance of two diffraction peaks around 2θ = 38.60 ± 0.008° and 43.30 ± 0.003°, ascribed to the (111) and (200) planes of fcc crystalline Au [32]. The peak positions are shifted to some extent from the bulk position which reflects the strain that may be caused by the geometry and the polymer–filler interfacial interaction. Two Gaussian peaks (red line in Figure 1b) were fitted to the data and the minimum crystalline size of a NP (D_{np}) was derived from the width obtained for the (111) peak in the diffraction pattern, after deducting the instrumental width using the Debye–Scherrer formula [31]:

$$D_{np} = \frac{0.94\lambda}{B \cos\theta}, \quad B = \sqrt{W_E^2 - W_{ins}^2} \qquad (1)$$

where λ is the wavelength of the X-rays and B is the full width at half maximum in radians. W_E and W_{ins} (0.35 rad.) are the experimental and instrumental width, respectively. The lattice constant nearly matches with the bulk value of Au (0.408 nm) which confirms crystalline gold NPs in the polymer matrix. Results obtained are given in Table 1. Diffraction plots of all three samples at room temperature are shown in the Figure S2d. As one can see from Table 1, the crystalline size in all three MLs is almost the same despite of the varying volume fraction of gold in ML. This is as per expectation, since the deposited in-plane mass thickness of the individual gold layers is the same (1.0 ± 0.1 nm) for all samples and the relative metal content is varied only by varying the thickness of the polymer layers.

Table 1. Au and polymer layer thicknesses, minimum crystalline size (111) (D_{np}), lattice constant (a), in-plane interparticle distance (ξ_H) of all three multilayers at room temperature. Errors indicated only in the first line hold for all values.

Sample	δ(Au) (nm)	δ(PPFC) (nm)	D_{np} (nm)	a (nm)	ξ_H (nm)
Au/PPFC_1	1 ± 0.1	19 ± 0.8	2.43 ± 0.004	0.403 ± 0.001	4.30 ± 0.05
Au/PPFC_2	1	9	2.46	0.404	4.34
Au/PPFC_3	1	5.6	2.45	0.406	4.32

3.3. GISAXS

In-plane cuts $I(q_y)$ were made at the Yoneda peak of Si (q_z = 0.236 nm^{-1}) and the DPDAK software package [33] was used to extract quantitative information from the data sequence. The last image taken at a particular temperature is illustrated in Figure 2. Please note that a small tilt can be observed in the measured 2D pattern which we associate to an initial tilt in the heater assembly on which the specimen was mounted (see Figure 1 for details of the setup). Tilt was accounted for when making the cuts, see indicated area (white block) in Figure 2, first left column.

The GISAXS pattern shows a broad side peak that emerges at large q_y values (Figure 1c). This peak is related to a maximum interference of scattered waves describing the NP correlation distance, often called interparticle distance ξ_H [16,25,34]. From the position of the side peak (horizontal) one can estimate the average interparticle distance (ξ_H) using the following formula [16],

$$\xi_H \approx \frac{2\pi}{q_y} \qquad (2)$$

where q_y is the position of the side peak in the y direction. In the Yoneda cut $I(q_y)$ at constant q_z = 0.236 nm^{-1}, the presence of a peak (Figure 1c) evidences that the island distribution is not completely random, but are separated by a preferential nearest neighbor distance ξ, (center-to-center), indicating short range ordering of Au NPs in PPFC. The side maxima observed for pristine MLs are dominated by the structure factor signals and indicate a substantial lateral ordering of the NPs.

The peak in the q_y range of 1.45–1.46 nm^{-1} corresponds to the interparticle distance $\xi_H \approx 4.3 \pm 0.05$ nm in real space, obtained by simulation (see SI for details). Simulation of Si/[Au(1 nm)/PPFC(19 nm)] was done using distorted wave born approximation (DWBA). The error in interparticle distance ($\xi_H \approx 4.3 \pm 0.05$ nm) is the fitting error of average size and distance. Indeed, the GISAXS pattern shows a broad distance variation of 25% which was also used in the simulation. Yoneda cuts of three pristine samples are shown in Figure S2e. The same interparticle distances ξ_H were obtained for all three samples (see Table 1), which corroborates a constant in-plane thickness of gold resulting in a uniform interlayer morphology. Similar to Equation (2), the vertical interlayer-particle distance ξ_V was derived using,

$$\xi_V \approx \frac{2\pi}{q_z} \tag{3}$$

where q_z denotes the position of the first order Bragg peak in the off-detector cut.

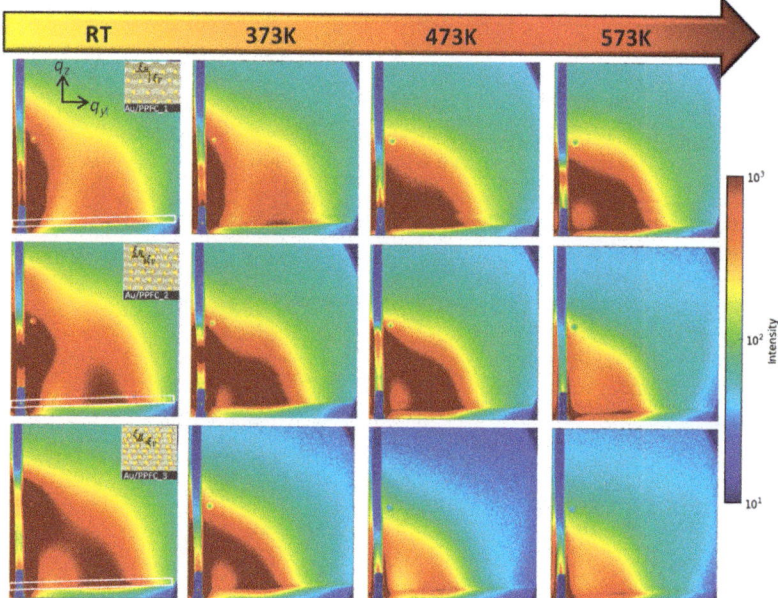

Figure 2. GISAXS pattern of MLs (**a**) Au/PPFC_1, (**b**) Au/PPFC_2, and (**c**) Au/PPFC_3 at the indicated temperatures. The white rectangle block in all three ML's room temperature patterns shows the projected area chosen to extract structural lateral features after applying 2° tilt correction. The inset of GISAXS room temperature images of three MLs shows the schematic layer arrangement and their horizontal (ξ_H) and vertical order (ξ_V).

The off-detector cut demonstrates the vertical ordering in MPNC, executed (along q_z) for 1.85 nm$^{-1} < q_y <$ 1.89 nm^{-1} (Figure S3d). In the vertical direction, an estimable interlayer interparticle correlation is observed due to the periodic structure of the MPNC ML (Table 2). As the first Au layer is deposited on the Si wafer, Au NPs are formed which exhibit in-plane local ordering of a granular nanoclusters assembly [16,35]. The next deposited PPFC amorphous layer perfectly covers the Au NPs layer. The successive alternate deposition of Au and PPFC layers leads to the appearance of small undulations/waviness with a period order given by the vertically interlayer-particle distance (ξ_V), see also Figure S2a. Layer-by-layer self-organized growth of NPs is the origin of this vertical order. The vertical order leads to well defined Bragg peaks in the scattering intensity from the MLs. The Bragg peaks in the vertical direction show the strong structural coherence between the individual layers of the MLs [36–38]. The off-detector cuts of three pristine MLs are given in Figure S2f. Vertically, the Bragg peak is shifted towards higher q values from sample Au/PPFC_1 to Au/PPFC_3. The bilayer period of Au/PPFC_1 to Au/PPFC_3 was varied in order to obtain different volume fractions of gold in the total ML stack of three MLs (Figure S2a–c). Thus, the shift in the Bragg peak position signals the fact that as the PPFC layer thickness decreases from Au/PPFC_1 to Au/PPFC_3, the separation of the Bragg peaks increases (due to reduce periodicity in successive MLs.; Figure S2f). The presence of a successive number of Bragg peaks in the GISAXS pattern of the MLs indicates that the NPs in the layers are well separated and the metal–polymer period is rather well established. In addition, the peaks broaden from sample Au/PPFC_1 to Au/PPFC_3 (Figure S2f). This can be explained as the interfaces are not sharp, but rather show a diffuse compositional profile which is evidence of a strong intermixing of metal and polymer at the interfaces in the pristine sample. This can be understood in terms of interface modulation [39] and stress in geometry of the three MLs according to the volume fraction of Au in PPFC. In this direction, Amarandei et al. reported that in MPNCs, the Au NPs interactions with the underlying layers must be considered [34,40]. For MLs Au/PPFC_2 and Au_PPFC3, the polymer thickness is comparable (low enough) to NPs size, which thermodynamically favors the interlayer interaction of NPs [41]. The undulated structure of the PPFC films increases with decreasing PPFC layer thickness because the thinner PPFC layer covers the waviness of the Au layers less efficiently (see schematic structures shown in Figure S2a–c). Consequently, the embedding of the Au inclusions in the intervening PPFC matrix increases and the vertical coherence length decreases [38]. We further note that the intensity of the first Bragg peak (indicated by arrow in Figure S2f) depends on the effective interdiffusivity/embedding in the MLs.

Table 2. Interlayer-particle distance (ξ_V) calculated from the first order Bragg peak of the multilayers at various temperatures. The errors only given in one row hold for all three MLs with varying temperature. n.q.: not quantifiable.

Multilayer	ξ_V (nm) 273 K	ξ_V (nm) 373 K	ξ_V (nm) 473 K	ξ_V (nm) 573 K
Au/PPFC_1	10.45 ± 0.03	9.79 ± 0.04	9.18 ± 0.07	5.13 ± 0.1
Au/PPFC_2	6.86 ± 0.03	6.61 ± 0.04	4.65 ± 0.07	n.q.
Au/PPFC_3	3.82 ± 0.03	7.02 ± 0.04	n.q.	n.q.

The interlayer-particle distance (ξ_V) decreased from ML Au/PPFC_1 to Au/PPFC_3 (from 10.5 to 3.8 nm) due to the decrease in the period in successive MLs. This is in conformity with the constant in-plane gold thickness and varying polymer thickness in the three successive MLs. From the off-detector cuts of pristine MLs (as given in Figure S2f), the structure and morphology of the NPs in a vertical stack can be estimated. For extracting the lateral distance ξ_H, we performed a simulation of the GISAXS signal in the Yoneda cut using the software IsGISAXS [42]. Details are described in Figure S4.

3.4. Structural Changes during and after Annealing

GIWAXS: Annealing of the nanocomposite thin films well above the glass transition temperature (T_g) reduces the interaction bond strength between metal NPs and polymer molecules. The ML structure undergoes reorganization because of the change in the physical structure of the polymer at higher temperatures. The mobility of the NPs increases due to weaker interface interaction, accompanied by a weaker entanglement density of the polymer chains. As a result, NPs start to diffuse through the polymer network. More and more NPs come in contact with each other, and eventually coagulation of these particles occurs [18,43]. This would increase the average crystalline size/NP size, interparticle distance, and may modify NPs morphology. As mentioned earlier, to study the effect of temperature on the nanocomposite properties, isochronal thermal annealing of the MLs was performed at various temperatures. A number of scattering patterns were recorded in a period of 20 to 30 min. with exposure time of 298 s at intervals of 5 min for statistical analysis. Thus, at each temperature, a number of scattering patterns were recorded as a function of time. To estimate the kinetic behavior of the NPs in the polymer matrix, the diffusion of NPs was estimated using the equation of diffusion [44,45],

$$D_{np}^2 = 2D_f t \tag{4}$$

where D_{np} denotes the minimum crystalline size (Equation (1)), D_f is the diffusion constant and t is the annealing time. Figure 3a shows the variation of minimum crystalline size in Au/PPFC_3 as a function of $\sqrt{t}$ at different annealing temperatures. One can see that the crystalline size increases linearly as a function of $\sqrt{t}$ at all investigated temperatures, suggesting diffusion-controlled growth of NPs in the polymer matrix at the varying temperatures. The slope of the straight line is proportional to the diffusivity of NPs in the polymer matrix. The variation in crystalline size was found to be systematic in all the samples (see Figure S5a,b). The obtained values of 'D_f' are used to find the activation energy (E_a) from Equation (5) [38,46]

$$D_f = D_0 \exp\left(\frac{-E_a}{RT}\right) \tag{5}$$

where R is the gas constant, D_0 is a pre-exponential factor, and T is the annealing temperature. From the slope of ln (D_f) vs. $10^3/T$ plot (shown in Figure 3b), the activation energy of Au NPs interdiffusion in Au/PPFC_3 is calculated as $E_a = (0.34 \pm 0.02$ eV). Arrhenius plots of ML Au/PPFC_1 and Au/PPFC_2 are given in Figure S5c,d.

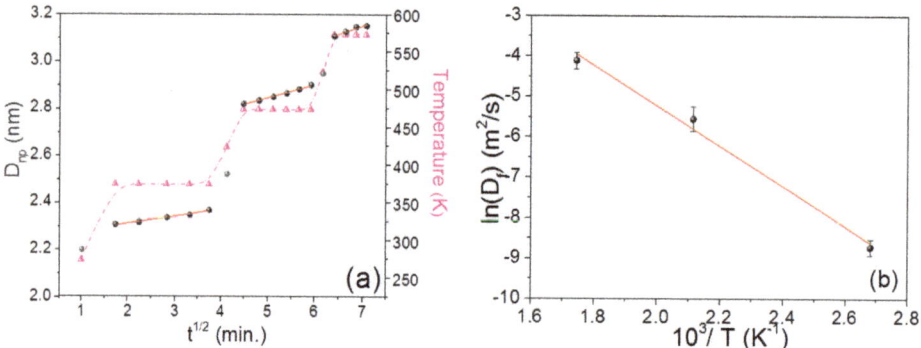

Figure 3. (a) Crystalline size (from GIWAXS) variation in Au/PPFC_3 as a function of annealing time $\left(\sqrt{t}\right)$ at varying temperature; the dotted (pink) curve represents the temperature profile of the sample with time and temperature (b) Arrhenius plot showing the relationship between ln(D_f) (where, D_f is the diffusion constant) and $10^3/T$ for the Au/PPFC_3 ML obtained from in situ temperature-dependent in-plane GIWAXS.

The glass transition temperature of bulk PTFE is around 394 K. It is interesting to see from Figure 3b that even at 373 K, which is below the glass transition temperature, the diffusion of NPs in the polymer matrix is governed by the same activation energy as evidenced by the Arrhenius plot. A change in viscosity across the glass transition temperature is expected to result in a significant change in the diffusion mechanism of NPs. The NP diffusion at all the three temperatures studied in the present work is governed by the same activation energy and suggests that no glass transition takes place in the temperature range investigated. The glass transition temperature is known to get reduced in the thin film form or with the inclusion of nanofillers [47,48]. Thus, it is quite likely that in the present system, the glass transition temperature has gone below 373 K (well above RT (273 K) but less than 394 K). The activation energy decreased from Au/PPFC_1 (1.03 ± 0.02) eV to Au/PPFC_3 (0.340 ± 0.02) eV [Au/PPFC_2 (0.429 ± 0.02)]. This is due to the fact that the polymer intermediate barrier layer between the two Au layers was successively decreased in the three MLs. Thus, the polymer layer mass density is in decreasing order, causing a decrease in activation energy.

GISAXS: The film morphology at various temperatures was accessed with GISAXS. The 2D GISAXS patterns recorded at various temperatures are shown in Figure 2. The Yoneda peak is accompanied by a broad side peak at higher q_y. Figure 4a–c gives the Yoneda cuts made for the three MLs at different temperatures The lateral correlation peak in Au/PPFC_1 and Au/PPFC_2 is continuously shifting towards lower q_y up to 473 K and then slightly shifts to higher q_y at 573 K signalizing first an increase in the interparticle distance of the NPs and then a slight decrease at 573 K. In Au/PPFC_3, the interparticle distance is continuously increasing with temperature up to 573 K. In the ML structure, the diffusion of NPs at lower temperature is mostly confined within the Au layers, i.e., parallel to the substrate. However, the NPs gain higher mobility in both horizontal and vertical directions with increasing temperature. At 573 K, the polymer is close to its molten state and the NPs thus gain more freedom in both directions. The maximum scattering intensity related to the NPs layer rapidly moves towards lower q_y at 573 K, indicating a change in the interparticle distance ξ_H. Thus the interparticle distance at higher temperature is due to the NPs mobility in both directions. However, Au/PPFC_3 has the lowest polymer layer thickness; due to this insufficient intermediate polymer thickness, it does not completely separating the Au NP layers. Hence, the NPs achieve the highest mobility at a lower temperature and the interparticle distance continuously increases as a collective response of the mobility of NPs in both vertical and horizontal directions in the ML. It is worth mentioning that at higher temperature, a higher order peak at larger q_y values (marked in Figure 4) occurs. This indicates a more ordered state of the structure due to the thermal annealing.

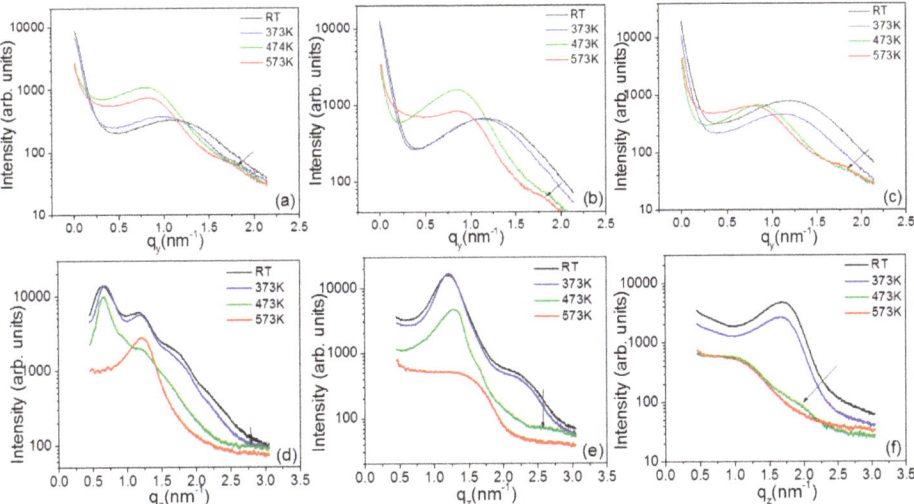

Figure 4. Directional cuts obtained from MLs at various temperatures. Yoneda cuts (horizontal) are shown in (**a–c**) and off-detector cuts (vertical) are shown in (**d–f**) for the MLs Au/PPFC_1, Au/PPFC_2, and Au/PPFC_3, respectively. In each plot, the arrow at higher q values denotes the presence of a higher order peak.

The in-plane average NPs radius R was derived from the effective layer thickness (δ) of gold and the interparticle distance based on the geometrical model by Schwartzkopf et al. [16] by applying it to spherical NPs. The geometrical model assumes that the effective deposited material is locally separated into a hexagonal array of spherical clusters in a distance ξ_H. According to this model assumption, the average radius of the supported clusters can be calculated by:

$$R = \sqrt[3]{\frac{3^{1.5}}{8\pi}\xi_H^2\delta} \qquad (6)$$

With effective thickness $\delta = 1.0 \pm 0.1$ nm of Au and interparticle distance ξ_H. The resulting sizes of the NPs for different temperatures are given in Table 3.

Table 3. NPs radius (R) and interparticle distance (ξ_H) variation in three MLs with varying temperature. The values for the errors given in row 1 hold for all values in the columns.

Temperature	MLs	ξ_H (nm)	R (nm)
RT	Au/PPFC_1	4.30 ± 0.05	1.56 ± 0.003
	Au/PPFC_2	4.34	1.57
	Au/PPFC_3	4.32	1.57
373 K	Au/PPFC_1	5.20	1.67
	Au/PPFC_2	5.81	1.91
	Au/PPFC_3	5.92	1.93
473 K	Au/PPFC_1	6.92	2.02
	Au/PPFC_2	6.03	1.96
	Au/PPFC_3	6.31	2.01
573 K	Au/PPFC_1	6.48	1.94
	Au/PPFC_2	5.89	1.93
	Au/PPFC_3	6.35	2.03

Particle radius slightly decreased in MLs Au/PFC_1 and Au_PPFC_2 from temperature 473 K to 573 K. This might be due to slight form change. Particle growth at various temperatures and their corresponding activation energy (0.109 ± 0.01 eV) for interdiffusion of adjacent NPs through the polymer matrix is shown in Figure 5a,b. One can observe a decrease in activation energy from Au/PPFC_1 (0.572 ± 0.01 eV) to Au/PPFC_2 (0.129 ± 0.01 eV) in Figure S6a,b. The activation energy is found to be lower in the analysis of particle growth by GISAXS than when using minimum crystalline size estimation via GIWAXS. The linear incremental trend in the activation energy plot is same in both calculated from crystalline size (GIWAXS) and particle size (GISAXS). The results are consistent, as the minimum crystalline size is lower than the particle size and it increases with temperature.

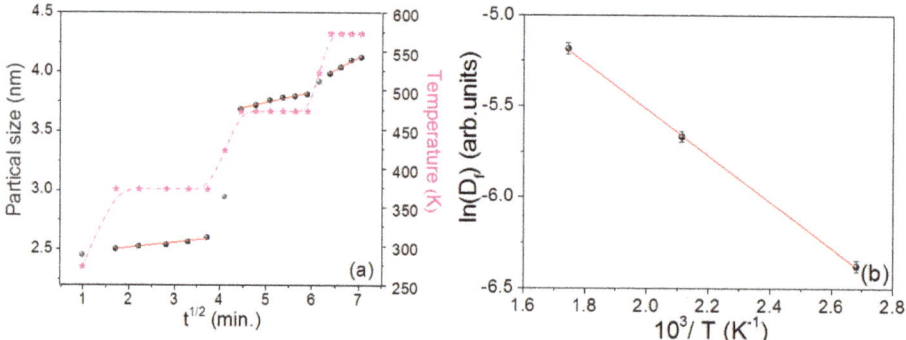

Figure 5. (a) Particle size variation in Au/PPFC_3 as a function of annealing time $\left(\sqrt{t}\right)$ at varying temperature; the dotted (pink) curve represents the temperature profile of the sample with time and temperature (b) Arrhenius plot showing the relationship between $\ln(D_f)$ and $10^3/T$ for the Au/PPFC_3 ML obtained from in situ temperature-dependent in-plane GISAXS; red line showing linear fit.

Off-detector cuts were also made from the data to more carefully analyze the changes occurring during annealing; they are shown in Figure 4d–f. One can notice that at 373 K there is no significant change in the Bragg peak position in all three MLs, showing that the ordering is preserved below the glass transition temperature (T_g) (one should expect change in morphology crossing glass transition temperature), although there is small decrease in intensity due to the gradual mixing at the interfaces.

However, a further increase in temperature to 473 K results in a significant change in the structure. At 473 K, the first Bragg peak in Au/PPFC_1 ML is slightly shifted towards lower q_z, while for other two MLs the Bragg peak is shifted towards higher q_z and the higher order Bragg peaks become suppressed. This might be because of a rearrangement of the NPs in PPFC leading to a more compact structure of the ML with higher intermixing at the metal–polymer interface. A further increase in temperature results in a rapid decrease in interlayer particle distance (ε_V). Although ε_V is decreased remarkably at 573 K, the presence of a broad Bragg peak or hump shows that the layered structure is sustained. The Bragg peak is increasingly broadening from ML Au/PPFC_1 to Au/PPFC_3 as a result of higher intermixing at elevated temperature. This transformation shows that the MLs are converted into a diffused metal–polymer structure at 573 K. The interlayer particle distance of the MLs at various temperatures is listed in Table 2.

3.5. Optical Properties at Room Temperature

Ordered NPs are promising structures that enable interconversion of the propagation of electromagnetic waves and thereby promote strongly enhanced local fields used for a number of practical applications such as photonics, optical sensors, etc. [43,49,50]. UV-Vis spectra of the three pristine MLs are shown in Figure 6. The spectra have been taken in off-specular reflection mode; dips in the spectra are due to absorption. One can notice characteristic features associated with NP assembly and their arrangement in the ML structure. In the wavelength range of 250 nm to 400 nm, some interference is seen [51] because of the ML structure. In the higher wavelength region, a surface plasmon is emerging in the spectrum, indicated by the diamond symbols in Figure 6. UV-Vis spectra of the nanocomposites show strong absorption in the range of 440 nm to around 520 nm. In the present case, we will refer to this as local surface plasmon resonance (LSPR) because in ML structures, NPs layers are buried in the polymer matrix. One can notice the presence of a single plasmon resonance in ML Au/PPFC_1 while for Au/PPFC_2 and Au/PPFC_3, the plasmon resonance splits in two.

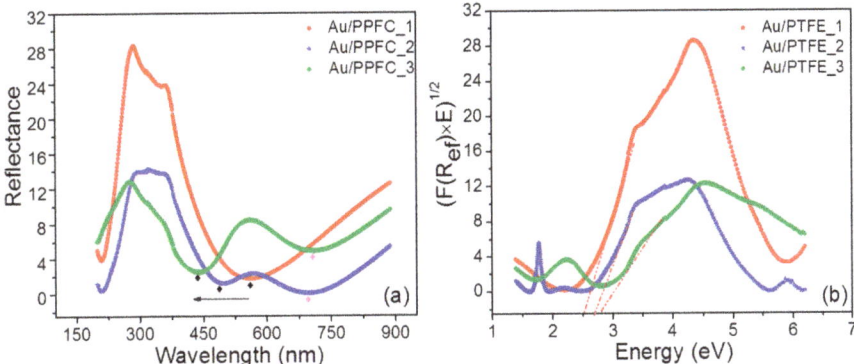

Figure 6. (a) UV-Vis spectra of three Au/PPFC MPNCs at room temperature; the local surface plasmon resonance (LSPR) in the three MLs are indicated by a diamond symbol (♦) and the arrow indicates the shift in LSPR. (b) Plot of $[F(R_{ef}) \times (h\nu)]$ vs. energy for indirect band gap calculation by applying linear approximations (fits are indicated by the red dash-dotted lines).

A single broad Gaussian plasmon dip in the reflection spectra can be an indication of spherical NPs [52,53]. The analysis has been done considering the spherical shape of the NPs, which was also confirmed by the GISAXS data. The polarizability is a function of the inclusion geometry (independent of volume) and orientation with respect to the applied electrical field, termed as the depolarization factor. The optical study shows that the electric field distribution across the surface of the spherical particle seems to be uniform, and thus all the free conduction electrons oscillate in phase, resulting in one plasmon resonance, regardless of the type of incident polarization [54,55]. The isotropic optical response of the spherical particles can be altered either by increasing their aspect ratio (changing the particles morphology) or coupling, i.e., placing them in close proximity of other particles [50,56–58]. In Au/PPFC_1, the periodic polymer thickness is 19 nm which ensures that the voids between NPs become completely filled with PPFC. The structurally continuous polymer barrier layer completely separates the metallic layer vertically, which in turn hinders the effective movement of NPs in the vertical direction (Figure S2a). Thus, an ultrafine constant thickness of 1 nm of gold ensures local monodispersion of in-plane particles with small size variation which supports the appearance of a single plasmon in Au/PPFC_1. Local monodisperse distribution of NPs has also been confirmed by GISAXS. In the other two MLs, because of the lower thicknesses of polymer (9 nm and 5.6 nm, respectively), the polymer film is not continuous, but PPFC grows preferentially within the voids between neighboring particles. The over layer PPFC thickness is not capable of complete isolation of metal NPs in a vertical stack. As a consequence, the NPs distance becomes small enough to give rise to plasmon coupling between the interacting particles due to close proximity, which changes the distribution of the induced surface dipoles and hence the electric field (Figure S2b,c). This yields a disturbance in the electronic oscillations which can result in a change in the optical response and thus lead to the excitation of more than one plasmon mode (PM) [56–58]. The broadening of plasmon resonance indicates a broad distribution of particle sizes in the MLs. The reflection amplitude increases with the metal filling factor, which shows that the NP density is directly proportional to the absorption intensity [59]. Sequels to this plasmon are shifted to lower wavelength (blue shift) from sample Au/PPFC_1 to Au/PPFC_3. There are a number of reasons that can lead to the observed blue plasmon shift. The optical properties of nanocomposites are highly sensitive to the morphology and the surrounding dielectric medium. The decrease in particle size or increase in interparticle distance can result in a blue shift of the plasmon resonance [3,60,61]. In the present case, the Au NP size and interparticle distance in the three pristine samples have been found to be almost consistent (Table 3) with deposition and self-assembly into NP of 1 nm Au per layer thickness, which confirms that in pristine MLs, NPs are still surrounded by polymer, thus blocking the possible mobility of NPs. The shape is also found to be spherical (confirmed by GISAXS & GIWAXS), so we exclude these reasons for the plasmon shift in our system. In this case, among the possible causes for the plasmon shift is a possible variation in the refractive index of the polymer in the vicinity of the metal particles [7,49]. In the present study, we have varied the polymer thickness in the MLs; this may induce variation of the mass density of the polymer. Due to this, the polymer–filler bonding would vary; this structural change can alter the refractive index of the polymer [62,63]. Given this information, we think that in pristine Au/PPFC_2 and Au/PPFC_3 MLs, the splitting of the surface plasmon resonance is the effect of close coupling and varied polymer refractive index. The LSPR of gold NPs can be used to assess the binding energy. To quantify this, the indirect band gape energy of the three MLs was calculated using the Kubelka–Munck radiative transfer model [64–66].

$$F(h\nu) = \left(F\left(R_{ef}\right) \times E\right)^{\frac{1}{2}}, \quad F(R_{ef}) = \frac{(1 - R_{ef})^2}{2R_{ef}} \qquad (7)$$

where R_{ef} is the reflectance. The model allows the calculation of the reflectance from a layer that both scatters and absorbs light. The linear fit through the LSPR yields the indirect band gap energy of the system. Figure 6b depicts the plot of $\left(F\left(R_{ef}\right) \times E\right)^{\frac{1}{2}}$ vs. energy for three MLs at room temperature and indicates a linear fit. The resulting extrapolated values are given in Table 4. From Table 4, one can see that the indirect band gap exhibits a systematic increase from sample Au/PPFC_1 to Au/PPFC_3. According to the Penn model [67], the dielectric constant of semiconductor materials [68,69] and other materials [70] varies inversely with the band gap energy. Thus, as one goes from Au/PPFC_1 to Au/PPFC_3, the dielectric constant is expected to decrease, which should result in blue-shift of the SPR, as observed in the present case, corroborating our hypothesis. These findings elucidate the nature of plasmon modes in this ML system, which involves strong light–matter coupling, and sets the level for the controlled bond formation by light excitation.

Table 4. Evaluated indirect band gap energy (in eV) for three multilayers at different temperatures.

Sample	RT	373 K	473 K	573 K
Au/PPFC_1	2.51 ± 0.04	2.61 ± 0.03	2.67 ± 0.05	3.18 ± 0.03
Au/PPFC_2	2.58 ± 0.04	2.82 ± 0.03	2.94 ± 0.05	3.70 ± 0.03
Au/PPFC_3	2.66 ± 0.04	2.85 ± 0.03	2.93 ± 0.05	2.95 ± 0.03

3.6. Temperature-Dependent Optical Response

The sensitivity of the nanocomposite film to the host medium was tested by monitoring the change in the Au-LSPR position as a function of temperature. The MLs were ex-situ annealed at 373 K, 473 K, and 573 K with the same annealing protocol used for GISAXS and GIWAXS measurements and their UV-Vis spectra were recorded. From Figure S7, one can see that in all the three cases the LSPR peak exhibits a blue shift with increasing annealing temperature. Furthermore, the magnitude of the shift decreases as one goes from sample Au/PPFC_1 to Au/PPFC_3 (Figure S8). Even a slight change from a nonspherical to a spherical shape can lead to a blue shift of the plasmon resonance [71,72]. The decrease in the magnitude of the blue shift in the three MLs is an indication of the reorganization of the NPs. The band width of the LSPR band of the annealed sample indicates only a small variation in the size distribution. In general, by annealing the samples well above the T_g of the polymer, the response to the dielectric environment significantly was enhanced. The increased sensitivity of the annealed sample is attributed to the increased mobility of both polymer chains and Au NPs in the close molten state of the polymer. The embedding of the Au NPs changed from sparse to a denser form due to a successive decrease in polymer thickness in the three MLs (Figure S2). In addition to this, the NP size increases with thermal annealing. Larger particles move slowly and this is why the magnitude of the shift is decreased from Au/PPFC_1 to Au/PPFC_3 (Figure S8). The results obtained are summarized in Table 4. Thus, the optical response of NPs during thermal treatment can possibly be explained in terms of the joint effect of effective refractive index variation and the depolarization factor as the particles come closer and there is a possibility of a slight change in shape (depolarization factor also contributes due to the small neck (rod-like shape) between two spherical particles) [73].

The result shows that, by using adequate postdeposition annealing, tailoring of the optical properties of such a system is possible and could be preferential for plasmatic-driven applications. From Table 4, one can see an increase in the band gap energy with increasing temperature. This shows a higher mismatch of the crystal momentum in the valence and conduction bands [74]. Qualitatively, according to the Penn model, this demonstrates the decrement of the dielectric constant (refractive index). An increase of the dielectric constant leads to a shift of the absorption maximum towards longer wavelengths [75,76]. However, the present results are in conformity with a decreasing dielectric constant, as the dielectric constant is supposed to decrease from Au/PPFC_1 to Au/PPFC_3, which leads to a blue shift of the plasmon resonance.

4. Conclusions

A simple, practical approach has been developed to make nanoengineered ML structures that can control the optical response of the NPs. A well-defined in-plane interparticle correlation is observed in all the three pristine MLs with an average distance of 4.30 ± 0.05 nm (Table 1), in conformity with simulations. Furthermore, a strong interlayer-particle correlation is observed in the vertical direction, in agreement with the bilayer period variation in the three MLs (Figure S2; Table 2). The reordering of NPs occurs as an effect of thermal annealing. The dependence of the NPs' size on the square root of the annealing time suggests diffusion-controlled growth of NPs. These findings indicate that the mobility of metal NPs can be affected by the volume fraction of metal in a polymer and the following annealing. Enhanced diffusion and intermixing is the main reason for the structural changes. It is worth noting that after annealing at 573 K, the interparticle correlation became more isotropic in both in-plane/horizontal and vertical directions. At the same time, NP size increases with increasing temperature and their shape becomes more spherical due to a reduction in the surface energy. The stability limit can be shifted to higher temperature by varying the intermediate polymer layer thickness, which can be useful in device applications. UV-Vis analysis shows that the LSPR frequency exhibits systematic variation with the volume fraction and thermal annealing. The variation in the dielectric constant of the material is found to be a possible factor affecting the LSPR frequency. The structural properties of the MLs are found to be in good agreement with their optical responses. Thus, the optical properties are tunable with appropriate choice of thermal treatment and volume fraction of metal in a favorable polymer matrix. This work opens the way to tune ML optical properties via controlling the growth of metal NPs.

Supplementary Materials: The following are available online at http://www.mdpi.com/2079-4991/9/9/1249/s1, Figure S1: Schematic diagram of alternate sputter deposition, Table S1: Filling factor, density of composite and volume fraction from the 3 MLs, Figure S2: Schematic structure presentations of the three MLs, the layer waviness is because of Au inclusion in PPFC matrix (a–c), 1D GIWAXS, GISAXS–yoneda and off detector cut, Figure S3: The sketch of GISAXS/GIWAXS experiment at Elettra BL 5.2 illustrates the angles and the distance used, Figure S4: Bilayer Simulation in ML Au/PPFC_1, Figure S5: Minimum crystalline size variation with time and temperature in MLs (a) Au/PPFC_1, (b) Au/PPFC_2 MLs and their corresponding Arrhenius-plot, Figure S6: Particle size variation with time and temperature in MLs a)Au/PPFC_1, b) Au/PPFC_2 MLs and their corresponding Arrhenius-plot, Figure S7: UV-Vis reflectance spectra of MLs Au/PPFC_1 (a), Au/PPFC_2 (b) and Au/PPFC_3 (c) at different temperatures, Figure S8: $F(Ref) \times (h^*v)^{1/2}$ as a function of photon energy for a)Au/PPFC_1, b)Au/PPFC_2 and c)Au/PPFC_3 at varying temperature.

Author Contributions: Conceptualization: P.P. and A.G., Methodology: P.P., A.G. and S.B., Investigation and & formal analysis P.P., A.G., M.S. and S.V.R., Writing-(Original draft preparation) P.P., Review and edition S.V.R., M.S., A.R., A.G., S.B. and P.P., Supervision: A.G.

Funding: This research received no external funding.

Conflicts of Interest: The authors declare no conflict of interest.

References

1. Faupel, F.; Zaporojtchenko, V.; Strunskus, T.; Elbahri, M. Faupel.pdf. *Adv. Eng. Mater.* **2010**, *12*, 1177–1190. [CrossRef]
2. Sanchez, C.; Belleville, P.; Popall, M.; Nicole, L. Applications of advanced hybrid organic–inorganic nanomaterials: From laboratory to market. *Chem. Soc. Rev.* **2011**, *40*, 696–753. [CrossRef] [PubMed]
3. Srivastava, S.; Haridas, M.; Basu, J.K. Optical properties of polymer nanocomposites. *Bull. Mater. Sci.* **2008**, *31*, 213–217. [CrossRef]
4. Lopes, W.A.; Jaeger, H.M. Hierarchical self-assembly of metal nanostrucures on diblock copolymer scaffolds. *Nature* **2001**, *414*, 735–738. [CrossRef] [PubMed]
5. Turković, A.; Dubček, P.; Juraić, K.; Drašner, A.; Bernstorff, S. SAXS Studies of Tio$_2$ nanoparticles in polymer electrolytes and in nanostructured films. *Materials* **2010**, *3*, 4979–4993. [CrossRef] [PubMed]
6. Schwartzkopf, M.; Santoro, G.; Brett, C.J.; Rothkirch, A.; Polonskyi, O.; Hinz, A.; Metwalli, E.; Yao, Y.; Strunskus, T.; Faupel, F.; et al. Real-time monitoring of morphology and optical properties during sputter deposition for tailoring metal–polymer interfaces. *ACS Appl. Mater. Interfaces* **2015**, *7*, 13547–13556. [CrossRef] [PubMed]
7. Alsawafta, M.; Badilescu, S.; Paneri, A.; Van Truong, V.; Packirisamy, M. Gold-poly(methyl methacrylate) nanocomposite films for plasmonic biosensing applications. *Polymers* **2011**, *3*, 1833–1848. [CrossRef]
8. Roth, S.V.; Santoro, G.; Risch, J.F.H.; Yu, S.; Schwartzkopf, M.; Boese, T.; Döhrmann, R.; Zhang, P.; Besner, B.; Bremer, P.; et al. Patterned diblock co-polymer thin films as templates for advanced anisotropic metal nanostructures. *ACS Appl. Mater. Interfaces* **2015**, *7*, 12470–12477. [CrossRef]
9. Hua, Y.; Chandra, K.; Dam, D.H.M.; Wiederrecht, G.P.; Odom, T.W. Shape-dependent nonlinear optical properties of anisotropic gold nanoparticles. *J. Phys. Chem. Lett.* **2015**, *6*, 4904–4908. [CrossRef]
10. Torrisi, V.; Ruffino, F. Metal–polymer nanocomposites: (Co-)evaporation/(Co)sputtering approaches and electrical properties. *Coatings* **2015**, *5*, 378–424. [CrossRef]

11. Ciprari, D.; Jacob, K.; Tannenbaum, R. Characterization of polymer nanocomposite interphase and its impact on mechanical properties. *Macromolecules* **2006**, *39*, 6565–6573. [CrossRef]
12. Li, S.; Meng Lin, M.; Toprak, M.S.; Kim, D.K.; Muhammed, M. Nanocomposites of polymer and inorganic nanoparticles for optical and magnetic applications. *Nano Rev.* **2010**, *1*, 5214. [CrossRef] [PubMed]
13. Shen, Y.; Lin, Y.; Nan, C.W. Interfacial effect on dielectric properties of polymer nanocomposites filled with core/shell-structured particles. *Adv. Funct. Mater.* **2007**, *17*, 2405–2410. [CrossRef]
14. Cao, G.Z. *Nanostructures Nanomaterials: Synthesis, Properties Applications*, 2nd ed.; Imperial College Press: London, UK, 2004; ISBN 1-86094-4159.
15. Takele, H.; Schürmann, U.; Greve, H.; Paretkar, D.; Zaporojtchenko, V.; Faupel, F. Controlled growth of Au nanoparticles in co-evaporated metal/polymer composite films and their optical and electrical properties. *Eur. Phys. J. Appl. Phys.* **2006**, *33*, 83–89. [CrossRef]
16. Schwartzkopf, M.; Buffet, A.; Körstgens, V.; Metwalli, E.; Schlage, K.; Benecke, G.; Perlich, J.; Rawolle, M.; Rothkirch, A.; Heidmann, B.; et al. From atoms to layers: In situ gold cluster growth kinetics during sputter deposition. *Nanoscale* **2013**, *5*, 5053. [CrossRef] [PubMed]
17. Kelly, K.L.; Coronado, E.; Zhao, L.L.; Schatz, G.C. The optical properties of metal nanoparticles: The influence of size, shape, and dielectric environment. *J. Phys. Chem. B* **2003**, *107*, 668–677. [CrossRef]
18. Heilmann, A. *Polymer Films with Embedded Metal Nanoparticles*; Springer: Berlin, Germany, 2003; ISBN 978-3540431510.
19. Zaporojtchenko, V.; Podschun, R.; Schürmann, U.; Kulkarni, A.; Faupel, F. Physico-chemical and antimicrobial properties of co-sputtered Ag-Au/PTFE nanocomposite coatings. *Nanotechnology* **2006**, *17*, 4904–4908. [CrossRef]
20. Kim, S.H.; Kim, M.; Lee, J.H.; Lee, S.-J. Self-cleaning transparent heat mirror with a plasma polymer fluorocarbon thin film fabricated by a continuous roll-to-roll sputtering process. *ACS Appl. Mater. Interfaces* **2018**, *10*, 10454–10460. [CrossRef]
21. Lahav, M.; Vaskevich, A.; Rubinstein, I. Biological sensing using transmission surface plasmon resonance spectroscopy. *Langmuir* **2004**, *20*, 7365–7367. [CrossRef]
22. Zekonyte, J.; Zaporojtchenko, V.; Faupel, F. Investigation of the drastic change in the sputter rate of polymers at low ion fluence. *Nucl. Instrum. Methods Phys. Res. B* **2005**, *236*, 241–248. [CrossRef]
23. Lei, Y.; Mehmood, F.; Lee, S.; Greeley, J.; Lee, B.; Seifert, S.; Winans, R.E.; Elam, J.W.; Meyer, R.J.; Redfern, P.C.; et al. Increased silver activity for direct. *Science* **2010**, *328*, 224–228. [CrossRef] [PubMed]
24. Vegso, K.; Siffalovic, P.; Benkovicova, M.; Jergel, M.; Luby, S.; Majkova, E.; Capek, I.; Kocsis, T.; Perlich, J.; Roth, S.V. GISAXS analysis of 3D nanoparticle assemblies—effect of vertical nanoparticle ordering. *Nanotechnology* **2012**, *23*, 045704. [CrossRef] [PubMed]
25. Müller-Buschbaum, P. Applications of Synchrotron light to scattering and diffraction in materials and life sciences: A basic introduction to grazing incidence small-angle X-ray scattering. *Lect. Notes Phys.* **2009**, *776*, 61–89.
26. Hexemer, A.; Müller-Buschbaum, P. Advanced grazing-incidence techniques for modern soft-matter materials analysis. *IUCrJ* **2015**, *2*, 106–125. [CrossRef] [PubMed]
27. Choi, J.; Choi, M.-J.; Yoo, J.-K.; Park, W.I.; Lee, J.H.; Lee, J.Y.; Jung, Y.S. Localized surface plasmon-enhanced nanosensor platform using dual-responsive polymer nanocomposites. *Nanoscale* **2013**, *5*, 7403. [CrossRef] [PubMed]
28. Novotny, L.; Hecht, B. Surface plasmon. In *Principles of Nano-Optics*; Cambridge University Press: Cambridge, UK, 2012; pp. 407–450. ISBN 9780511813535.
29. Pandit, P.; Banerjee, M.; Pandey, K.K.; Sharma, S.M.; Gupta, A. Role of substrate in melting behavior of Langmuir-Blodgett films. *Colloids Surf. A Physicochem. Eng. Asp.* **2015**, *471*, 159–163. [CrossRef]
30. Amenitsch, H.; Bernstorff, S.; Kriechbaum, M.; Lombardo, D.; Mio, H.; Rappolt, M.; Laggner, P. Performance and first results of the ELETTRA high-flux beamline for small-angle X-ray scattering. *J. Appl. Crystallogr.* **1997**, *30*, 872–876. [CrossRef]
31. Bushroa, A.R.; Rahbari, R.G.; Masjuki, H.H.; Muhamad, M.R. Approximation of crystallite size and microstrain via XRD line broadening analysis in TiSiN thin films. *Vacuum* **2012**, *86*, 1107–1112. [CrossRef]

32. Davey, W.P. Precision measurements of the lattice constants of twelve common metals. *Phys. Rev.* **1925**, *538*, 753–761. [CrossRef]
33. Benecke, G.; Wagermaier, W.; Li, C.; Schwartzkopf, M.; Flucke, G.; Hoerth, R.; Zizak, I.; Burghammer, M.; Metwalli, E.; Müller-Buschbaum, P.; et al. A customizable software for fast reduction and analysis of large X-ray scattering data sets: Applications of the new DPDAK package to small-angle X-ray scattering and grazing-incidence small-angle X-ray scattering. *J. Appl. Crystallogr.* **2014**, *47*, 1797–1803. [CrossRef]
34. Levine, J.R.; Cohen, J.B.; Chung, Y.W. Thin film island growth kinetics: A grazing incidence small angle X-ray scattering study of gold on glass. *Surf. Sci.* **1991**, *248*, 215–224. [CrossRef]
35. Babonneau, D.; Petroff, F.; Maurice, J.L.; Fettar, F.; Vaurès, A.; Naudon, A. Evidence for a self-organized growth in granular Co/Al$_2$O$_3$ multilayers. *Appl. Phys. Lett.* **2000**, *76*, 2892–2894. [CrossRef]
36. Fullerton, E.E.; Schuller, I.K.; Vanderstraeten, H.; Bruynseraede, Y. Structural refinement of superlattices from X-ray diffraction. *Phys. Rev. B* **1992**, *45*, 9292–9310. [CrossRef]
37. Fullerton Eric, E.; Kumar, S.; Grimsditch, M.; Kelly David, M.; Schuller Ivan, K. X-ray-diffraction characterization and sound-velocity measurements of W/Ni multilayers. *Phys. Rev. B* **1993**, *48*, 2560–2567. [CrossRef] [PubMed]
38. Reddy Raghavendra, V.; Gupta, A.; Gome, A.; Leitenberger, W.; Pietsch, U. In situ X-ray reflectivity and grazing incidence X-ray diffraction study of L 10 ordering in 57Fe/Pt multilayers. *J. Phys. Condens. Matter* **2009**, *21*, 186002. [CrossRef] [PubMed]
39. Shia, D.; Hui, C.Y.; Burnside, S.D.; Giannelis, E.P. An interface model for the prediction of Young's modulus of layered silicate-elastomer nanocomposites. *Polym. Compos.* **1998**, *19*, 608–617. [CrossRef]
40. Amarandei, G.; O'Dwyer, C.; Arshak, A.; Corcoran, D. The stability of thin polymer films as controlled by changes in uniformly sputtered gold. *Soft Matter* **2013**, *9*, 2695. [CrossRef]
41. Amarandei, G.; Clancy, I.; O'Dwyer, C.; Arshak, A.; Corcoran, D. Stability of ultrathin nanocomposite polymer films controlled by the embedding of gold nanoparticles. *ACS Appl. Mater. Interfaces* **2014**, *6*, 20758–20767. [CrossRef]
42. Lazzari, R. IsGISAXS: A program for grazing-incidence small-angle X-ray scattering analysis of supported islands. *J. Appl. Crystallogr.* **2002**, *35*, 406–421. [CrossRef]
43. Etrich, C.; Fahr, S.; Hedayati, M.; Faupel, F.; Elbahri, M.; Rockstuhl, C. Effective optical properties of plasmonic nanocomposites. *Materials* **2014**, *7*, 727–741. [CrossRef]
44. Pradell, T.; Crespo, D.; Clavaguera, N.; Clavaguer-Mora, M.T. Diffusion controlled grain growth in primary crystallization: Avrami exponents revisited. *J. Phys. Condens. Matter* **1998**, *10*, 3833–3844. [CrossRef]
45. Gupt, P.; Gupta, A.; Shukl, A.; Ganguli, T.; Sinha, A.K.; Principi, G.; Maddalena, A. Structural evolution and the kinetics of Cu clustering in the amorphous phase of Fe-Cu-Nb-Si-B alloy. *J. Appl. Phys.* **2011**, *110*, 033573. [CrossRef]
46. Gupta, A.; Gupta, M.; Chakravarty, S.; Wille, H.; Leupold, O. Fe diffusion in amorphous and nanocrystalline alloys studied using nuclear resonance reflectivity. *Phys. Rev. B* **2005**, *72*, 1–8. [CrossRef]
47. Bansal, A.; Yang, H.; Li, C.; Cho, K.; Benicewicz, B.C.; Kumar, S.K.; Schadler, L.S. Quantitative equivalence between polymer nanocomposites and thin polymer films. *Nat. Mater.* **2005**, *4*, 693–698. [CrossRef] [PubMed]
48. Forrest, J.A.; Mattsson, J. Reductions of the glass transition temperature in thin polymer films: Probing the length scale of cooperative dynamics. *Phys. Rev. E* **2000**, *61*, 53–56. [CrossRef] [PubMed]
49. Garcia, M.A. Surface plasmons in metallic nanoparticles: Fundamentals and applications. *J. Phys. D Appl. Phys.* **2011**, *44*, 283001. [CrossRef]
50. Reinhard, B.M.; Siu, M.; Agarwal, H.; Alivisatos, A.P.; Liphardt, J. Calibration of dynamic molecular rulers based on plasmon coupling between gold nanoparticles. *Nano Lett.* **2005**, *5*, 2246–2252. [CrossRef]
51. Kats, M.A.; Capasso, F. Optical absorbers based on strong interference in ultra-thin films. *Laser Photonics Rev.* **2016**, *10*, 735–749. [CrossRef]
52. Felidj, N.; Auberd, J.; Levi, G. Discrete dipole approximation for ultraviolet-visible extinction spectra simulation of silver and gold colloids. *J. Chem. Phys.* **1999**, *111*, 1195–1208. [CrossRef]
53. Mahmoud, M.A.; Chamanzar, M.; Adibi, A.; El-Sayed, M.A. Effect of the dielectric constant of the surrounding medium and the substrate on the surface plasmon resonance spectrum and sensitivity factor of highly symmetric systems. *J. Am. Chem. Soc.* **2012**, *134*, 6434–6442. [CrossRef]

54. Jensen, T.R.; Schatz, G.C.; Van Duyne, R.P. Nanosphere lithography: surface plasmon resonance spectrum of a periodic array of silver nanoparticles by ultraviolet–visible extinction spectroscopy and electrodynamic modeling. *J. Phys. Chem. B* **1999**, *103*, 2394–2401. [CrossRef]
55. Hulteen, J.C.; Van Duyne, R.P. Nanosphere lithography: A materials general fabrication process for periodic particle array surfaces. *J. Vac. Sci. Technol. A Vac. Surf. Films* **1995**, *13*, 1553–1558. [CrossRef]
56. Voshchinnikov, N.V.; Farafonov, V.G. Optical properties of spheroidal particles. *Astrophys. Space Sci.* **1993**, *204*, 19–86. [CrossRef]
57. Chu, Y.; Banaee, M.G.; Crozier, K.B. Double-resonance plasmon substrates for surface-enhanced Raman scattering with enhancement at excitation and stokes frequencies. *ACS Nano* **2010**, *4*, 2804–2810. [CrossRef] [PubMed]
58. Zhou, Y.; Li, X.; Ren, X.; Yang, L.; Liu, J. Designing and fabricating double resonance substrate with metallic nanoparticles–metallic grating coupling system for highly intensified surface-enhanced Raman spectroscopy. *Analyst* **2014**, *139*, 4799–4805. [CrossRef] [PubMed]
59. Hedayati, M.K.; Fahr, S.; Etrich, C.; Faupel, F.; Rockstuhl, C.; Elbahri, M. The hybrid concept for realization of an ultra-thin plasmonic metamaterial antireflection coating and plasmonic rainbow. *Nanoscale* **2014**, *6*, 6037–6045. [CrossRef] [PubMed]
60. El-Brolossy, T.A.; Abdallah, T.; Mohamed, M.B.; Abdallah, S.; Easawi, K.; Negm, S.; Talaat, H. Shape and size dependence of the surface plasmon resonance of gold nanoparticles studied by Photoacoustic technique. *Eur. Phys. J. Spec. Top.* **2008**, *153*, 361–364. [CrossRef]
61. Ahmad, T.; Wani, I.A.; Ahmed, J.; Al-Hartomy, O.A. Effect of gold ion concentration on size and properties of gold nanoparticles in TritonX-100 based inverse microemulsions. *Appl. Nanosci.* **2014**, *4*, 491–498. [CrossRef]
62. Ghanipour, M.; Dorranian, D. Effect of Ag-nanoparticles doped in polyvinyl alcohol on the structural and optical properties of PVA films. *J. Nanomater.* **2013**, *2013*, 1–10. [CrossRef]
63. Scaffardi, L.B.; Tocho, J.O. Size dependence of refractive index of gold nanoparticles. *Nanotechnology* **2006**, *17*, 1309–1315. [CrossRef]
64. Hecht, H.G. The interpretation of diffuse reflectance spectra. *J. Res. NBS A Phys. Chem.* **1976**, *80*, 567–583. [CrossRef]
65. Murphy, A. Band-gap determination from diffuse reflectance measurements of semiconductor films, and application to photoelectrochemical water-splitting. *Sol. Energy Mater. Sol. Cells* **2007**, *91*, 1326–1337. [CrossRef]
66. R, L.; R, G. Band-gap energy estimation from diffuse reflectance measurements on sol–gel and commercial TiO_2: A comparative study. *J. Sol-Gel Sci. Technol.* **2012**, *61*, 1–7.
67. Penn, D.R. Wave-number-dependent dielectric function of semiconductors. *Phys. Rev.* **1962**, *128*, 2093–2097. [CrossRef]
68. Ravindra, N.M.; Ganapathy, P.; Choi, J. Energy gap-refractive index relations in semiconductors-An overview. *Infrared Phys. Technol.* **2007**, *50*, 21–29. [CrossRef]
69. Aziz, S.B.; Rasheed, M.A.; Ahmed, H.M. Synthesis of polymer nanocomposites based on [methyl cellulose](1−x):(CuS)x (0.02 M ≤ x ≤ 0.08 M) with desired optical band gaps. *Polymers* **2017**, *9*, 194. [CrossRef] [PubMed]
70. Gupta, K.S.; Singh, J.; Akhtar, J. Materials and processing for gate dielectrics on silicon carbide (SiC) surface. In *Physics and Technology of Silicon Carbide Devices*; Intech Open: London, UK, 2013; pp. 207–234, ISBN 978-953-51-0917-4.
71. Portalès, H.; Pinna, N.; Pileni, M.P. Optical response of ultrafine spherical silver nanoparticles arranged in hexagonal planar arrays studied by the DDA method. *J. Phys. Chem. A* **2009**, *113*, 4094–4099. [CrossRef]
72. Pileni, M.P. Optical properties of nanosized particles dispersed in colloidal solutions or arranged in 2D or 3D superlattices. *New J. Chem.* **1998**, *22*, 693–702. [CrossRef]
73. Lim, T.H.; McCarthy, D.; Hendy, S.C.; Stevens, K.J.; Brown, S.A.; Tilley, R.D. Real-Time TEM and kinetic monte carlo studies of the coalescence of decahedral gold nanoparticles. *ACS Nano* **2009**, *3*, 3809–3813. [CrossRef]
74. Trainer, D.J.; Putilov, A.V.; Di Giorgio, C.; Saari, T.; Wang, B.; Wolak, M.; Chandrasena, R.U.; Lane, C.; Chang, T.R.; Jeng, H.T.; et al. Interlayer coupling induced valence band edge shift in mono-to few-layer MoS_2. *Sci. Rep.* **2017**, *7*, 40559. [CrossRef]

75. Hilger, A.; Tenfelde, M.; Kreibig, U. Silver nanoparticles deposited on dielectric surfaces. *Appl. Phys. B Lasers Opt.* **2001**, *73*, 361–372. [CrossRef]
76. Berg, K.J.; Berger, A.; Hofmeister, H. Small silver particles in glass surface layers produced by sodium-silver ion exchange—Their concentration and size depth profile. *Z. Phys. D Atoms Mol. Clust.* **1991**, *20*, 309–311. [CrossRef]

© 2019 by the authors. Licensee MDPI, Basel, Switzerland. This article is an open access article distributed under the terms and conditions of the Creative Commons Attribution (CC BY) license (http://creativecommons.org/licenses/by/4.0/).

Article

Achieving Secondary Dispersion of Modified Nanoparticles by Hot-Stretching to Enhance Dielectric and Mechanical Properties of Polyarylene Ether Nitrile Composites

Yong You, Ling Tu, Yajie Wang, Lifen Tong, Renbo Wei * and Xiaobo Liu *

Research Branch of Advanced Functional Materials, School of Materials and Energy, University of Electronic Science and Technology of China, Chengdu 610054, China
* Correspondence: weirb10@uestc.edu.cn (R.W.); liuxb@uestc.edu.cn (X.L.); Tel.: +86-028-8320-7326 (X.L.)

Received: 5 June 2019; Accepted: 10 July 2019; Published: 12 July 2019

Abstract: Enhanced dielectric and mechanical properties of polyarylene ether nitrile (PEN) are obtained through secondary dispersion of polyaniline functionalized barium titanate (PANI-*f*-BT) by hot-stretching. PANI-*f*-BT nanoparticles with different PANI content are successfully prepared via in-situ aniline polymerization technology. The transmission electron microscopy (TEM), fourier transform infrared spectroscopy (FTIR), X-ray photoelectron spectroscopic instrument (XPS) and Thermogravimetric analysis (TGA) results confirm that the PANI layers uniformly enclose on the surface of $BaTiO_3$ nanoparticles. These nanoparticles are used as functional fillers to compound with PEN (PEN/PANI-*f*-BT) for studying its effect on the mechanical and dielectric performance of the obtained composites. In addition, the nanocomposites are uniaxial hot-stretched by 50% and 100% at 280 °C to obtain the oriented nanocomposite films. The results exhibit that the PANI-*f*-BT nanoparticles present good compatibility and dispersion in the PEN matrix, and the hot-stretching endows the second dispersion of PANI-*f*-BT in PEN resulting in enhanced mechanical properties, crystallinity and permittivity-temperature stability of the nanocomposites. The excellent performances of the nanocomposites indicate that a new approach for preparing high-temperature-resistant dielectric films is provided.

Keywords: nanocomposites; surface-functionalization; secondary dispersion; hot-stretching

---

1. Introduction

With the increasing requirements of modern microelectronic components, the miniaturized and flexible dielectric materials are attracting more and more attention for various applications [1,2]. However, up to now, a single component material has been unable to meet these demands. Although the traditional inorganic ceramic dielectrics are widely used owing to their high dielectric constant, their inherent characteristics of heavy weight, difficult processing and brittleness fail to meet the current practical application [3,4]. In comparison, polymeric materials have exhibited the advantages of being lightweight and flexible, but their low dielectric constant also limits their application to a great extent [5–7]. Therefore, combining the two-component materials is an effective way to overcome these limitations [8–10].

In recent years, polymer-based nanocomposites have proved to be an important dielectric material by virtue of the high dielectric permittivity, flexibility and excellent thermal stability for widely using as dielectrics in the electronic system [11–15]. It is mainly because the nanocomposites can absorb the dominants of polymer matrix and inorganic nanofillers. Nevertheless, the prerequisite for the nanocomposites demonstrating excellent properties is to realize good compatibility between nanoparticles and polymer matrix [16,17]. In general, modifying the micro-interface of the nanoparticles can greatly improve the compatibility of nanoparticles with matrix, which can also regulate the dielectric permittivity and thermal stability of composites [18–20]. Thus, it is important to design and fabricate surface functionalized nanoparticles on the basis of maintaining the properties of nanoparticles while improving the compatibility. Polyaniline (PANI), as a conducting polymer, has been widely used as the filler to compound with polymer matrix [21] or as the surface agent to modify the nanoparticles [22,23] by virtue of its excellent conductivity after doping.

Although surface-functionalized nanofillers can effectively improve their compatibility with the matrix enhancing the dispersion of nanofillers in the matrix, they will inevitably agglomerate at a high filler level, resulting in reduced mechanical properties and permittivity-temperature stability. Therefore, it is crucial to find a technology that can achieve secondary dispersion of fillers and improve the overall performance of composites [24]. Uniaxial hot-stretching, a method to achieve high orientation of polymer materials under the action of external force, realizes a secondary uniform dispersion of the fillers along the orientation direction. In addition, the hot-stretching technology can effectively promote the regularity of polymer molecular chains and improve their crystallinity [23–28].

In this paper, novel PANI-*functionalized*-nanoparticles of different polymer content via in-situ polymerization technology are fabricated and characterized in detail. Also, these surface functionalized nanoparticles are used as functional fillers to promote the performances of polyarylene ether nitrile (PEN). In addition, the PEN-based nanocomposites are uniaxial hot-stretched by 50% and 100% at 280 °C. The corresponding properties of the oriented nanocomposite films are investigated in detail.

2. Experimental

2.1. Materials

$BaTiO_3$ (~60 nm, cubic) was bought from TPL Co., Dallas, Texas, USA. Potassium carbonate (K_2CO_3), 2, 6-dichlorobenzonitrile (DCBN), biphenol (BP), ammonium persulfate ((NH_4)$_2S_2O_8$), aniline (C_6H_7N), hydrochloric acid and alcohol were supplied by Chengdu KeLong chemicals, Chengdu, China. N-methyl-2-pyrrolodone (NMP) was bought from Chengdu Changzheng chemicals, Chengdu, China.

2.2. Preparation of PEN

PEN was prepared by 2, 6-dichlorobenzonitrile and biphenol in our laboratory through the previously reported method [28].

2.3. Preparation of PANI-f-BT Nanoparticles

The PANI-*f*-BT nanoparticles were fabricated by in-situ aniline polymerization method [29]. The specific steps are shown in Figure 1a. Firstly, $BaTiO_3$ (~60 nm, 1.0 g) was added into 100 mL deionized water and ultrasonicated for 1 h. Then the dispersion was cooled to 0–5 °C with an ice bath (step A). At the same time, a certain amount of aniline dissolved in 50 mL HCl (0.1 M) was also cooled in another ice bath. Next, the aniline solution was quickly dripped into the $BaTiO_3$ dispersion under nitrogen atmosphere (step B). After stirring for 1 h, pre-cooled ammonium persulfate dissolving in deionized water was added into the mixture of $BaTiO_3$/aniline for oxidative polymerization for 18 h (step C and D). Finally, the PANI-*f*-BT nanoparticle was obtained through filtration and drying (step E). In addition, the preparation diagram of PANI-*f*-BT nanoparticles is shown in Figure 1b. In this system, the molar ratio of ammonium persulfate to aniline is controlled to be 1.2:1, and the amounts of aniline are 0.1, 0.2 and 0.3 mL for PANI-*f*-BT-a, PANI-*f*-BT-b and PANI-*f*-BT-c, respectively.

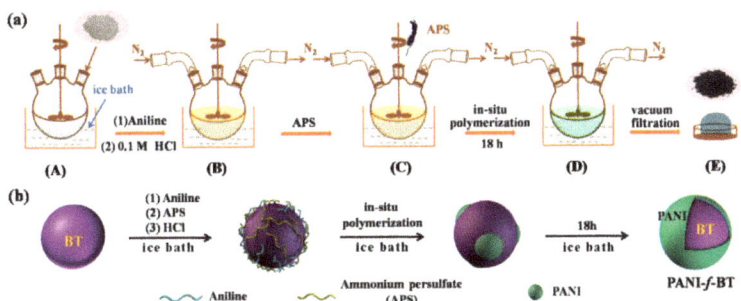

Figure 1. Experimental steps (**a**) and schematic diagram (**b**) of the polyaniline functionalized barium titanate (PANI-*f*-BT) nanoparticles.

2.4. Preparation of Nanocomposites

PEN-based nanocomposite films with 40 wt% pure BT and PANI-*f*-BT nanoparticles were prepared by solution casting method [30], which were named PEN/BT, PEN/PANI-*f*-BT-a, PEN/PANI-*f*-BT-b, PEN/PANI-*f*-BT-c, respectively.

2.5. Preparation of Oriented Nanocomposites by Hot-Stretching

The orientation process of nanocomposite film was carried out in a 280 °C oven by uniaxial hot-stretching method according to the previously reported paper [25]. All the nanocomposite films were stretched by 50% and 100%, respectively. The detailed steps were as follows: first of all, the two ends of the films (10 mm × 150 mm) were fixed by clamps, where the distance was controlled to be 100 mm; next, one clamp was suspended at the top of the oven, and the other one was hung with a 200 g balancing weight. The distance between the bottom of the oven and the balancing weight was adjusted to be 50 mm and 100 mm, corresponding to the stretching ratios of 50% and 100%. After the films were stretched to the required length, they were quickly removed and quenched in cold water. For comparison, the un-stretched nanocomposite films were also treated at 280 °C for the same time.

2.6. Characterization

The chemical structure of PANI-*f*-BT nanoparticles was characterized on a fourier transform infrared spectroscopy (FTIR, 8400S, Shimadzu, Japan) in the transmission mode between 4000 and 500 cm^{-1} by incorporating PANI-f-BT in the KBr. The elemental analysis was tested on an X-ray photoelectron spectroscopic instrument (XPS, ESCA 2000, VG Microtech, UK) using a monochromic Al Kα (h_v = 1486.6 eV) X-ray source. The micro-structures of PANI-f-BT were carried out on the transmission electron microscopy (TEM, JEM-2100F, JEOL, Japan) at 200 kV by dispersing PANI-f-BT on the copper network. The micro-structures of the PANI-f-BT nanocomposites were also obtained using a scanning electron microscopy (SEM, 6490LV, JSM, Japan) at 20 kV by sputtering gold on the fractured surface of PANI-f-BT. The crystalline structure of PANI-f-BT was characterized by X-ray diffractometer (XRD, RINT2400, Rigaku, Japan) with Cu Kα radiation. Thermal properties of samples were tested under N_2 atmosphere by differential scanning calorimetry (DSC, Q100, TA Instruments, New Castle, USA) from 40 °C to 380 °C with a heating rate of 10 °C/min and Thermogravimetric analysis (TGA, Q50, TA Instruments, New Castle, USA) from 50 °C to 800 °C with a heating rate of 20 °C/min. The mechanical properties of the compounds were measured by Universal Testing Machine (SANS CMT6104, China) with a stretching speed of 5 mm/min. All the films were cut into standard strips (10 mm × 150 mm), and the reported data is the average value obtained by testing five samples. The Dielectric properties of the polymeric compounds were tested on a precision LCR meter (TH 2819A, Tonghui, China). The films were cut into regular pieces (10 mm × 10 mm) and both sides were coated with the conductive silver paste to form a plate capacitor.

3. Results and Discussion

3.1. Microstructure and Morphology of PANI-f-BT

In this work, enhanced dielectric and mechanical properties of PEN are obtained by hot-stretching. PANI-f-BT nanoparticles with different polymer content are fabricated via in-situ aniline polymerization technology by controlling the content of aniline, and then incorporated into PEN matrix by ultrasonication achieving the first dispersion of the fillers in PEN matrix. In addition, the PEN-based nanocomposites are uniaxial hot-stretched by 50% and 100% at 280 °C, obtaining the second dispersion of the fillers. Resulting from the excellent compatibility between PANI-f-BT and PEN and the second dispersion of PANI-f-BT in PEN by hot-stretching, the obtained composites demonstrate enhanced crystallinity, mechanical and dielectric properties.

In order to characterize the microstructure of the functionalized nanoparticles, BT and PANI-f-BT are characterized by TEM, as shown in Figure 2a,b. It can be seen from Figure 2a that BT shows a smooth surface without distinct interface at its periphery. In comparison, Figure 2b shows different interfaces at the edges of PANI-f-BT. A layer of polymer corona is uniformly coated around the BT, indicating that the surface of BT is wrapped with a compact polyaniline layer [31]. The chemical structure of PANI-f-BT is characterized by FTIR (Figure 2c). It is clear that the strong band at 567 cm^{-1} spectra of pristine BT and PANI-f-BT is corresponding to the vibration of Ti-O [32]. Besides, obvious absorption bands at 1586 and 1497 cm^{-1} can be found on the FTIR spectrum of PANI-f-BT, which belong to the skeleton vibration of benzene rings from polyaniline [33]. Compared to BT, the additional characteristic absorption peaks at 3428 and 1189 cm^{-1} from the spectrum of PANI-f-BT are the absorption peaks of N-H and Ar-N vibration, which proves that polyaniline exists in the PANI-f-BT [33].

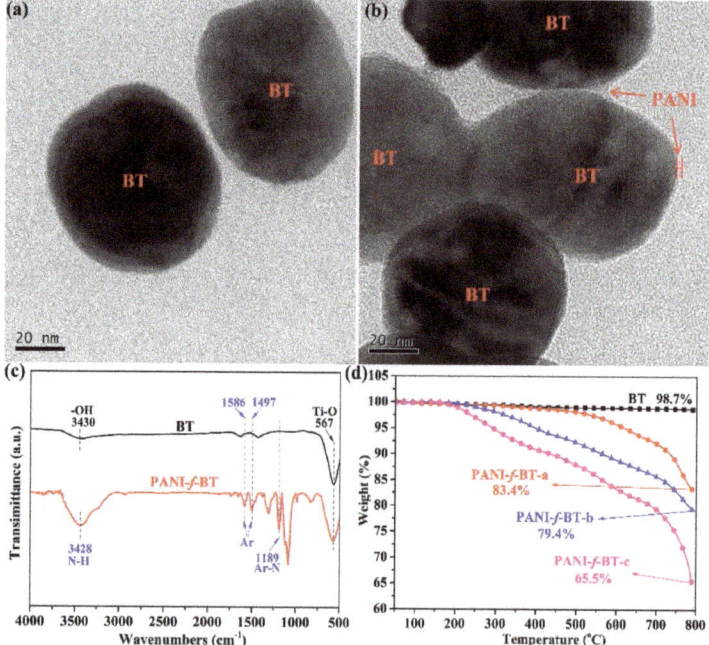

Figure 2. Transmission electron microscopy (TEM) images of (**a**) barium titanate (BT) and (**b**) polyaniline functionalized barium titanate (PANI-f-BT-b); (**c**) Fourier transform infrared spectroscopy (FTIR) spectrum of BT and PANI-f-BT-b; (**d**) Thermogravimetric analysis (TGA) curves of the nanofillers.

In addition, the chemical composition of the obtained PANI-f-BT is further characterized by XPS measurement (Figure 3). As shown in Figure 3a, it is obvious that the Ba3d, Ba4d, Ba4p, Ti2p and O1s peaks can be observed on the full scanned XPS spectrum of PANI-f-BT indicating the existence of BT. In addition, two peaks at 286 eV and 402 eV on the spectrum of PANI-f-BT are corresponding to C1s and N1s from polyaniline. Ba3d spectrum of PANI-f-BT presents two peaks at 779.6 and 794.9 eV which belong to Ba3d$_{5/2}$ and Ba3d$_{3/2}$, respectively (Figure 3b) [34]. The Ti2p spectrum of PANI-f-BT also shows two peaks at 457.9 eV (Ti2p$_{3/2}$) and 463.9 eV (Ti2p$_{1/2}$) (Figure 3c) [32]. What is more, the N1s spectrum of PANI-f-BT can be differentiated into three peaks: 398.1 eV (–N=), 398.9 eV (–NH–) and 400.1 eV (N$^+$), respectively [35] as shown in Figure 3d.

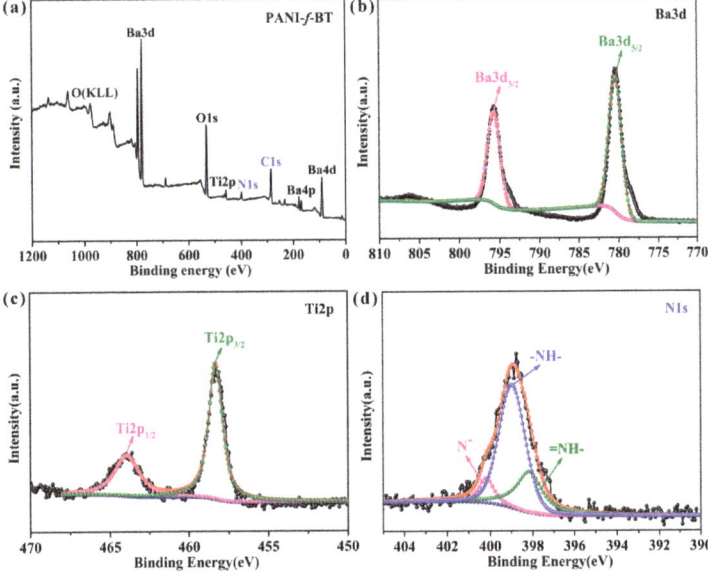

Figure 3. The X-ray photoelectron spectroscopic instrument (XPS) spectrum of PANI-f-BT-b: (a) full scanned spectrum; (b) Ba3d; (c) Ti2p; (d) N1s.

Beside the characterization of PANI-f-BT, the contents of polyaniline in PANI-f-BT are determined by TGA test. As shown in Figure 2d, it is clear that BT nanoparticle does not demonstrate any weight loss, even when heated to 800 °C. In comparison, the residue of PANI-f-BT-a, PANI-f-BT-b and PANI-f-BT-c is 83.4%, 79.4% and 65.5% at 800 °C, respectively. The decrement of the residue indicates the existence of PANI in PANI-f-BT. Simultaneously, the less residue of the PANI-f-BT nanoparticles, the more PANI in the PANI-f-BT nanoparticles. Therefore, all these results including TEM, FTIR, XPS and TGA suggest that PANI is successfully grown on the surface of BT after the in-situ polymerization procedure.

3.2. Morphology of PEN/PANI-f-BT Composites

After the fabrication and characterization of the PANI-f-BT nanoparticles, they are introduced into the PEN matrix to prepare the PEN/PANI-f-BT composites. The miscibility between PANI-f-BT and PEN matrix, which is one of the most important factors affecting the properties of the nanocomposites, is firstly investigated by SEM measurement. Figure 4a shows the microstructures of PEN/BT nanocomposites, which exhibit a poor interfacial adhesion. In addition, a large number of BT nanoparticles are observed at the cross-section of the PEN matrix with serious spherical agglomeration. This phenomenon is mainly due to the high content of BT as well as the poor miscibility between PEN and BT [36]. Contrarily, the PANI-*f*-BT nanoparticles reveal a homogeneous dispersion in PEN matrix without agglomeration (Figure 4b). This result is mainly caused by that the PANI layer on PANI-*f*-BT nanoparticles which improves compatibility with PEN. Therefore, the modification of BT with PANI can effectively improve the compatibility between BT and PEN [36].

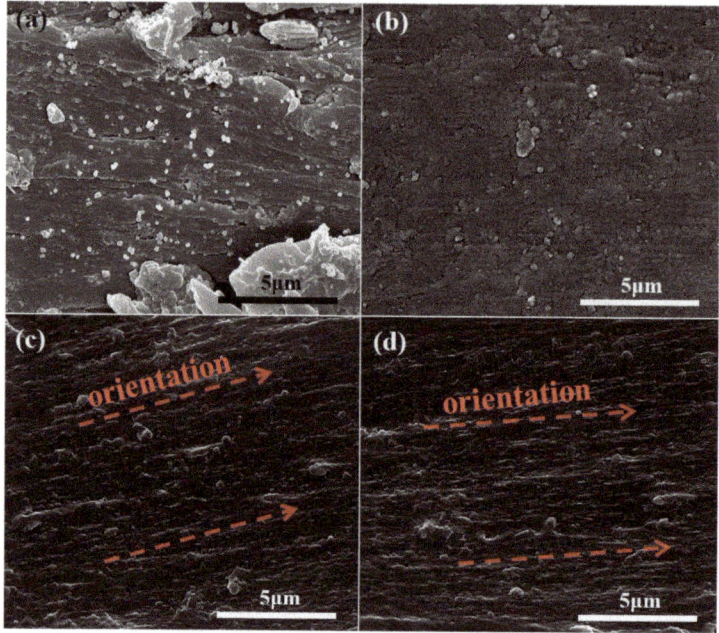

Figure 4. Cross-sectional scanning electron microscopy (SEM) images of (**a**) PEN/BT; (**b**) PEN/PANI-*f*-BT-b; (**c**) PEN/PANI-*f*-BT-b hot-stretched by 50%; (**d**) PEN/PANI-*f*-BT-b hot-stretched by 100%.

To achieve the secondary dispersion of PANI-*f*-BT in PEN, the PEN/PANI-*f*-BT composites are hot-stretched in a home-made oven. Figure 4c,d are the cross-sectional SEM images of PEN/PANI-*f*-BT-b after hot-stretching at a stretching ratio of 50% and 100%, from which obvious orientation of the sample caused by the directional arrangement of polymer molecular chains under the action of external forces is observed. After hot-stretching, the PANI-*f*-BT nanoparticles are secondarily dispersed in the PEN matrix along the orientation direction. A schematic model of the evolution process of PANI-*f*-BT in the polymer matrix during uniaxial stretching is presented in Figure 5. Before the hot-stretching, the spherical PANI-*f*-BT is isotropically dispersed in the PEN matrix. With the commencement of the hot-stretching, the PANI-*f*-BT nanoparticles rearrange along the orientation direction of the stretching. Finally, the enhanced dispersion of PANI-*f*-BT in PEN matrix is obtained after the hot-stretching. Combining the improved compatibility between PANI-*f*-BT and PEN and the secondarily dispersion of PANI-*f*-BT in PEN matrix induced by hot-stretching, enhanced properties of the PEN/PANI-*f*-BT composites can be imaged.

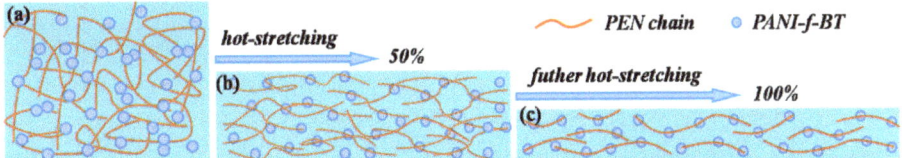

Figure 5. The theoretical model of the evolution process of inner network during uniaxial stretching: (a) the original composite film, the composite film hot-stretched by (b) 50% and (c) 100%.

3.3. Thermal Properties and Crystallization of PEN/PANI-f-BT Composites

As a crystalline polymer, the crystallization behavior of PEN is another important factor affecting the properties of the nanocomposites. The crystallization behavior of PEN/PANI-f-BT composites is studied by DSC and XRD. Figure 6 shows the DSC curves of the PEN/BT (Figure 6a) and PEN/PANI-f-BT (Figure 6b–d) nanocomposites before and after hot-stretching. It clearly shows that the melting peaks of all nanocomposites are not observed before hot-stretching, while they appear after hot-stretching. The melting enthalpy (ΔH_m) of PEN/PANI-f-BT-b at the stretching ratios of 0%, 50% and 100% is 0, 6.4 and 9.1 J/g, respectively (Table 1). With the increase of stretching ratios from 50% to 100%, the ΔH_m of these composites increases gradually, meaning that the crystallinity of the nanocomposites increases [25,37]. This would be due to the rearrangement of PEN molecular chains during the hot-stretching resulting the transitions of the samples from amorphous regions to crystalline regions and from irregular crystals to regular crystals [28]. Besides, the half peak width of the melting peaks also shows the crystalline information of the samples. Generally, the perfect crystals exhibit smaller half peak width than the imperfect crystals. The half peak width is 9.34, 6.55, 5.68 and 5.71 °C for PEN/BT, PEN/PANI-f-BT-a, PEN/PANI-f-BT-b and PEN/PANI-f-BT-c respectively, at the stretching ratio of 100%. The widest half peak width of PEN/BT at 100% stretching ratio is mainly owing to the worst compatibility and dispersion of BT in the PEN matrix as confirmed by the SEM observation (Figure 4a). In addition, DSC curves also demonstrate that the glass transition (T_g) of the all nanocomposites increases slightly during the hot-stretching process. For instance, the T_g of PEN/PANI-f-BT-b increases from 218.6 °C to 221.7 °C as stretching ratios increases from 0% to 100% (Figure 6c and Table 1). This result can also be explained by the arrangement of the PEN chains after hot-stretching, leading to the harder movement of them. Moreover, the increase of crystallinity further limits the movement of the chain segments [38]. Furthermore, the DSC curves for both cooling and heating are shown in Figure S1, as can be seen from the figure that during the first cooling scan and the second heating scan, the T_gs obtained from the curves are lower than the one obtained from the first heating scan due to the supercooling effect [39]. What is more, the melting point disappears during the first cooling scan and the second heating scan due to the semi-crystalline property of the polymers and slow crystalline rate of the polymers [40]. In addition, we also characterize the crystal structure state of the samples after treatment at 200 °C, which is shown in Figure S2. It can be clearly seen from the figure that the PEN/PANI-f-BT-b nanocomposites with a 50% and 100% stretching ratio still show an obvious melting peak in the second heating curve after treating them at 200 °C for 10 min. It is indicated that the crystal structure of the PEN/PANI-f-BT-b nanocomposites is thermodynamic stable when used at 200 °C. More importantly, as shown in Figure S2d, it is clear that the melting peak is also maintained in the second heating curve of the PEN/PANI-f-BT-b nanocomposites with a 100% stretching ratio after treating at 300 °C for 10 min. Therefore, all these results confirm that the crystal structure of the hot-stretched PEN/PANI-f-BT nanocomposite films can still maintain good thermal stability during the practical application (<200 °C).

Table 1. Thermal and mechanical properties of nanocomposites at different stretching ratios.

Samples	T_g (°C)	ΔH_m (J/g)	Tensile Strength (MPa)	Tensile Modulus (GPa)
PEN/BT 0% 0%	218.8	-	67.2 ± 3.9	1.93 ± 0.10
PEN/BT 50%	219.5	4.2	90.3 ± 5.7	2.25 ± 0.11
PEN/BT 100%	221.9	7.5	119.7 ± 6.3	2.75 ± 0.15
PEN/PANI-f-BT-a 0% 0%	218.5	-	78.5 ± 3.8	1.99 ± 0.09
PEN/PANI-f-BT-a 50%	219.4	5.9	108.6 ± 4.2	2.41 ± 0.13
PEN/PANI-f-BT-a 100%	221.7	8.4	141.2 ± 5.9	2.94 ± 0.12
PEN/PANI-f-BT-b 0% 0%	218.6	-	83.8 ± 3.6	2.11 ± 0.08
PEN/PANI-f-BT-b 50%	219.4	6.4	126.4 ± 4.6	2.77 ± 0.10
PEN/PANI-f-BT-b 100%	221.7	9.1	161.1 ± 5.3	3.37 ± 0.14
PEN/PANI-f-BT-c 0% 0%	218.1	-	79.6 ± 3.9	2.01 ± 0.10
PEN/PANI-f-BT-c 50% 50%	219.0	4.7	109.1 ± 5.1	2.57 ± 0.11
PEN/PANI-f-BT-c 100% 100%	221.4	7.6	144.3 ± 5.7	3.05 ± 0.13

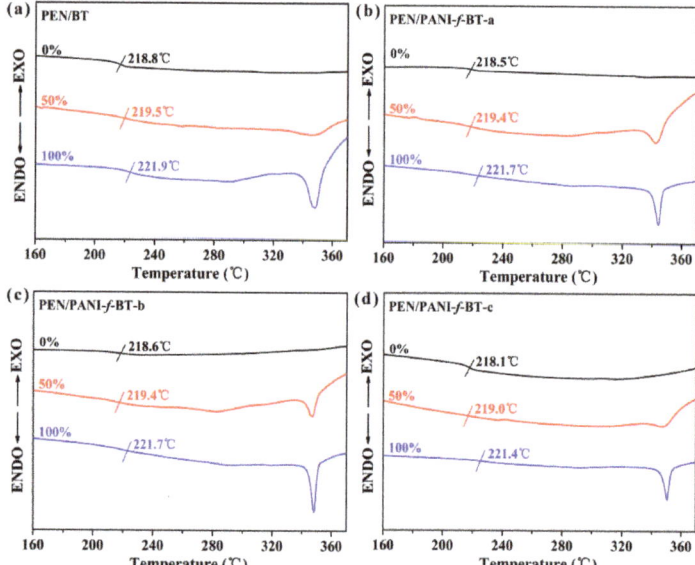

Figure 6. The differential scanning calorimetry (DSC) curves of nanocomposites with different stretching ratios: (a) PEN/BT; (b) PEN/PANI-f-BT-a; (c) PEN/PANI-f-BT-b; (d) PEN/PANI-f-BT-c.

XRD is usually employed to study the crystals and crystallinity of samples. Figure 7 typically shows the XRD patterns of PEN/PANI-f-BT-b at different stretching ratios. As shown in the figure, the diffraction peaks at around 32°, 38° and 45° are observed from all three samples which are coming from the diffractions of (110), (111) and (200) of BT. As for PEN, no crystalline peak is observed before hot-stretching. In comparison, two diffraction peaks at 17° and 23° which are coming from the diffractions of (111) and (112) of PEN are observed after the hot-stretching. The crystallinities of PEN/PANI-f-BT-b with the stretching ratios from 0% to 100% are 0%, 11.2% and 16.4%, which are calculated from wide-angle XRD spectrogram by using Jade 6 software [41].

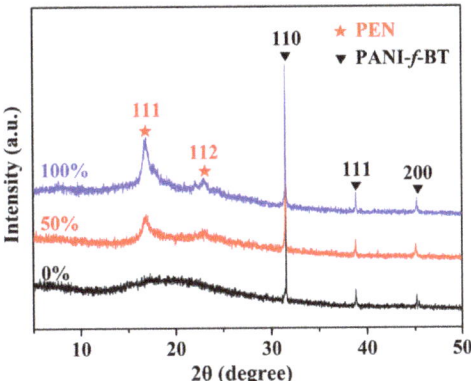

Figure 7. The wide-angle XRD patterns of PEN/PANI-*f*-BT-b at different stretching ratios.

3.4. Mechanical Properties of PEN/PANI-f-BT Composites

Resulting from the improved compatibility and crystalline of the composite, their enhanced properties are further investigated. Tensile strength and tensile modulus, two most important mechanical properties of the high-performance engineering plastics, are typically studied. The tensile properties of the PEN-based nanocomposites at different stretching ratios are shown in Figure 8. The tensile strength of PEN/BT nanocomposites is 67.2, 90.3 and 119.7 MPa, at the stretching ratio 0%, 50% and 100%, respectively. As expected, the tensile strengths of all nanocomposites demonstrate a significant increase after modifying the surface of BT and hot-stretching, which are shown in Figure 8a. Moreover, tensile strength of PEN/PANI-*f*-BT-b nanocomposite is 83.8 MPa without stretching. What is more, it increases to 161.1 MPa when the stretching ratio is 100%, with an increment of 90%. The detailed mechanical data of the nanocomposites are listed in Table 1. It is obvious that the tensile strengths of PEN/PANI-*f*-BT nanocomposites are higher than that of PEN/BT nanocomposites, which are attributed to the better dispersion and compatibility between PEN and PANI-*f*-BT. In addition, it can be concluded that a substantial increase in mechanical property of the nanocomposites is contributed by the high orientation of the molecular chains and the newly formed oriented crystals after hot-stretching process [25,41]. The tensile modulus of the nanocomposites exhibits a similar tendency as that of tensile strength, which results from the same reasons as mentioned above [25].

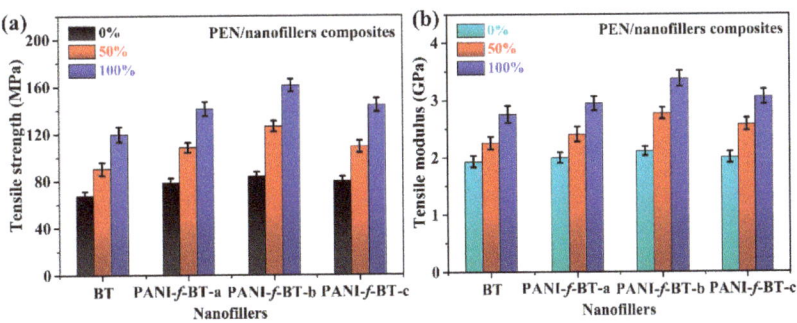

Figure 8. The mechanical properties of nanocomposite films at different stretching ratios: (a) tensile strength and (b) tensile modulus.

3.5. Dielectric Properties of PEN/PANI-f-BT Composites

PEN has shown prospective application in film capacitors and other electronic devices. However, the permittivity of PEN is relatively low (~4.0 at 1 kHz) which failed to meet the high permittivity requirement for film capacitors. Herein, PANI-f-BT nanoparticles are used as a filler to mix with PEN for preparing PEN/PANI-f-BT nanocomposites. The dependence between dielectric properties of the obtained nanocomposites and varying frequency (100 Hz to 1 MHz) is shown in Figure 9a. Compared with PEN/BT, the dielectric constants of all PEN/PANI-f-BT nanocomposites are slightly lower than those of PEN/BT. The decreasing of dielectric constant of PEN/PANI-f-BT nanocomposites is caused by the PANI layer on the surface of BT nanoparticles which will hinder the charge movements from BT to PEN matrix. This usually leads to a decrease in interfacial polarization of the system [42]. In addition, it is clear that the permittivity of PEN/PANI-f-BT nanocomposites is more stable than that of PEN/BT nanocomposite with the change of frequency (100 Hz to 1 MHz). This is due to that the introduction of organic shell layer which enhances the compatibility between nanofillers and PEN matrix which depresses the Maxwell-Wagner polarization [30]. The dielectric loss of the PEN/PANI-f-BT nanocomposites is shown in Figure 9b. Although the content of nanofillers is up to 40 wt%, the dielectric loss of PEN/PANI-f-BT nanocomposites is still below 0.028 (1 kHz). This phenomenon also resulted from the improved compatibility between PANI-f-BT and PEN [30]. What is more, the dielectric loss of the PEN/PANI-f-BT nanocomposites demonstrates a similar trend with the changing of frequency and fillers as that of their permittivity. Furthermore, the electrical conductivity of the studied samples is shown in Figure S3. It can be seen from the figure that the electrical conductivity of PEN/PANI-f-BT nanocomposites is almost the same (10^{-10} S cm^{-1}) at 100 Hz (Figure S3a), which indicates that all the nanocomposite films are insulators.

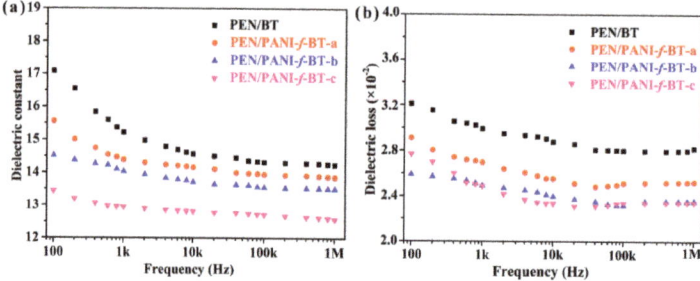

Figure 9. (a) Dielectric constant and (b) dielectric loss of the nanocomposites with the changing of frequency.

Compared with the commonly used biaxially oriented polypropylene (BOPP) and Poly(vinylidene fluoride) (PVDF), PEN, which is a kind of special thermoplastic engineering material, has demonstrated its potential application as high temperature dielectrics. Therefore, the permittivity of all samples at different temperature is further researched. As shown in Figure 10a, the dielectric constant of nanocomposites is measured at 1 kHz in the range of 25 to 250 °C. It is clear that the dielectric constant of all nanocomposites is stable before their T_g, and it increases hastily and obviously when the temperature exceeds their T_g. This is due to the molecular chains of the PEN that are frozen at a temperature lower than its T_g. However, as the temperature increases, the macromolecular chains are thawed which strengthens the mobility of electrons and enhances polarization in the system [43]. As a result, the T_g of the polymeric dielectrics can be obtained from their permittivity-temperature curves. As shown in Figure 10a, the T_g of PEN/PANI-f-BT-c obtained from the permittivity-temperature curve is 218 °C which is the same as that obtained from its DSC curve. The T_g of the other composites is also around 218 °C. The high T_g of the composites ensures the potential application of them at high temperature. Beside the qualitative description of the stable permittivity of the composites before their T_g, the quantitative value of dielectric constant with the change of temperature (temperature coefficients of dielectric constant) is calculated according to Equation (1) [44]:

$$\tau_\varepsilon = \frac{\varepsilon_{T2} - \varepsilon_{T1}}{\varepsilon_{T0}(T_2 - T_1)} \tag{1}$$

where τ_ε is the temperature coefficient of dielectric constant, ε_{T0} is the dielectric constant of room temperature, ε_{T1} is the dielectric constant of initial temperature and ε_{T2} is the dielectric constant of the final temperature. T_1 and T_2 are the initial and final temperatures, respectively. According to Equation (1), the calculated results of the temperature coefficients of dielectric constant of all nanocomposites are shown in Figure 10b. The temperature coefficients of dielectric constant of the composites are lower than 5×10^{-4} °C^{-1} within the temperature range from 25 to 100 °C. They are still lower than 3×10^{-3} °C^{-1} even though in the temperature range from 25 to 200 °C, indicating that they are extraordinarily stable, even at temperature up to 200 °C. In addition, it is notable that the temperature coefficients of dielectric constant of the composites decrease with the increase of the PANI content in the PANI-f-BT. This result is also largely due to the introduction of organic shell layer can improve the compatibility between nanoparticles and PEN matrix, which can effectively reduce the charge accumulation and interfacial polarization between inorganic nanoparticles and matrix [30].

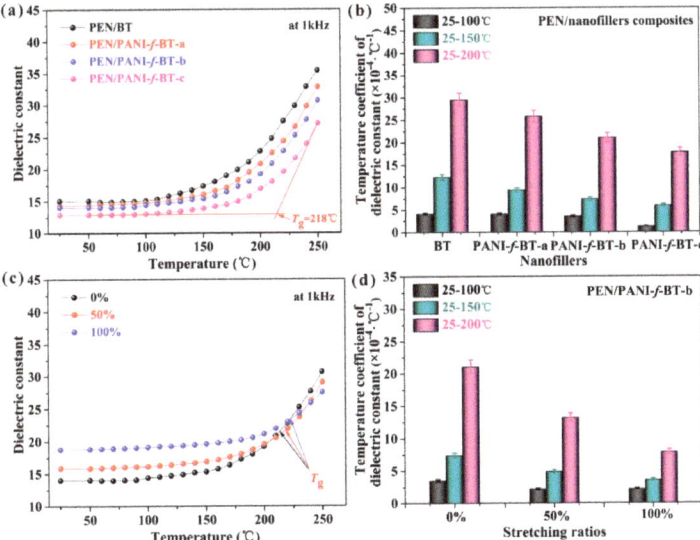

Figure 10. (a) Permittivity-temperature dependences and (b) temperature coefficients of dielectric constant of the nanocomposites; (c) permittivity-temperature dependences and (d) temperature coefficients of dielectric constant of PEN/PANI-f-BT-b at different stretching ratios.

Moreover, to make a clear view of the properties of the PEN/PANI-f-BT nanocomposites, a comparison of dielectric constant, dielectric loss and working temperature at room temperature and 1 kHz of reported polymer-based composites are summarized in Table 2. As can be seen from the table, the dielectric constant and loss of PEN/PANI-f-BT are 14 and 0.025 when the content of PANI-f-BT nanoparticles is 40 wt%. Although the dielectric constant is lower than that of the most widely used PVDF based composites, it is comparable to the P(VDF-HFP)/BT-OPA, PES/BT-CuPc and PAEN/BT@CPAEN system. In addition, the dielectric loss of PEN/PANI-f-BT nanocomposites maintains at a relatively low level and is lower than that of most PVDF based composites. More importantly, the inherent low temperature resistance of PVDF (<120 °C) will limit its application in high temperature environment.

In comparison, in this work, the PEN/PANI-*f*-BT nanocomposites can be used as flexible dielectric films at around 200 °C, which provides a new approach for preparing high-temperature-resistant dielectric films.

Table 2. Dielectric constant at 1 kHz and 25 °C, dielectric loss at 1 kHz and 25 °C, working temperature of typical polymer-based composites.

Samples	Content	Dielectric Constant (1 kHz, 25 °C)	Dielectric Loss (1 kHz, 25 °C)	Working Temperature (°C)	Ref.
PVDF/BT	60 vol%	95	~0.04	<120	[45]
PVDF/BT-PDOPA	50 vol%	56.8	0.04	<120	[46]
PVDF/BT-TDPA	40 vol%	48	0.03	<120	[47]
PVDF/BT-SiO$_2$	10 vol%	14.7	0.02	<120	[48]
hydantoin/BT-P(VDF-HFP)	50 vol%	48.9	0.06	120	[49]
P(VDF-HFP)/BT-OPA	30 vol%	~15	0.08	120	[50]
PES/BT-CuPc	40 vol%	~17	~0.12	~160	[51]
PAEN/BT@CPAEN	40 wt%	13	0.023	~180	[30]
PEN/PANI-*f*-BT	40 wt%	14	0.025	~200	This work

The dielectric properties of the PEN/PANI-*f*-BT nanocomposites are further improved by hot-stretching. As shown in Figure 10c, the dielectric constant of PEN/PANI-*f*-BT-b is 14.0 at 1 kHz without stretching (at 25 °C). After hot-stretching, its dielectric constant increases to 15.9 and 18.7 when the stretching ratio is 50% and 100% respectively (at 25 °C). It is well-known that the micro-capacitor networks are often formed in the nanocomposites [28]. During hot-stretching, the disordered nanoparticles are realigned to form the oriented micro-capacitors in the polymer matrix along the orientation direction (Figure 5), contributing to the enhancement of the dielectric constant of nanocomposites. In addition, the electrical conductivity of PEN/PANI-*f*-BT-b nanocomposites presents a slight increase with the increase of stretching ratio (Figure S3b). This is a result of the second dispersion of PANI-*f*-BT-b in the system. What is more, the permittivity of the PEN/PANI-*f*-BT-b nanocomposites after hot-stretching at different temperatures is also investigated in this work, as shown in Figure 10c. It can be seen that the T_g of PEN/PANI-*f*-BT-b obtained from the permittivity-temperature curve increases to 219 °C and 221°C at the 50% and 100% respectively, indicating that the composites can be used at higher temperature after hot-stretching. It is well-known that the service temperature of BOPP is higher than PP due to the stretching of the sample. Furthermore, the temperature coefficients of dielectric constant of PEN/PANI-*f*-BT-b nanocomposites at different stretching ratios are shown in Figure 10d. As can be seen, the temperature coefficients of dielectric constant of PEN/PANI-*f*-BT-b nanocomposites exhibit a gradual downward trend. More importantly, the temperature coefficients of dielectric constant of PEN/PANI-*f*-BT-b are lower than 5×10^{-4} °C, which are very important for its application at high temperature. This is because the high orientation of the molecular chain and the increase of crystallinity limit the movement of the chain segments, which is consistent with the conclusion of the thermal properties [52]. The results reveal that the modification of the filler and hot-stretching method can effectively increase the dielectric constant and the stability of the nanocomposites at different temperatures, which presents great potential for it to be used as a high performance dielectric film in harsh environments.

4. Conclusions

In conclusion, enhanced dielectric and mechanical properties of polyarylene ether nitrile are obtained through secondary dispersion of polyaniline functionalized barium titanate (PANI-*f*-BT) by hot-stretching. PANI-*f*-BT nanoparticles with different PANI contents are prepared via in-situ aniline polymerization technology, and then characterized by TEM, XPS, FTIR and TGA. The results confirm that the polymer layers have uniformly enclosed on the surface of BaTiO$_3$ nanoparticles. The obtained PANI-*f*-BT nanoparticles are used as functional fillers to compound with PEN for preparing the PEN/PANI-*f*-BT nanocomposites. In addition, these PEN-based nanocomposites are uniaxial hot-stretched by 50% and 100% at 280 °C to obtain the oriented nanocomposite films. The results show that the PANI-*f*-BT nanoparticles present well compatibility and dispersion in the PEN matrix, and the hot-stretching can achieve a second dispersion of the PANI-*f*-BT nanoparticles in PEN, which can effectively enhance the comprehensive properties of nanocomposites like mechanical properties, crystallinity, dielectric constant and so on. When the stretching ratios increase from 0% to 100%, the tensile strengths of PEN/PANI-*f*-BT-b nanocomposite film increase from 83.8 to 161.1 MPa, and the crystallinities increase from 0% to 16.4%. Most importantly, the permittivity temperature dependences of nanocomposites after hot-stretching are more stable than that of original nanocomposites. The excellent performances of the stretched composites indicate that these samples present a great potential to be used as a high-performance dielectric film in high-temperature environments.

Supplementary Materials: The following are available online at http://www.mdpi.com/2079-4991/9/7/1006/s1, Figure S1: The DSC curves of PEN/PANI-*f*-BT nanocomposite films with different stretching ratios. Figure S2: The DSC curves of PEN/PANI-*f*-BT-b nanocomposite films with different stretching ratios. Figure S3: The electrical conductivity of PEN/PANI-*f*-BT nanocomposites.

Author Contributions: Y.Y. and X.L. conceived and designed the experiments; Y.Y. performed the experiments; L.T. (Ling Tu), Y.W. and L.T. (Lifen Tong) analyzed the data; Y.Y. wrote the paper. R.W. and X.L. edited and revised manuscript. All the authors approved the final version of the manuscript.

Funding: This research received no external funding.

Acknowledgments: The authors wish to give thanks for the financial supports from the National Natural Science Foundation of China (No. 51773028 and 51603029), China Postdoctoral Science Foundation (2017M623001) and National Postdoctoral Program for Innovative Talents (BX201700044).

Conflicts of Interest: The authors declare no conflict of interest.

References

1. Dang, Z.M.; Yuan, J.K.; Yao, S.H.; Liao, R.J. Flexible nanodielectric materials with high permittivity for power energy storage. *Adv. Mater.* **2013**, *25*, 6334–6365. [CrossRef]
2. Chu, B.J.; Zhou, X.; Ren, K.L.; Neese, B.; Lin, M.R.; Wang, Q.; Bauer, F.; Zhang, Q.M. A dielectric polymer with high electric energy density and fast discharge speed. *Science* **2006**, *313*, 334–336. [CrossRef] [PubMed]
3. Hao, Y.N.; Wang, X.H.; O'Brien, S.; Lombardi, J.; Li, L.T. Flexible BaTiO$_3$/PVDF gradated multilayer nanocomposite film with enhanced dielectric strength and high energy density. *J. Mater. Chem. C* **2015**, *3*, 9740–9747. [CrossRef]
4. Kim, P.; Jones, S.C.; Hotchkiss, P.J.; Haddock, J.N.; Kippelen, B.; Marder, S.R.; Perry, J.W. Phosphonic acid-modified barium titanate polymer nanocomposites with high permittivity and dielectric strength. *Adv Mater.* **2007**, *19*, 1001–1005. [CrossRef]
5. Maier, G. Low dielectric constant polymers for microelectronics. *Prog. Polym. Sci.* **2001**, *26*, 3–65. [CrossRef]
6. Qi, L.; Lee, B.I.; Chen, S.; Samuels, W.D.; Exarhos, G.J. High-dielectric-constant silver-epoxy composites as embedded dielectrics. *Adv. Mater.* **2005**, *17*, 1777–1781. [CrossRef]
7. Bi, K.; Bi, M.; Hao, Y.; Luo, W.; Cai, Z.; Wang, X.; Huang, Y. Ultrafine core-shell BaTiO$_3$@SiO$_2$ structures for nanocomposite capacitors with high energy density. *Nano Energy* **2018**, *51*, 513–523. [CrossRef]
8. Maliakal, A.; Katz, H.; Cotts, P.M.; Subramoney, S.; Mirau, P. Inorganic oxide core, polymer shell nanocomposite as a high-*k* gate dielectric for flexible electronics applications. *J. Am. Chem. Soc.* **2005**, *127*, 14655–14662. [CrossRef]
9. You, Y.; Zhan, C.H.; Tu, L.; Wang, Y.J.; Hu, W.B.; Wei, R.B.; Liu, X.B. Polyarylene ether nitrile-based high-*k* composites for dielectric applications. *Int. J. Polym. Sci.* **2018**, 5161908. [CrossRef]
10. Zotti, A.; Zuppolini, S.; Borriello, A.; Zarrelli, M. Thermal properties and fracture toughness of epoxy nanocomposites loaded with hyperbranched-polymers-based core/shell nanoparticles. *Nanomaterials* **2019**, *9*, 418. [CrossRef]
11. Li, J.; Seok, S.I.; Chu, B.; Dogan, F.; Zhang, Q.; Wang, Q. Nanocomposites of ferroelectric polymers with TiO$_2$ nanoparticles exhibiting significantly enhanced electrical energy density. *Adv. Mater.* **2009**, *21*, 217–221. [CrossRef]

12. Wang, Y.J.; Tong, L.F.; You, Y.; Tu, L.; Zhou, M.R.; Liu, X.B. Polyethylenimine assisted bio-inspired surface functionalization of hexagonal boron nitride for enhancing the crystallization and the properties of poly(arylene ether nitrile). *Nanomaterials* **2019**, *9*, 760. [CrossRef] [PubMed]
13. Wei, R.B.; Wang, J.L.; Zhang, H.X.; Han, W.H.; Liu, X.B. Crosslinked polyarylene ether nitrile interpenetrating with zinc ion bridged graphene sheet and carbon nanotube network. *Polymers* **2017**, *9*, 342. [CrossRef] [PubMed]
14. Wang, Z.D.; Liu, J.Y.; Cheng, Y.H.; Chen, S.Y.; Yang, M.M.; Huang, J.L.; Wang, H.K.; Wu, G.L.; Wu, H.J. Alignment of boron nitride nanofibers in epoxy composite films for thermal conductivity and dielectric breakdown strength improvement. *Nanomaterials* **2018**, *8*, 242. [CrossRef] [PubMed]
15. Zhi, C.; Bando, Y.; Terao, T.; Tang, C.; Kuwahara, H.; Golberg, D. Boron nanotube–polymer composites: Towards thermoconductive, electrically insulating polymeric composites with boron nitride nanotubes as fillers. *Adv. Funct. Mater.* **2009**, *19*, 1857–1862. [CrossRef]
16. Xu, M.Z.; Lei, Y.X.; Ren, D.X.; Chen, S.J.; Chen, L.; Liu, X.B. Synergistic effects of functional CNTs and h-BN on enhanced thermal conductivity of epoxy/cyanate matrix composites. *Nanomaterials* **2018**, *8*, 997. [CrossRef] [PubMed]
17. Yang, M.; Hu, C.; Zhao, H.; Haghi-Ashtiani, P.; He, D.; Yang, Y.; Yuan, J.; Bai, J. Core@double-shells nanowires strategy for simultaneously improving dielectric constants and suppressing losses of poly (vinylidene fluoride) nanocomposites. *Carbon* **2018**, *132*, 152–156. [CrossRef]
18. Ma, J.; Azhar, U.; Zong, C.; Zhang, Y.; Xu, A.; Zhai, C.; Zhang, L.; Zhang, S. Core-shell structured PVDF@BT nanoparticles for dielectric materials: A novel composite to prove the dependence of dielectric properties on ferroelectric shell. *Mater. Design* **2019**, *164*, 107556. [CrossRef]
19. Song, Y.; Shen, Y.; Liu, H.; Lin, Y.; Li, M.; Nan, C.W. Improving the dielectric constants and breakdown strength of polymer composites: Effects of the shape of the BaTiO$_3$ nanoinclusions, surface modification and polymer matrix. *J. Mater. Chem.* **2012**, *22*, 16491–16498. [CrossRef]
20. Niu, Y.; Bai, Y.; Yu, K.; Wang, Y.; Xiang, F.; Wang, H. Effect of the modifier structure on the performance of barium titanate/poly(vinylidene fluoride) nanocomposites for energy storage applications. *ACS Appl. Mater. Interfaces* **2015**, *7*, 24168–24176. [CrossRef]
21. Wei, R.; Li, K.; Ma, J.; Zhang, H.; Liu, X. Improving dielectric properties of polyarylene ether nitrile with conducting polyaniline. *J. Mater. Sci. Mater. Electron.* **2016**, *27*, 9565–9571. [CrossRef]
22. Yu, S.; Qin, F.; Wang, G. Improving the dielectric properties of poly (vinylidene fluoride) composites by using poly (vinyl pyrrolidone)-encapsulated polyaniline nanorods. *J. Mater. Chem. C* **2016**, *4*, 1504–1510. [CrossRef]
23. Zhang, X.; He, Q.L.; Gu, H.B.; Wei, S.Y.; Guo, Z.H. Polyaniline stabilized barium titanate nanoparticles reinforced epoxy nanocomposites with high dielectric permittivity and reduced flammability. *J. Mater. Chem. C* **2013**, *1*, 2886–2899. [CrossRef]
24. Li, Z.M.; Yang, M.B.; Lu, A.; Feng, J.M.; Huang, R. Tensile properties of poly(ethylene terephthalate) and polyethylene in-situ microfiber reinforced composite formed via slit die extrusion and hot-stretching. *Mater. Lett.* **2002**, *56*, 756–762. [CrossRef]
25. You, Y.; Huang, X.; Pu, Z.; Jia, K.; Liu, X. Enhanced crystallinity, mechanical and dielectric properties of biphenyl polyarylene ether nitriles by unidirectional hot-stretching. *J. Polym. Res.* **2015**, *22*, 221. [CrossRef]
26. Li, L.; Zhou, T.; Liu, J.Z.; Ran, Q.P.; Ye, G.D.; Zhang, J.H.; Liu, P.Q.; Zhang, A.; Yang, Z.Q.; Xu, D.G.; et al. Formation of a large-scale shish-kebab structure of polyoxymethylene in the melt spinning and the crystalline morphology evolution after hot stretching. *Polym. Adv. Technol.* **2015**, *26*, 77–84. [CrossRef]
27. Tian, Y.; Zhu, C.Z.; Gong, J.H.; Yang, S.L.; Ma, J.H.; Xu, J. Lamellae break induced formation of shish-kebab during hot stretching of ultra-high molecular weight polyethylene precursor fibers investigated by in situ small angle X-ray scattering. *Polymer* **2014**, *55*, 4299–4306. [CrossRef]
28. You, Y.; Du, X.; Mao, H.; Tang, X.; Wei, R.; Liu, X. Synergistic enhancement of mechanical, crystalline and dielectric properties of polyarylene ether nitrile-based nanocomposites by unidirectional hot stretching-quenching. *Polym. Int.* **2017**, *66*, 1151–1158. [CrossRef]
29. Tian, C.; Du, Y.; Xu, P.; Qiang, R.; Wang, Y.; Ding, D.; Xue, J.; Ma, J.; Zhao, H.; Han, X. Constructing uniform core-shell PPy@PANI composites with tunable shell thickness toward enhancement in microwave absorption. *ACS Appl. Mater. Interfaces* **2015**, *7*, 20090–20099. [CrossRef]

30. Tang, H.L.; Wang, P.; Zheng, P.L.; Liu, X.B. Core-shell structured BaTiO$_3$@polymer hybrid nanofiller for poly(arylene ether nitrile) nanocomposites with enhanced dielectric properties and high thermal stability. *Compos. Sci. Technol.* **2016**, *123*, 134–142. [CrossRef]
31. Kim, J.; Kim, D.; Kim, J.; Kim, Y.; Hui, K.N.; Lee, H. Selective substitution and tetragonality by co-doping of dysprosium and thulium on dielectric properties of barium titanate ceramics. *Electron. Mater. Lett.* **2011**, *7*, 155–159. [CrossRef]
32. Li, Y.; Li, J.; Gao, X.; Qi, S.; Ma, J.; Zhu, J. Synthesis of stabilized dispersion covalently-jointed SiO$_2$@polyaniline with core-shell structure and anticorrosion performance of its hydrophobic coating for Mg-Li alloy. *Appl. Surf. Sci.* **2018**, *462*, 362–372. [CrossRef]
33. You, Y.; Wang, Y.; Tu, L.; Tong, L.; Wei, R.; Liu, X. Interface modulation of core-shell structured BaTiO$_3$@polyaniline for novel dielectric materials from its nanocomposite with polyarylene ether nitrile. *Polymers* **2018**, *10*, 1378. [CrossRef] [PubMed]
34. Yang, Y.; Wang, X.H.; Sun, C.K.; Li, L.T. Structure study of single crystal BaTiO$_3$ nanotube arrays produced by the hydrothermal method. *Nanotechnology* **2009**, *20*, 055709. [CrossRef] [PubMed]
35. Qaiser, A.A.; Hyland, M.M.; Patterson, D.A. Surface and charge transport characterization of polyaniline-cellulose acetate composite membranes. *J. Phys. Chem. B* **2011**, *115*, 1652–1661. [CrossRef] [PubMed]
36. Wei, R.; Yang, R.; Xiong, Z.; Xiao, Q.; Li, K.; Liu, X. Enhanced dielectric properties of polyarylene ether nitriles filled with core-shell structured PbZrO$_3$ around BaTiO$_3$ nanoparticles. *J. Electron. Mater.* **2018**, *47*, 6177–6184. [CrossRef]
37. You, Y.; Wei, R.; Yang, R.; Yang, W.; Hua, X.; Liu, X. Crystallization behaviors of polyarylene ether nitrile filled in multi-walled carbon nanotubes. *RSC Adv.* **2016**, *6*, 70877–70883. [CrossRef]
38. Tu, L.; You, Y.; Tong, L.; Wang, Y.; Hu, W.; Wei, R.; Liu, X. Crystallinity of poly(arylene ether nitrile) copolymers containing hydroquinone and bisphenol A segments. *J. Appl. Polym. Sci.* **2018**, *135*, 46412. [CrossRef]
39. Mollova, A.; Androsch, R.; Mileva, D.; Schick, C.; Benhamida, A. Effect of supercooling on crystallization of polyamide 11. *Macromolecules* **2013**, *46*, 828–835. [CrossRef]
40. Mao, M.; Das, S.; Turner, S.R. Synthesis and characterization of poly(aryl ether sulfone) copolymers containing terphenyl groups in the backbone. *Polymer* **2007**, *48*, 6241–6245. [CrossRef]
41. Wei, R.; Tu, L.; You, Y.; Zhan, C.; Wang, Y.; Liu, X. Fabrication of crosslinked single-component polyarylene ether nitrile composite with enhanced dielectric properties. *Polymer* **2019**, *161*, 162–169. [CrossRef]
42. Huang, X.; Jiang, P. Core-shell structured high-k polymer nanocomposites for energy storage and dielectric applications. *Adv. Mater.* **2014**, *27*, 546–554. [CrossRef] [PubMed]
43. Li, W.; Elzatahry, A.; Aldhayan, D.; Zhao, D.Y. Core-shell structured titanium dioxide nanomaterials for solar energy utilization. *Chem. Soc. Rev.* **2018**, *47*, 8203–8237. [CrossRef] [PubMed]
44. Yang, R.; Wei, R.; Tong, L.; Jia, K.; Liu, X.; Li, K. Crosslinked polyarylene ether nitrile film as flexible dielectric materials with ultrahigh thermal stability. *Sci. Rep.* **2016**, *6*, 36434. [CrossRef] [PubMed]
45. Prateek; Thakur, V.K.; Gupta, R.K. Recent progress on ferroelectric polymer-based nanocomposites for high energy density capacitors: Synthesis, dielectric properties, and future aspects. *Chem. Rev.* **2016**, *116*, 4260–4317. [CrossRef] [PubMed]
46. Lin, M.F.; Thakur, V.K.; Tan, E.J.; Lee, P.S. Surface functionalization of BaTiO$_3$ nanoparticles and improved electrical properties of BaTiO$_3$/polyvinylidene fluoride composite. *RSC Adv.* **2011**, *1*, 576–578. [CrossRef]
47. Ye, H.J.; Shao, W.Z.; Zhen, L. Tetradecylphosphonic acid modified BaTiO$_3$ nanoparticles and its nanocomposite. *Colloids Surf. A* **2013**, *427*, 19–25. [CrossRef]
48. Yu, K.; Niu, Y.; Bai, Y.; Zhou, Y.; Wang, H. Poly(vinylidene fluoride) polymer based nanocomposites with significantly reduced energy loss by filling with core-shell structured BaTiO$_3$/SiO$_2$ nanoparticles. *Appl. Phys. Lett.* **2013**, *102*, 102903. [CrossRef]
49. Luo, H.; Zhang, D.; Jiang, C.; Yuan, X.; Chen, C.; Zhou, K.C. Improved dielectric properties and energy storage density of poly(vinylidene fluoride-co-hexafluoropropylene) nanocomposite with hydantoin epoxy resin coated BaTiO$_3$. *ACS Appl. Mater. Interfaces* **2015**, *7*, 8061–8069. [CrossRef] [PubMed]
50. Ehrhardt, C.; Fettkenhauer, C.; Glenneberg, J.; Münchgesang, W.; Pientschke, C.; Großmann, T.; Zenkner, M.; Wagner, G.; Leipner, H.S.; Buchsteiner, A.; et al. BaTiO$_3$-P(VDF-HFP) nanocomposite dielectrics-influence of surface modification and dispersion additives. *Mater. Sci. Eng. B* **2013**, *178*, 881–888. [CrossRef]

51. Xu, W.H.; Yang, G.; Jin, L.; Liu, J.; Zhang, Y.H.; Zhang, Z.C.; Jiang, Z.H. High-*k* polymer nanocomposites filled with hyperbranched phthalocyanine-coated BaTiO$_3$ for high-temperature and elevated field applications. *ACS Appl. Mater. Interfaces* **2018**, *10*, 11233–11241. [CrossRef] [PubMed]
52. Song, Z.Y.; Hou, X.X.; Zhang, L.Q.; Wu, S.Z. Enhancing crystallinity and orientation by hot-stretching to improve the mechanical properties of electrospun partially aligned polyacrylonitrile (PAN) nanocomposites. *Materials* **2011**, *4*, 621–632. [CrossRef] [PubMed]

© 2019 by the authors. Licensee MDPI, Basel, Switzerland. This article is an open access article distributed under the terms and conditions of the Creative Commons Attribution (CC BY) license (http://creativecommons.org/licenses/by/4.0/).

Article

Rapid Self-Assembly of Metal/Polymer Nanocomposite Particles as Nanoreactors and Their Kinetic Characterization

Andrew Harrison [1], Tien T. Vuong [1], Michael P. Zeevi [1], Benjamin J. Hittel [1], Sungsool Wi [2] and Christina Tang [1,*]

1. Department of Chemical and Life Sciences Engineering, Virginia Commonwealth University, Richmond, VA 23284-3028, USA; harrisona3@vcu.edu (A.H.); vuongtt@vcu.edu (T.T.V.); zeevimp@vcu.edu (M.P.Z.); hittelbj@vcu.edu (B.J.H.)
2. The National High Magnetic Field Laboratory, Florida State University, Tallahassee, FL 32310, USA; sungsool@magnet.fsu.edu
* Correspondence: ctang2@vcu.edu

Received: 24 January 2019; Accepted: 18 February 2019; Published: 28 February 2019

Abstract: Self-assembled metal nanoparticle-polymer nanocomposite particles as nanoreactors are a promising approach for performing liquid phase reactions using water as a bulk solvent. In this work, we demonstrate rapid, scalable self-assembly of metal nanoparticle catalyst-polymer nanocomposite particles via Flash NanoPrecipitation. The catalyst loading and size of the nanocomposite particles can be tuned independently. Using nanocomposite particles as nanoreactors and the reduction of 4-nitrophenol as a model reaction, we study the fundamental interplay of reaction and diffusion. The induction time is affected by the sequence of reagent addition, time between additions, and reagent concentration. Combined, our experiments indicate the induction time is most influenced by diffusion of sodium borohydride. Following the induction time, scaling analysis and effective diffusivity measured using NMR indicate that the observed reaction rate are reaction- rather than diffusion-limited. Furthermore, the intrinsic kinetics are comparable to ligand-free gold nanoparticles. This result indicates that the polymer microenvironment does not de-activate or block the catalyst active sites.

Keywords: nanoreactor; catalyst confinement; Flash Nanoprecipitation; diffusion

1. Introduction

Self-assembled amphiphilic molecules, both small molecules and macromolecules, that confine catalysts to micelle, vesicle, and Janus particle nanoreactor systems have proven to offer an efficient approach to perform organic reactions using water as a bulk solvent [1–3]. Using Janus particles, catalysts can be incorporated into a portion of the nanocomposite particle and the other portion imparts stability to the system. Asymmetric catalyst loading can facilitate particle motion driven by a chemical reaction [3]. In other nanoreactor systems, surfactant micelles that incorporate organic, metal (homogeneous), and metal nanoparticle catalysts have been used for a wide range of coupling reactions in water [4–6]. Confining catalyzed organic reactions to the nanoreactor environment can be leveraged to speed up various chemical reactions [7,8]. Improved yield and selectivity when compared to traditional organic solvents has been reported [6,9].

Core-shell polymer systems have also been considered. Polymeric micelles have been used for several reactions such as asymmetric aldol reactions catalyzed by L-proline [10], acylation [11], hydroaminomethylation of octane catalyzed by Ru-based nanoparticles [12], etc., with extensive reviews available elsewhere [13]. Another approach has been to immobilize metal nanoparticles within polyelectrolyte-brushes synthesized on a polystyrene core [14]. The polymer microenvironment of these systems can lead to increased local concentrations of reactants, which can accelerate reactions, facilitate reactions of otherwise non-reactive species [9,10,15–17], confer temperature or pH dependent catalytic activity [18], and/or provide specificity based on hydrophobicity [15].

Generally, these promising approaches have involved design and synthesis of amphiphiles, block copolymers, or polyelectrolytes that contain catalyst or ligand for covalent attachment of the catalyst. Additionally, nanoreactor properties, such as catalyst loading and nanoreactor size, are related to the molecular properties of the synthesized material. Thus, varying the nanoreactor properties would require additional syntheses. Approaches to metal nanoparticle catalyst-polymer nanocomposite particle fabrication that would facilitate (1) modular material (off-the-shelf polymer, catalyst) selection, (2) tunable properties (size and catalyst loading), and (3) rapid, scalable production would be beneficial to expanding their potential application.

Flash NanoPrecipitation (FNP) is a rapid, scalable method of polymer self-assembly that may be useful for producing nanoreactors. In Flash NanoPrecipitation, an amphiphilic block copolymer and hydrophobic core material are dissolved in a water miscible organic solvent and rapidly mixed against water using a confined impinging jet mixer. Upon mixing, the rapid decrease in solvent quality causes the hydrophobic core material to precipitate and the block copolymer to micellize directing formation of the overall nanocomposite particle. This particle assembly ends when the hydrophobic block of the block copolymer adsorbs on the precipitating core material preventing further growth, while the hydrophilic block sterically stabilizes the nanoparticle. Given the molecular weight of the block copolymer, dynamic exchange of the block copolymer does not occur [9,19,20], so the resulting structure is kinetically-trapped.

Hydrophobic, inorganic nanoparticles have been incorporated into nanocomposite particles by dispersing the nanoparticles with the dissolved block copolymer and then mixing with confined impinging jets. Upon mixing, colloidal aggregation and block copolymer self-assembly occur due to the decrease in solvent quality. Nanocomposite particle assembly is complete when sufficient hydrophobic blocks of the block copolymer adsorb to the nanoparticle clusters to prevent further colloidal aggregation. For example, Gindy et al. demonstrated fabrication of polymer nanostructures containing colloidal gold using Flash NanoPrecipitation [21]. More recently, Pinkerton et al. encapsulated iron oxide nanoparticles for medical imaging applications [22]. For medical imaging, ~100 nm composite nanostructures with tunable inorganic nanoparticle loading were achieved. These studies suggest that Flash NanoPrecipitation is a suitable method for nanoreactor fabrication. However, the ability to independently tune inorganic nanoparticle loading and nanocomposite particle size has yet to be demonstrated.

Other important considerations when using the nanocomposite particles as nanoreactors are the reaction and diffusion within the system. In small molecule micelle systems that are thermodynamically stable, there is constant molecular exchange between the bulk solvent, and the confined hydrophobic mesophase facilitates reaction [6,9]. In the kinetically-trapped systems produced by Flash Nanoprecipitation, reactants and products reach the catalyst by partitioning from the bulk and diffusing through the nanoreactor structure [17]. The potential mass transfer limitations and the effect of incorporation into the nanocomposite particle on reactivity of the catalyst need to be established.

In this work, we use Flash NanoPrecipitation for rapid and scalable self-assembly of hybrid metal nanoparticle catalyst-polymer nanocomposite nanoreactors. Independently tuning the nanoreactor properties, namely size and gold loading, is investigated. We focus on fundamental understanding of reaction and diffusion using the reduction of 4-nitrophenol as a model reaction. Kinetic and scaling analysis following the induction time are also discussed.

2. Materials and Methods

2.1. Materials

Citrate stabilized 5 nm gold nanoparticles were purchased from Ted Pella. Polystyrene (PS, M_W 800–5000 g/mol) was purchased from Polysciences, Inc. Sodium borohydride and 4-nitrophenol were purchased from Sigma Aldrich (St. Louis, MO, USA). Dodecanethiol (DDT) stabilized 5 nm nanoparticles, tetrahydrofuran (THF), HPLC grade), ethanol (ACS reagent grade), and diethyl ether (ACS reagent grade) were purchased from Fisher Scientific (Fairmont, NJ, USA). Environmental Grade Hydrochloric Acid 30-38% and Environmental Grade Nitric Acid 70% were purchased from GFS Chemicals (Columbus, OH, USA). The ^{1}H-NMR solvent D_2O with 4,4-dimethyl-4-silapentane-1-sulfonic acid DSS as an internal standard was purchased from Cambridge Isotope Lab, Inc (Andover, MA, USA). These chemicals and materials were used as received. Polystyrene-b-polyethylene glycol (PS-b-PEG, PS_m-b-PEG_n where m = 1600 g/mol and n = 5000 g/mol) was obtained from Polymer Source (Product No. P13141-SEO). Prior to use, PS-b-PEG was dissolved in THF (500 mg/mL) and precipitated in ether (~1:20 v/v THF:ether). The PS-b-PEG was recovered by centrifuging, decanting, and drying under vacuum at room temperature for 2 days.

2.2. Nanoreactor Assembly

For self-assembly, the gold nanoparticles need to be dispersed in a water miscible solvent such as THF with molecularly dissolved block copolymer. To disperse the gold nanoparticles in THF, the as-received dodecanethiol stabilized gold nanoparticles in toluene (1 mL) were precipitated into ethanol (45 mL) and filtered using a Buchner funnel. The filtered nanoparticles from the filter cake were resuspended in THF and concentrated via evaporation at room temperature overnight to achieve a nominal concentration of around 20 mg/mL. The final concentration was confirmed by inductively coupled plasma optical emission spectroscopy using an Agilent 5110 (ICP-OES, Santa Clara, CA, USA). UV spectra collected on an Ocean Optics FLAME-S-UV-VIS with a HL-2000-FHSA light source (Largo, FL, USA) were compared before and after the solvent switch to confirm processing did not significantly affect gold nanoparticle size.

For nanoreactor self-assembly, typically, PS-b-PEG (6 mg), dodecanethiol stabilized 5 nm gold nanoparticles (0.5 mg), and PS homopolymer (co-precipitate, 5.5 mg) were added to 0.5 mL of tetrahydrofuran (THF) and sonicated at 55 °C for 30 min. Using a manually operated confined impinging jet mixer with dilution (CIJ-D) [23,24] with achievable Reynolds' numbers >1300, the resulting THF mixture was rapidly mixed against 0.5 mL of water into a stirring vial of water (4 mL). The resulting dispersion (5 mL total) was stored at room temperature for further characterization and analysis without purification. The nanocomposite particle properties were tuned by adjusting the total solids concentration or the relative amounts of gold nanoparticles and the co-precipitate at a constant total mass or a constant total core volume based on the bulk density of gold and co-precipitate.

2.3. Nanoreactor Characterization

Nanoreactor size was measured after mixing using a Malvern Zetasizer Nano ZS (Westborough, MA, USA) with a backscatter detection angle of 173°. Size distributions are reported using the average of four measurements of the intensity weight distributed with normal resolution. The reported size is the peak 1 mean intensity. The polydispersity index (PDI) is defined from the moment of the cumulant fit of the autocorrelation function calculated by the instrument software (appropriate for samples with PDI < 0.3) and is reported as a measure of particle size distribution. UV absorbance spectra (300 to 1200 nm) of the nanoparticle dispersions were measured at room temperature with an Ocean Optics FLAME-S-UV-VIS with a HL-2000-FHSA light source (Largo, FL, USA). For visualization by TEM, samples were submerged in a dilute dispersion of nanoreactors (10-fold dilution with water) for one hour and dried at ambient conditions overnight. Samples were imaged using a Zeiss Libra 120 TEM (Oberkochen, Germany) using an accelerating voltage of 120 kV. To determine the gold nanoparticle concentration, nanoreactor dispersions were dissolved in THF and digested in aqua regia (1:3 nitric acid:hydrochloric acid by volume) and diluted to 5% v/v aqua regia. Gold concentration of the digested sample was measured using inductively coupled plasma optical emission spectroscopy measurements with an Agilent 5110 (Santa Clara, CA, USA).

2.4. Kinetic Analysis

The catalytic performance of the nanoreactors was evaluated using the reduction of 4-nitrophenol with sodium borohydride as a model reaction using well established procedures [25,26]. Briefly, the nanoreactors were diluted with 4-nitrophenol (aq.) and aqueous sodium borohydride (within 5 min of preparation) and the reduction of 4-nitrophenol was monitored using UV spectroscopy (Ocean Optics FLAME-S-VIS-NIR-ES, Largo, FL, USA, with a HL-2000-FHSA light source (300–1200 nm) with a CUV-UV cuvette holder placed on a stir plate). The final reaction mixture contained less than 0.01 vol% THF. The induction time and apparent reaction rate (k_{app}) were determined from tracking the absorbance at 425 nm as a function of time. The values of k_{app} and induction time are the averages (± standard deviations) of at least 3 trials of each experiment. Detailed procedures are provided in the Supporting Information.

2.5. Langmuir-Hinshelwood Kinetics

For more detailed kinetic analysis, we performed full kinetic analysis considering the two-step reaction mechanism previously established [26]. Full kinetic analysis is described by the reaction rate of each step and the Langmuir adsorption constants of 4-nitrophenol, borohydride, and the stable intermediate. We determined the rate constants for both steps by solving the coupled rate equations using the numerical method previously described and fitting the experimental data (average of three experimental trials) [26].

2.6. NMR Measurements

To evaluate effective transport of the 4-nitrophenol, ^{1}H-NMR spectroscopy and pulsed field gradient (PFG) NMR, combined with saturated transfer difference (STD) spectroscopy, using a Bruker 800 MHz cryo-probe (Billerica, MA, USA) was performed in accordance with the methods described in the Supplemental Information. Briefly, 4-nitrophenol molecules in close proximity to the nanoreactor core were analyzed based on spin diffusion of selectively saturated polystyrene, in conjunction with an applied magnetic field gradient. Relevant intensities were analyzed as a function of gradient strength to determine the diffusion coefficient of the molecules. Since nanoreactors diffuse in free solution at least 3 orders of magnitude slower than molecules, the measured diffusion coefficient was considered the effective diffusion coefficient of the solute within the nanoreactor [27–29].

3. Results and Discussion

3.1. Nanoreactor Self-Assembly

To perform Flash NanoPrecipitation, dodecanethiol stabilized 5 nm gold nanoparticles were dispersed in THF with the molecular dissolved, PS and PS-b-PEG, and rapidly mixed with water using a hand-operated confined impinging jet mixer. The entire formation process was accomplished in less than a second; further, the process can be performed continuously at large scales [24,30,31]. Due to their hydrophobic nature and particle aggregation during assembly, the gold nanoparticles are expected to be in the hydrophobic core of the nanoreactor [3,21], forming a nanoparticle-macromolecular system [32]. Due to the high molecular weight of the polystyrene block, no dynamic exchange of the block copolymer is expected [20]. The resulting nanoreactors were ~130 nm indicated by a single Gaussian peak with PDI <0.2 measured by dynamic light scattering (DLS). The dispersions were stable when stored at room temperature for at least 2 months as there was no significant change in size or size distribution by DLS (Supporting Info, Figure S5), and no macroscopic precipitation of unencapsulated gold was observed.

We further characterized the nanoreactors using UV-Vis spectroscopy. Prior to Flash NanoPrecipitation, the dodecanethiol-stabilized nanoparticles dispersed in toluene showed a peak absorbance at 495 nm (as received and after switching solvents). The nanoreactors showed a peak absorbance of 520 nm (Figure 1b). The peak shift could occur due to differences in hydrophobicity of the surrounding environment [33]. Since the polystyrene microenvironment should have similar hydrophobicity as toluene, we attribute the red-shift to plasmonic coupling due to close proximity of the encapsulated gold nanoparticles, which has been previously observed with polymer-gold nanocomposite particles [34].

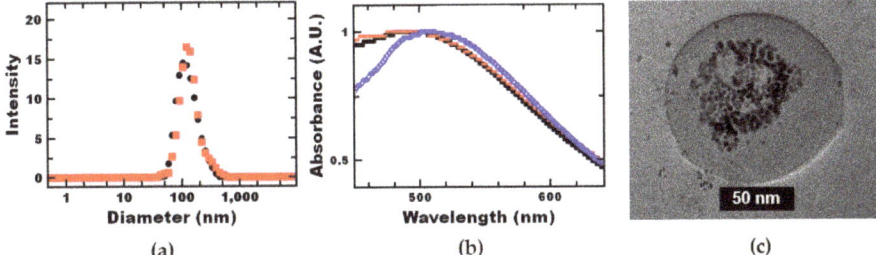

Figure 1. Polymer nanoreactors were fabricated via self-directed assembly. (a) DLS confirms the uniform size distribution of the ~130 nm self-assembled polymer nanoreactors (black circles) and confirms that the size is the same after the reduction of 4-nitrophenol (red squares). (b) UV-vis analysis shows that the absorbance of the gold nanoparticle remains unchanged through the solvent switch from toluene (black filled circles) to tetrahydrofuran (THF) (red open circles). A red-shift is seen upon encapsulation within polymer nanoreactors (blue open diamonds) due to close proximity of the encapsulated gold nanoparticles. (c) TEM imaging demonstrates that multiple gold nanoparticles were encapsulated within the core of the nanoreactors.

The structure of the nanocomposite particles was visualized using TEM. Based on TEM imaging, clustering of the gold nanoparticles during assembly resulted in multiple catalytic gold nanoparticles per nanoreactor. The majority of the gold nanoparticles appear to be in the nanoreactor core, although multiple polymer layers are not visible on TEM due to low electron density. This result is consistent with previous reports of encapsulated gold nanoparticles via Flash NanoPrecipitation (FNP) [21,35]. Based on TEM, some of the gold may also be associated with the PEG-layer of the nanoreactors whereas unassociated gold would be expected to precipitate out of the dispersion as well as affect the size distribution measured by DLS. Since we do not observe gold precipitate from the dispersion, and the size of the TEM size is consistent with DLS with PDI <0.2, we assume all the gold in the dispersion is associated with the nanoreactors. Finally, we confirmed the amount of gold by ICP-OES. We found the polystyrene nanoreactors retained 74% of the gold from the THF-gold nanoparticle solution and the loss can be attributed to the hold-up volume during mixing.

Next, we aimed to independently tune the nanoreactor properties, size and gold loading, using formulation parameters. Nanoreactor assembly depends on the relative time scales of block copolymer micellization, gold nanoparticle clustering, and co-precipitate nucleation and growth. Therefore, the overall nanoreactor size can be affected by the ratio of core material to block copolymer, as well as the total concentration of components in the organic stream [22].

Varying the ratio of block copolymer to core materials has been an effective method for tuning nanostructure size via Flash NanoPrecipitation [23,36]. To vary nanoreactor size, the amount of block copolymer concentration can be increased (Supporting Information, Figure S6), but the gold loading is also affected. In order to vary the nanoreactor size while holding the gold loading constant, we varied the total solids concentration holding the mass ratio of gold to polystyrene co-precipitate constant. As expected, the nanoreactor size increased with increasing total solids concentration. This effect has been attributed to an increase in the rate of particle core relative to nucleation [36,37]. Using this approach, the nanoreactor size could be tuned between 100 and 200 nm with nominal gold loading of 4 wt % (Figure 2a). This level of gold loading is comparable other polymer nanocomposite systems with low volume additions of inorganic nanoparticles that demonstrate enhanced functional performance [32].

Next, we aimed to vary the gold loading independently of nanoparticle size. Holding the total core material mass constant and varying the ratio of gold to polymer resulted in a decrease in nanoparticle size with increasing gold concentration. In contrast, with gold nanoparticles and block copolymer without a co-precipitate, Gindy et al. observed that increasing the gold loading results in an increase in nanocomposite particle size that is attributed to the increase in the amount of gold core relative to the block copolymer [21]. The difference is our use of a co-precipitate. We attribute the trend observed in this case to the increase in the number density of gold nanoparticles that act as nucleating agents that seed particle growth via heterogeneous nucleation [37,38].

To guide nanoreactor formulation, the Smoluchowski diffusion limited aggregation model has previously been used to formulate inorganic nanoparticle-polymer nanocomposite particles via Flash NanoPrecipitation [36]. Based on the model, nanoreactor size can be predicted using:

$$R = \left(K \frac{k_B T c_{core}^{5/3}}{\pi \mu \rho c_{BCP}} \right)^{1/3} \tag{1}$$

where R is the aggregate radius, K is a constant of proportionality for formation time, k_B is Boltzmann's constant, T is the absolute temperature, c_{core} is the concentration of core material, c_{BCP} is the concentration of block copolymer, μ is the solvent viscosity, and ρ is the core material density. This model suggests that the nanoreactor size is affected by the volume more than the mass of the core. Thus, as an alternative to holding the mass of the core constant, we held the volume of the core constant, according to:

$$V_{cm} = \frac{m_{AuNP}}{\rho_{AuNP}} + \frac{m_{PS}}{\rho_{PS}} \tag{2}$$

where V_{cm} is the total volume of the core materials, m_{AuNP} and m_{PS} are the masses of the gold nanoparticles and polystyrene core materials, respectively, finally ρ_{AuNP} and ρ_{PS} are the densities of the gold nanoparticles and polystyrene core materials, respectively. The core volume was selected from the standard formulation, a nominal gold loading of 4% and nanoreactor concentration of 2.4 mg/mL. Using the density of bulk gold and polystyrene, which are 19.32 g/mL and 1.04 g/mL, respectively, the core material volume was found to be 5.33 µL. Using the approach of constant volume, the gold loading was tuned between 4 and 50 nominal wt % at a nanoreactor size of ~130 nm (Figure 2b).

Overall, nanoreactors were assembled in a rapid, scalable, single-step method using Flash NanoPrecipitation. Nanoreactor size could be tuned independently of gold nanoparticle loading by varying the total solids concentration at a constant ratio of gold to polystyrene. Interestingly, the gold nanoparticle loading was tuned independently of nanoreactor size by varying the ratio of gold to polystyrene at constant total core volume. The constant core volume approach may be useful for formulations of multiple components with disparate densities e.g., inorganic particle-polymer nanocomposite particles.

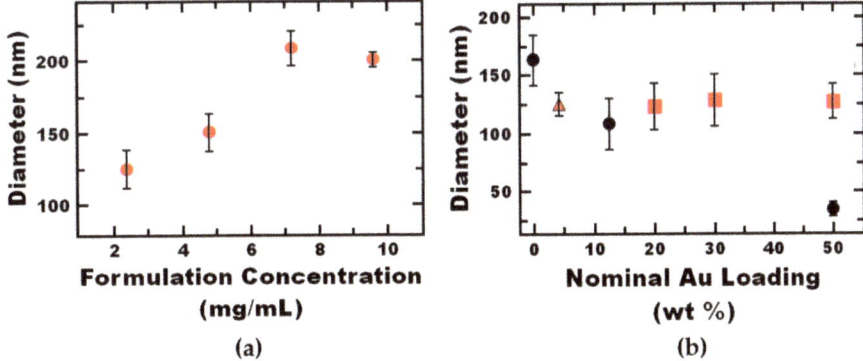

Figure 2. Hydrodynamic diameter of polystyrene nanoreactors measured by DLS with varying total nanoreactor material concentration in the formulation. (**a**) By varying the total material concentration with constant ratio of components tunable nanoreactor size between 100–200 nm. (**b**) By varying the gold to polystyrene co-precipitate ratio at a constant nanoreactor core volume (red squares), as opposed to constant mass ratio (black circles), the nominal gold loading of polystyrene nanoreactors can be tuned at constant nanoreactor size (~130 nm). The standard formulation (4 wt % nominal gold loading, 2.4 mg/mL) is shown by the red triangle.

3.2. Initial Characterization of Nanoreactor Performance

To evaluate the catalytic performance of the nanoreactor, the reduction of 4-nitrophenol by sodium borohydride was used as a model reaction [39]. First, we confirmed the nanoreactors remained intact following the reaction; no significant change in size or polydispersity was observed by DLS (Figure 1a). Further, no macroscopic precipitation of gold nanoparticles was observed following the reaction.

In these initial studies, we assume all of the gold nanoparticles included in the formulation are associated with the nanoreactor and contribute to the observed catalytic activity. From TEM (Figure 1c), the gold nanoparticles may be associated with the hydrophobic core or hydrophilic shell or may be unencapsulated. Unencapsulated gold was not observed precipitating from the nanoreactors and would not contribute to the observed activity (Table 1, dodecanethiol-stabilized gold nanoparticles (DDT)). This is likely due to the lack of solubility as other hydrophobic inorganic nanoparticles have shown activity in water:solvent reaction mixtures [40]. If the dispersions contained trace amounts of unencapsulated gold, the reported values for k_1 would be slightly underestimated. The conversion of 4-nitrophenol confirmed the gold nanoparticles associated with the nanoreactors were catalytically active (Supporting Information, Figure S4a). The apparent reaction rate constant per surface area of gold, k_1, for the nanoreactors was 0.414 ± 0.095 L m^{-2}s^{-1}, which is comparable to the citrate-stabilized, 5 nm gold particles.

Table 1. Rate constants and induction times for various gold nanoparticles. PS, polystyrene; DDT dodecanethiol-stabilized gold nanoparticles.

Support	Diameter (nm)	k_1 (L m^{-2} s^{-1})	Induction Time (s)	Reference
PS	5	0.414 ± 0.095	229 ± 21	This Paper
DDT	5	Undetected	N/A	This Paper
Citrate	5	0.173 ± 0.026	5 ± 1	This Paper
Ligand-Free	7	0.17	N/A	[41]

Comparing the performance of the nanoreactors with other metal nanoparticle-polymer systems using the reaction rate considering the amount of gold catalyst (e.g., k_1 in Table 2), the nanoreactors demonstrate over 110-fold better catalytic activity than gold within polymer (PNIPAM-b-P4VP) micelles, despite a larger overall nanoreactor size. This difference may be attributed to P4VP-gold interactions that affect availability of active sites. Thus, the use of non-interacting co-precipitates and Flash NanoPrecipitation may provide an advantage to other polymer micelle systems that rely on gold-polymer interactions for self-assembly.

Further, the induction time and kinetics are similar to immobilized gold nanoparticles within polyelectrolyte brush shell on polystyrene core particle systems [26]. Specifically, the kinetics of the nanoreactors we report with 5 nm gold are comparable to polyelectrolyte brushes with 2.2 nm gold nanoparticles at the surface of the core-shell nanostructures, which are expected to have similar activities [42]. This result suggests that association of the catalyst with the nanoreactor does not sacrifice reactivity.

Table 2. Rate constants for various metal/polymer nanocomposite nanoreactors.

Support	AuNP Diameter (nm)	k_1 (L m^{-2} s^{-1})	Reference
Polystyrene nanoreactors	5	$(4.14 \pm 0.95) \times 10^{-1}$	This Paper
PNIPAM-b-P4VP Micelles	3.3	3.70×10^{-3}	[43]
Polyelectrolyte brush	2.2	2.70×10^{-1}	[14]

3.3. Probing Potential Mass Transfer Limitations

3.3.1. Induction Time

Notably, the induction time of the encapsulated gold nanoparticles is ~50-fold longer than citrate-stabilized nanoparticles (Table 1). This relatively long induction time has been previously observed with gold-nanoparticle-polymer nanoreactor systems. It may be attributed, in part, to slow surface restructuring upon encapsulation within the hydrophobic polystyrene microenvironment [14]. Additional factors that may increase induction time include: poisoning of the active sites when encapsulated within the nanoreactor core, reduction of the dissolved oxygen present in the reaction dispersion, and/or diffusion limitations [44,45].

To further understand the nature of the induction time in the nanoreactor system, we investigated both the sequence of addition and the time between adding the reactants (Figure 3). Under standard model reaction conditions, 4-nitrophenol was added first and allowed to equilibrate for 1 min, followed by the addition of the sodium borohydride. To probe potential diffusion limitations, we increased the time between adding the 4-nitrophenol and sodium borohydride 10-fold, and no significant change in induction time was observed. This result suggests that the induction time is not related to diffusion of 4-nitrophenol.

Moreover, switching the sequence to adding sodium borohydride first, followed by 4-nitrophenol after 1 min of equilibration did not significantly affect the induction time. Interestingly, when the equilibration time was increased in this case, the induction time was reduced by two orders of magnitude. This ~5 s induction time is comparable to the value measured for citrate-capped gold nanoparticles. This result indicates the long induction times relative to citrate stabilized gold nanoparticles may be attributed to diffusion of sodium borohydride. Further examining the effect of equilibration time, the induction time decreased from ~100 to 5 s when increasing the equilibration time from 1 to 3 min (Figure 4). Further increasing the equilibration time beyond 3 min did not significantly impact the induction time. Thus, it appears that it takes ~3 min for sufficient sodium borohydride to partition into the nanoreactor for the reaction to progress. This required equilibration time can be reduced by increasing the concentration of the borohydride (constant ratio of borohydride to 4-nitrophenol) (Supporting Information, Figure S7) which further indicates the relatively long induction time of the nanoreactors relative to citrate stabilized gold nanoparticles can be attributed to diffusion of the borohydride.

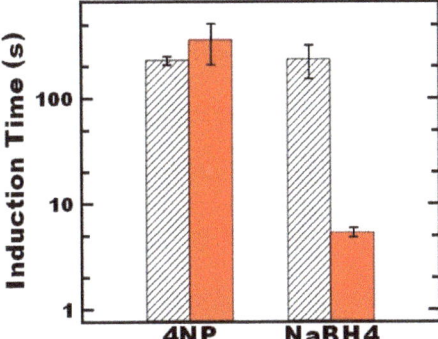

Figure 3. The effect of the sequence of reagent addition on the induction time of the 4-nitrophenol reaction. In all experiments, the 4-nitrophenol and sodium borohydride concentration followed standard conditions of 0.01 mM and 0.01 M, respectively. The indicated reagent was the first to be added, after which the reagent was allowed to equilibrate in the solution for either 1 min (black striped bars) or 10 min (red solid bars). The end of the equilibration period was the addition of the second reagent, at which point the reaction could progress.

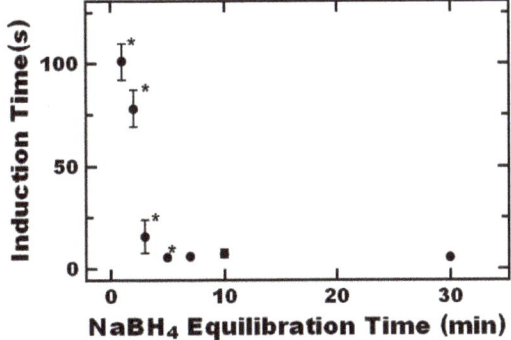

Figure 4. The effect of sodium borohydride equilibration on the induction time of the 4-nitrophenol reaction. Standard reagent concentrations of 0.01 mM and 0.01 M were used for 4-nitrophenol and sodium borohydride, respectively. Data points marked with an asterisk (*) are significantly different than each other ($p < 0.1$).

3.3.2. Reaction Rate

Next, we further investigated potential mass transfer limitations on the observed reaction rate following the induction time. A useful tool for determination of diffusion limitations is the second Damköhler number (DaII), which is a ratio of the reaction rate to the diffusion rate given by:

$$DaII = \frac{k_{app} C^{n-1}}{\beta a} \qquad (3)$$

where n is the reaction order, β is the mass transport coefficient (which is a quotient of the diffusion coefficient and the characteristic length of the system), and a is the interfacial area. To calculate DaII for a 130 nm diameter particle, the interfacial area (nanoreactor area per unit volume of nanoreactor dispersion) was estimated to be 2×10^4 m^{-1} based on the number of nanoreactors estimated using the aggregation number of the block copolymer previously reported [46,47]. The diffusion coefficient for 4-nitrophenol in the nanoreactor system was experimentally determined by NMR. Using PFG-NMR in conjunction with the STD spectroscopy, the effective diffusion coefficient of the 4-nitrophenol within the nanoreactors was determined to be $1.91 \pm 0.01 \times 10^{-8}$ m^2/s (Figure S2). Using this experimentally determined effective diffusion coefficient, the DaII is on the order of 10^{-6} indicating the reaction is significantly slower than diffusion; therefore, the apparent kinetics are reaction-limited.

A complementary approach was to consider the bimolecular reaction between 4-nitrophenol and nanoparticle catalyst using the Smoluchowski diffusion limited reaction model [48,49]. We varied the gold concentration by (1) varying the nanoreactor concentration to probe potential external diffusion limitations, and (2) varying the gold loading at constant nanoreactor concentration to examine potential internal diffusion limitations (Figure S8). When the nanoreactor concentration or the gold loading was increased, k_{app} increased; the 2nd order rate constant was on the order of 10^6 M^{-1}s^{-1}. These values are much lower than the $k_{bm} \sim 10^8$ M^{-1}s^{-1}, indicating that neither internal nor external diffusion from the bulk solution to the nanoreactor limited the apparent reaction kinetics.

Since there were no indications of diffusion limitations associated with the reaction following the induction time, we further characterized the reaction kinetics using Langmuir-Hinshelwood kinetics. Based on the previously established two-step reaction model [26], and fitting the measured concentration of 4-nitrophenol as a function of time (normalized after the induction time for conversions up to 30%) [26], the kinetics were comparable to other gold nanoparticle-polymer nanoreactor systems (Table 3 with plot, Figure S3, and full fit parameters, Table S1, in the Supporting Information). Interestingly, k_a and k_b observed for the gold encapsulated within the nanoreactors are comparable to ligand-free gold nanoparticles. This result suggests that the reactivity of the gold nanoparticle surface is not significantly affected by self-assembly and their incorporation into the nanoreactors.

Table 3. Langmuir-Hinshelwood rate constants obtained from fits to experimental data.

Reactor	k_a (10^4 mol/m^2 s)	k_b (10^5 mol/m^2 s)	Reference
Polystyrene Nanoreactors	4.32 ± 0.14	4.3 ± 0.5	This Study
Ligand-Free	5.8 ± 3.1	5.4 ± 2.0	[41]
Brush Shell	9.7 ± 2.9	7.8 ± 1.7	[26]

Overall, diffusion and partitioning of sodium borohydride into the polymer nanoreactor affect the induction time for the reaction. Sufficient equilibration time between adding the sodium borohydride and the 4-nitrophenol (~3 min) for the borohydride to partition and diffuse minimizes induction time. Notably, mass transfer effects are not observed after the induction time and the intrinsic kinetics are comparable to ligand-free gold nanoparticles.

4. Conclusions

Overall, we have presented rapid, scalable self-assembly of metal nanoparticle catalyst-polymer nanocomposite particles as nanoreactors. The size and gold loading of the nanoreactors can be tuned independently, with sizes and nominal loadings ranging from 100–200 nm and 4–50 wt% respectively. Using the 4-nitrophenol reduction as a model reaction, the induction time is affected by sequence or reagent addition, time between addition, and reagent concentration. Combined, our experiments indicate that the induction time is most influenced by diffusion of sodium borohydride. Scaling analysis and effective diffusivity measured using NMR, the observed reaction rate after the induction time are reaction- rather than diffusion-limited. Finally, the intrinsic reaction kinetics of gold associated with the polymer were comparable to ligand-free particles indicating the self-assembly process and resulting polymer microenvironment did not de-activate or block the catalyst active sites. Building on this foundational study, practical considerations such as nanoreactor recycling will be considered in future work.

Supplementary Materials: The following are available online at http://www.mdpi.com/2079-4991/9/3/318/s1, Figure S1: Representative Kinetic Data Analysis, Figure S2: PFG-NMR with STD, Figure S3: Langmuir-Hinshelwood Model Fitting, Figure S4: Confirmation of Nanoreactor Activity, Figure S5: Nanoreactor Stability, Figure S6: Nanoreactor Size Tuning with Block Copolymer Concentration, Figure S7: Effect of Sodium Borohydride Equilibration on Induction Time, Figure S8: External and Internal Mass Transfer, Table S1: Langmuir-Hinshelwood Fitting Parameters, Table S2: Rate Constant After Recycling.

Author Contributions: Conceptualization, A.H., S.W. and C.T.; Formal analysis, A.H., T.T.V., M.P.Z., B.J.H., S.W. and C.T.; Funding acquisition, S.W. and C.T.; Investigation, A.H., T.T.V., M.P.Z., B.J.H. and S.W.; Methodology, A.H., S.W. and C.T.; Project administration, C.T.; Supervision, C.T.; Validation, A.H. and C.T.; Visualization, A.H., S.W. and C.T.; Writing—original draft, A.H., S.W. and C.T.; Writing—review & editing, A.H., T.T.V., M.P.Z., B.J.H., S.W. and C.T.

Funding: This research was partially supported by startup funding at Virginia Commonwealth University, and NSF (Award number CMMI-1651957). A portion of this work was performed at the National High Magnetic Field Laboratory, which is supported by the National Science Foundation Cooperative Agreement No. DMR-1157490 and DMR-1644779 as well as the State of Florida.

Acknowledgments: The authors gratefully acknowledge funding for this research which was partially supported by startup funding at Virginia Commonwealth University, and NSF (Award number CMMI-1651957). This work was also supported by the National High Magnetic Field Laboratory through the NSF (DMR-1157490 and DMR-1644779) and by the State of Florida. We would also like to thank Joseph Turner of VCU and Christine Lacy of University of Richmond for their technical support.

Conflicts of Interest: The authors declare no conflict of interest. The founding sponsors had no role in the design of the study; in the collection, analyses, or interpretation of data; in the writing of the manuscript, and in the decision to publish the results.

References

1. Zhang, X.; Cardozo, A.F.; Chen, S.; Zhang, W.; Julcour, C.; Lansalot, M.; Blanco, J.F.; Gayet, F.; Delmas, H.; Charleux, B.; et al. Core-Shell Nanoreactors for Efficient Aqueous Biphasic Catalysis. *Chem. A Eur. J.* **2014**, *20*, 15505–15517. [CrossRef] [PubMed]
2. Cotanda, P.; Petzetakis, N.; O'reilly, R.K. Catalytic Polymeric Nanoreactors: More than a Solid Supported Catalyst. *MRS Commun.* **2012**, *2*, 119–126. [CrossRef]
3. Walther, A.; Muller, A.H.E. Janus Particles: Synthesis, Self-Assembly, Physical Properties, and Applications. *Chem. Rev.* **2013**, *113*, 5194–5261. [CrossRef] [PubMed]
4. Lipshutz, B.H.; Ghorai, S. "Designer"-Surfactant-Enabled Cross-Couplings in Water at Room Temperature. *Aldrichim. Acta* **2012**, *45*, 3–16. [CrossRef]
5. Lipshutz, B.H.; Ghorai, S. Transitioning Organic Synthesis from Organic Solvents to Water. What's Your E-Factor? *Green Chem.* **2014**, *16*, 3660–3679. [CrossRef] [PubMed]
6. La Sorella, G.; Strukul, G.; Scarso, A. Recent Advances in Catalysis in Micellar Media. *Green Chem.* **2015**, *17*, 644–683. [CrossRef]
7. Petrosko, S.H.; Johnson, R.; White, H.; Mirkin, C.A. Nanoreactors: Small Spaces, Big Implications in Chemistry. *J. Am. Chem. Soc.* **2016**, *138*, 7443–7445. [CrossRef] [PubMed]
8. Vriezema, D.M.; Aragone, M.C.; Elemans, J.A.A.W.; Cornelissen, J.J.L.M.; Rowan, A.E.; Nolte, R.J.M. Self-Assembled Nanoreactors. *Chem. Rev.* **2005**, *105*, 1445–1489. [CrossRef] [PubMed]
9. Lu, J.; Dimroth, J.; Weck, M. Compartmentalization of Incompatible Catalytic Transformations for Tandem Catalysis. *J. Am. Chem. Soc.* **2015**, *137*, 12984–12989. [CrossRef] [PubMed]
10. Lu, A.; Moatsou, D.; Hands-Portman, I.; Longbottom, D.A.; O'Reilly, R.K. Recyclable L-Proline Functional Nanoreactors with Temperature-Tuned Activity Based on Core-Shell Nanogels. *ACS Macro Lett.* **2014**, *3*, 1235–1239. [CrossRef]
11. Cotanda, P.; Lu, A.; Patterson, J.P.; Petzetakis, N.; Reilly, R.K.O. Functionalized Organocatalytic Nanoreactors: Hydrophobic Pockets for Acylation Reactions in Water. *Macromolecules* **2012**, *45*, 2377–2384. [CrossRef]
12. Gall, B.; Bortenschlager, M.; Nuyken, O.; Weberskirch, R. Cascade Reactions in Polymeric Nanoreactors: Mono (Rh)—And Bimetallic (Rh/Ir) Micellar Catalysis in the Hydroaminomethylation of 1-Octene. *Macromol. Chem. Phys.* **2008**, *209*, 1152–1159. [CrossRef]
13. De Martino, M.T.; Abdelmohsen, L.K.E.A.; Rutjes, F.P.J.T.; Van Hest, J.C.M. Nanoreactors for Green Catalysis. *Beilstein J. Org. Chem.* **2018**, *14*, 716–733. [CrossRef] [PubMed]

14. Wunder, S.; Lu, Y.; Albrecht, M.; Ballauff, M. Catalytic Activity of Faceted Gold Nanoparticles Studied by a Model Reaction: Evidence for Substrate-Induced Surface Restructuring. *ACS Catal.* **2011**, *1*, 908–916. [CrossRef]
15. Cotanda, P.; O'Reilly, R.K. Molecular Recognition Driven Catalysis Using Polymeric Nanoreactors. *Chem. Commun.* **2012**, *48*, 10280–10282. [CrossRef] [PubMed]
16. Lan, Y.; Yang, L.; Zhang, M.; Zhang, W.; Wang, S. Microreactor of Pd Nanoparticles Immobilized Hollow Microspheres for Catalytic Hydrodechlorination of Chlorophenols in Water. *ACS Appl. Mater. Interfaces* **2010**, *2*, 127–133. [CrossRef] [PubMed]
17. Angioletti-Uberti, S.; Lu, Y.; Ballauff, M.; Dzubiella, J. Theory of Solvation-Controlled Reactions in Stimuli-Responsive Nanoreactors. *J. Phys. Chem. C* **2015**, *119*, 15723–15730. [CrossRef]
18. Peter, N.; Tan, B.; Lee, C.H.; Li, P. Green Synthesis of Smart Metal/Polymer Nanocomposite Particles and Their Tuneable Catalytic Activities. *Polymers* **2016**, *8*, 105. [CrossRef]
19. Lempke, L.; Ernst, A.; Kahl, F.; Weberskirch, R.; Krause, N. Sustainable Micellar Gold Catalysis—Poly(2-Oxazolines) as Versatile Amphiphiles. *Adv. Synth. Catal.* **2016**, *358*, 1491–1499. [CrossRef]
20. Choi, S.H.; Bates, F.S.; Lodge, T.P. Molecular Exchange in Ordered Diblock Copolymer Micelles. *Macromolecules* **2011**, *44*, 3594–3604. [CrossRef]
21. Gindy, M.E.; Panagiotopoulos, A.Z.; Prud'Homme, R.K. Composite Block Copolymer Stabilized Nanoparticles: Simultaneous Encapsulation of Organic Actives and Inorganic Nanostructures. *Langmuir* **2008**, *24*, 83–90. [CrossRef] [PubMed]
22. Pinkerton, N.M.; Gindy, M.E.; Calero-Ddelc, V.L.; Wolfson, T.; Pagels, R.F.; Adler, D.; Gao, D.; Li, S.; Wang, R.; Zevon, M.; et al. Single-Step Assembly of Multimodal Imaging Nanocarriers: MRI and Long-Wavelength Fluorescence Imaging. *Adv. Healthc. Mater.* **2015**, *4*, 1376–1385. [CrossRef] [PubMed]
23. Tang, C.; Amin, D.; Messersmith, P.B.; Anthony, J.E.; Prud'homme, R.K. Polymer Directed Self-Assembly of pH-Responsive Antioxidant Nanoparticles. *Langmuir* **2015**, *31*, 3612–3620. [CrossRef] [PubMed]
24. Han, J.; Zhu, Z.; Gian, H.; Wohl, A.R.; Beaman, C.J.; Hoye, T.R.; Macosko, C.W. A Simple Confined Impingement Jets Mixer for Flash Nanoprecipitation. *J. Pharm. Sci.* **2012**, *101*, 4018–4023. [CrossRef] [PubMed]
25. Hervés, P.; Pérez-Lorenzo, M.; Liz-Marzán, L.M.; Dzubiella, J.; Lu, Y.; Ballauff, M. Catalysis by Metallic Nanoparticles in Aqueous Solution: Model Reactions. *Chem. Soc. Rev.* **2012**, *41*, 5577. [CrossRef] [PubMed]
26. Gu, S.; Wunder, S.; Lu, Y.; Ballauff, M.; Fenger, R.; Rademann, K.; Jaquet, B.; Zaccone, A. Kinetic Analysis of the Catalytic Reduction of 4-Nitrophenol by Metallic Nanoparticles. *J. Phys. Chem. C* **2014**, *118*, 18618–18625. [CrossRef]
27. Momot, K.I.; Kuchel, P.W. Pulsed Field Gradient Nuclear Magnetic Resonance as a Tool for Studying Drug Delivery Systems. *Concepts Magn. Reson.* **2003**, *19A*, 51–64. [CrossRef]
28. Mun, E.A.; Hannell, C.; Rogers, S.E.; Hole, P.; Williams, A.C.; Khutoryanskiy, V.V. On the Role of Specific Interactions in the Diffusion of Nanoparticles in Aqueous Polymer Solutions. *Langmuir* **2014**, *30*, 308–317. [CrossRef] [PubMed]
29. Wang, J.H. Self-Diffusion Coefficients of Water. *J. Phys. Chem.* **1965**, *69*, 4412. [CrossRef]
30. Johnson, B.K.; Prud, R.K. Chemical Processing and Micromixing in Confined Impinging Jets. *AIChE J.* **2003**, *49*, 2264–2282. [CrossRef]
31. Liu, Y.; Kathan, K.; Saad, W.; Prud, R.K. Ostwald Ripening of β-Carotene Nanoparticles. *Phys. Rev. Lett.* **2007**, *36102*, 8–11. [CrossRef]
32. Vaia, R.A.; Maguire, J.F. Polymer Nanocomposites with Prescribed Morphology: Going beyond Nanoparticle-Filled Polymers. *Chem. Mater.* **2007**, *19*, 2736–2751. [CrossRef]
33. Ghosh, S.K.; Nath, S.; Kundu, S.; Esumi, K.; Pal, T. Solvent and Ligand Effects on the Localized Surface Plasmon Resonance (LSPR) of Gold Colloids. *J. Phys. Chem. B* **2004**, *108*, 13963–13971. [CrossRef]
34. Lange, H.; Juárez, B.H.; Carl, A.; Richter, M.; Bastús, N.G.; Weller, H.; Thomsen, C.; Von Klitzing, R.; Knorr, A. Tunable Plasmon Coupling in Distance-Controlled Gold Nanoparticles. *Langmuir* **2012**, *28*, 8862–8866. [CrossRef] [PubMed]
35. Tang, C.; Sosa, C.L.; Pagels, R.F.; Priestley, R.D.; Prud'homme, R.K. Efficient Preparation of Size Tunable PEGylated Gold Nanoparticles. *J. Mater. Chem. B* **2016**, *4*, 4813–4817. [CrossRef]

36. Pagels, R.F.; Edelstein, J.; Tang, C.; Prud'homme, R.K. Controlling and Predicting Nanoparticle Formation by Block Copolymer Directed Rapid Precipitations. *Nano Lett.* **2018**, *18*, 1139–1144. [CrossRef] [PubMed]
37. D'Addio, S.M.; Prud'homme, R.K. Controlling Drug Nanoparticle Formation by Rapid Precipitation. *Adv. Drug Deliv. Rev.* **2011**, *63*, 417–426. [CrossRef] [PubMed]
38. Tang, C.; Prud'homme, R.K. Targeted Theragnostic Nanoparticles Via Flash Nanoprecipitation: Principles of Material Selection. In *Polymer Nanoparticles for Nanomedicines*; Springer: Cham, Switzerland, 2016; pp. 55–85.
39. Zhao, P.; Feng, X.; Huang, D.; Yang, G.; Astruc, D. Basic Concepts and Recent Advances in Nitrophenol Reduction by Gold- and Other Transition Metal Nanoparticles. *Coord. Chem. Rev.* **2015**, *287*, 114–136. [CrossRef]
40. Tadele, K.; Verma, S.; Nadagouda, M.N.; Gonzalez, M.A.; Varma, R.S. A Rapid Flow Strategy for the Oxidative Cyanation of Secondary and Tertiary Amines via C-H Activation. *Sci. Rep.* **2017**, *7*, 6–10. [CrossRef] [PubMed]
41. Gu, S.; Kaiser, J.; Marzun, G.; Ott, A.; Lu, Y.; Ballauff, M.; Zaccone, A.; Barcikowski, S.; Wagener, P. Ligand-Free Gold Nanoparticles as a Reference Material for Kinetic Modelling of Catalytic Reduction of 4-Nitrophenol. *Catal. Lett.* **2015**, *145*, 1105–1112. [CrossRef]
42. Fenger, R.; Fertitta, E.; Kirmse, H.; Thünemann, A.F.; Rademann, K. Size Dependent Catalysis with CTAB-Stabilized Gold Nanoparticles. *Phys. Chem. Chem. Phys.* **2012**, *14*, 9343–9349. [CrossRef] [PubMed]
43. Zhang, M.; Liu, L.; Wu, C.; Fu, G.; Zhao, H.; He, B. Synthesis, Characterization and Application of Well-Defined Environmentally Responsive Polymer Brushes on the Surface of Colloid Particles. *Polymer* **2007**, *48*, 1989–1997. [CrossRef]
44. Nigra, M.M.; Ha, J.M.; Katz, A. Identification of Site Requirements for Reduction of 4-Nitrophenol Using Gold Nanoparticle Catalysts. *Catal. Sci. Technol.* **2013**, *3*, 2976–2983. [CrossRef]
45. Menumerov, E.; Hughes, R.A.; Neretina, S. Catalytic Reduction of 4-Nitrophenol: A Quantitative Assessment of the Role of Dissolved Oxygen in Determining the Induction Time. *Nano Lett.* **2016**, *16*, 7791–7797. [CrossRef] [PubMed]
46. D'Addio, S.M.; Saad, W.; Ansell, S.M.; Squiers, J.J.; Adamson, D.H.; Herrera-Alonso, M.; Wohl, A.R.; Hoye, T.R.; MacOsko, C.W.; Mayer, L.D.; et al. Effects of Block Copolymer Properties on Nanocarrier Protection from in Vivo Clearance. *J. Control. Release* **2012**, *162*, 208–217. [CrossRef] [PubMed]
47. Budijono, S.J.; Russ, B.; Saad, W.; Adamson, D.H.; Prud'homme, R.K. Block Copolymer Surface Coverage on Nanoparticles. *Colloids Surf. A Physicochem. Eng. Asp.* **2010**, *360*, 105–110. [CrossRef]
48. Chang, Y.C.; Chen, D.H. Catalytic Reduction of 4-Nitrophenol by Magnetically Recoverable Au Nanocatalyst. *J. Hazard. Mater.* **2009**, *165*, 664–669. [CrossRef] [PubMed]
49. Ncube, P.; Bingwa, N.; Baloyi, H.; Meijboom, R. Catalytic Activity of Palladium and Gold Dendrimer-Encapsulated Nanoparticles for Methylene Blue Reduction: A Kinetic Analysis. *Appl. Catal. A Gen.* **2015**, *495*, 63–71. [CrossRef]

© 2019 by the authors. Licensee MDPI, Basel, Switzerland. This article is an open access article distributed under the terms and conditions of the Creative Commons Attribution (CC BY) license (http://creativecommons.org/licenses/by/4.0/).

Article

Comparative Analysis of Properties of PVA Composites with Various Nanofillers: Pristine Clay, Organoclay, and Functionalized Graphene

Jin-Hae Chang

Department of Polymer Science and Engineering, Kumoh National Institute of Technology, Gumi 39177, Korea; changjinhae@hanmail.net

Received: 7 January 2019; Accepted: 18 February 2019; Published: 1 March 2019

Abstract: Poly(vinyl alcohol) (PVA) nanocomposites containing three different nanofillers are prepared and compared in terms of their thermal properties, morphologies, and oxygen permeabilities. Specifically, pristine saponite (SPT) clay, hydrophilic organically modified bentonite (OMB), and hexadecylamine-functionalized graphene sheets (HDA-GSs) are utilized as nanofillers to fabricate PVA nanocomposite films. The hybrid films are fabricated from blended solutions of PVA and the three different nanofillers. The content of each filler with respect to PVA is varied from 0 to 10 wt%, and the changes in the properties of the PVA matrices as a function of the filler content are discussed. With respect to the hybrid containing 5 wt% of SPT, OMB, and HDA-GS, each layer in the polymer matrix consists of well-dispersed individual nanofiller layers. However, the fillers are mainly aggregated in the polymer matrix in a manner similar to the case for the hybrid material containing 10 wt% of fillers. In the thermal properties, SPT and OMB are most effective when the filler corresponds to 5 wt% and 7 wt% for HDA-GS, respectively, and the gas barrier is most effective with respect to 5 wt% content in all fillers. Among the three types of nanofillers that are investigated, OMB exhibits optimal results in terms of thermal stability and the gas barrier effect.

Keywords: poly(vinyl alcohol); nanocomposite; nanofiller; film

---

1. Introduction

Poly(vinyl alcohol) (PVA) is a water-soluble synthetic polymer with high hydrophilicity, biocompatibility, and non-toxicity [1,2]. The high capacity of PVA to simultaneously form both intra- and inter-chain hydrogen bonds make it a unique polymer that can interact with nanofillers such as clay, graphene, and functionalized-graphene [3,4]. For example, incorporation of graphene oxide (GO) into PVA facilitates good dispersion and interfacial interaction due to the presence of OH-bonds at the end due to the interaction of OH bonding. The content of clay can readily affect the thermo-mechanical properties and gas permeability [5,6].

Clay and graphene exhibit high aspect and high load transfer results in the agglomeration of layers in polymer matrix, and this makes dispersion difficult. In order to avoid this practical challenge, it can be exfoliated or organically modified to achieve the desired properties by dispersing nanofillers in compatible solvent via sonification [7].

In the wet state or especially after mild drying, clay layers are distributed and embedded in the PVA gel to yield a true nanoscale hybrid material. However, drying in vacuo can cause the re-aggregation of the clay layers. The steric constraints created by the PVA matrix impede the re-aggregation of the clay layers, and, thus, a few clay layers remain in the dispersed state. Ideally, useful nanocomposites are fabricated to create amorphous domains with uniformly distributed mineral layers. However, the preparation of PVA/clay nanocomposite materials from a solution is challenging due to the re-aggregation of the layers [8].

Clays with sandwich-type structures that typically consist of an octahedral Al sheet and multiple tetrahedral Si sheets are referred to as phyllosilicates [9,10]. There are several types of phyllosilicates including kaolinite, montmorillonite, hectorite, saponite, bentonite, and synthetic mica. In the present study, we selected saponite (SPT) [11,12] and hydrophilic organically modified bentonite (OMB) [13,14] as clays for the synthesis of clay/PVA polymer nanocomposites. Specifically, SPT and bentonite consist of stacked silicate 1-nm-thick sheets with lengths of approximately 165 nm and 68 nm, respectively. Currently, SPT and bentonite are widely employed as reinforcing fillers in polymeric matrices due to their excellent mechanical, electrical, and thermal properties and their low cost [15–17].

Generally, with the exception of SPT, several types of pristine clays are not compatible with most polymers, and, thus, require organic treatments to render them as organophilic. A common method for this type of organic treatment is based on the ion exchange of the cations within the clay with organic ammonium cations [18–20]. Thus, we selected hydrophilic organo-clay OMB as nano-filler for the synthesis of hybrid polymer films.

Graphene tends to aggregate or restack due to its strong stacking tendency and high cohesive energy. Additionally, it is insoluble in a variety of organic solvents due to its significantly hydrophobic nature and high specific surface area of graphene. Therefore, a key challenge in the preparation and processing of graphene-based composites corresponds to the prevention of aggregation. The functionalization of the graphene surface can introduce reactive moieties that disrupt the bundle structure and can potentially obtain individual sheets [21–23]. This type of functionalization involves the attachment of functional moieties to the open ends and walls of graphene to improve the solubility and dispersibility of graphene sheets (GSs) [24]. Hence, an optimal method to achieve a homogeneous graphene dispersion throughout a polymer matrix is the use of functionalized graphene sheets (FGSs) [25] that exhibit improved dispersion in solvents and polymers. Furthermore, covalent functionalization can provide the means to engineer the GS/polymer interface and, thereby, optimize the properties of the composite material.

Typically, traditional composite structures typically contain a significant content ($\approx$ 40 wt%) of a filler bound within a polymer matrix. However, significant changes in the properties of the materials are also possible at low loadings (< 10 wt%) of nano-fillers, such as exfoliated pristine clays, organoclays, and FGSs, in the hybrid materials [26,27]. The improvements in material performance are achieved as a result of the inherent properties of the nanofillers and also by optimizing the dispersion, interface chemistry, and nanoscale morphology. This is completed to utilize the advantages of the tremendous surface area per unit volume exhibited by nano-fillers (the theoretical limits correspond to 760 m^2/g for clay [28] and 2,630 m^2/g for graphene [29]).

In the present study, we prepared hybrid films containing PVA and an appropriate amount of filler ($\leq$ 10 wt%) and examined their properties as a function of the filler content and type. We examine and compare the thermal properties, morphologies, and oxygen permeation capabilities of PVA nanocomposites containing three different nanofillers, such as pristine clay SPT, hydrophilic organoclay OMB, and hexadecylamine-functionalized GSs (HDA-GSs). The thermal and oxygen barrier properties of the hybrids are also examined as a function of the nanofiller type and content in the PVA polymer matrix. Lastly, we investigate the effects of filler loadings on the morphologies of the PVA hybrid films.

2. Materials and Methods

2.1. Materials

The source clays, SPT, and OMB were obtained from Kunimine Ind. Co. (Tokyo, Japan) and Nanomer Co. (Seoul, Korea), respectively. The clays were passed through a 325-mesh sieve to remove impurities to yield an SPT clay with a cationic exchange capacity of 100 meq/100 g and bentonite with a cationic exchange capacity of 145 meq/100 g. Additionally, PVA with >99% saponification (M_w = approximately 89,000–98,000), graphite, and HDA were purchased from Aldrich Chem. Co. These materials were used in an as-received condition. Commercially available solvents were purified via distillation, and common reagents were used without further purification.

2.2. Synthesis of HDA-GS

Graphene oxide (GO) was synthesized from natural graphite using a multi-step route known as the Hummers method [30]. Furthermore, HDA-GS was synthesized from hexadecylamine (HDA) and GO based on the following procedure: GO (1 g) was dissolved in 1.5 L of distilled water. HDA (2.00 g; 8.28 × 10^{-3} mol) was added to 25 mL of ethanol, and the mixture was stirred at 25 °C under a steady stream of N_2, and subsequently added to the GO/water mixture. The resulting mixture was heated for 12 h at 25 °C under a steady stream of N_2, cooled to 25 °C, washed twice with a mixture of distilled water and ethanol (1:1, v/v), and dried under vacuum at 70 °C for 24 h to obtain HDA-GS. The synthetic route for HDA-GS is shown in Scheme 1.

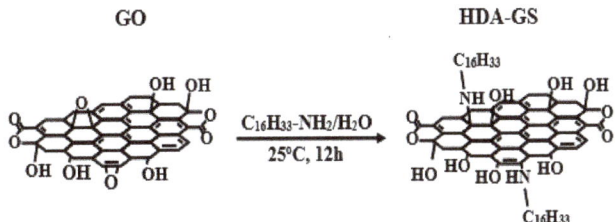

Scheme 1. Chemical structures of GO and HDA-GS.

2.3. Preparation of PVA Hybrid Films

The synthetic procedures used to produce the polymer hybrids were identical for all filler contents used in this experiment. Therefore, the preparation of 5-wt% SPT/PVA is detailed in this paper as a representative example. Specifically, SPT (0.1 g) was added to distilled water (20 mL) in a 100-mL beaker, and the mixture was stirred at 80 °C for 1 h. The resulting mixture was subjected to ultrasonication three times for 5 min to obtain a homogeneously dispersed clay solution. In a separate beaker, PVA (1.9 g) and distilled water (140 mL) were mixed at 80 °C for 3 h and subsequently added dropwise to the SPT/water system with vigorous stirring for 3 h to obtain a homogeneously dispersed system. The solution was cast on poly(ethylene terephthalate) (PET) films and evaporated in a vacuum oven at 35 °C for two days. After the removal of the solvent, the hybrid film was dried for a second time in a vacuum oven at 80 °C for a day.

2.4. Characterization

Fourier-transform infrared (FT-IR) spectra were obtained via an FT-IR 460 (JASCO, Tokyo, Japan) instrument in the range of 4000 to 600 cm^{-1} with KBr pellets. Wide-angle X-ray diffraction (XRD) measurements were performed at room temperature via a Rigaku (D/Max-IIIB, Tokyo, Japan) X-ray diffractometer with Ni-filtered CoK$_\alpha$ radiation. The scan rate corresponded to 2°/min over the 2θ range of 2° to 10°. Differential scanning calorimetry (DSC 200F3, Berlin, Germany) was performed on a NETZSCH instrument, and a thermogravimetric analyzer (AutoTGA 1000, New Castle, USA)) was employed as the thermogravimetric analysis instrument with a heating rate of 20 °C/min under the flow of N_2.

Atomic force microscopy (AFM, Multimode, NanoScope III, Digital instruments Inc. NY, USA) images were obtained on an AutoProbe CP/MT scanning probe microscope. The GO samples were dispersed in water and HDA-GS samples in toluene. The suspensions were ultrasonicated for 3 h and subsequently spin-coated at 5000 rpm on silicon wafers.

The morphologies of the fractured surfaces of the extrusion samples were investigated via a Hitachi S-2400 scanning electron microscope (SEM). In order to enhance the conductivity, the fractured surfaces were sputter-coated with gold via an SPI sputter coater. Transmission electron microscopy (TEM) samples were prepared by placing the PVA hybrid films on epoxy capsules and curing them at 70 °C for 24 h under vacuum. The cured epoxy capsules containing the PVA hybrids were microtomed into 90-nm thick slices and positioned on a 200-mesh copper net, and a layer of carbon (approximately 3-nm thick) was deposited on each slice. The TEM images of the ultrathin sections of the polymer hybrid samples were obtained via an EM 912 OMEGA TEM instrument with an acceleration voltage corresponding to 120 kV.

The O_2 transmission rates (O_2TRs) of the films were measured based on ASTM E96 with a Mocon DL 100 instrument. The O_2TRs were obtained at 23 °C, 0% relative humidity, and 1 atm pressure.

3. Results and Discussion

3.1. FT-IR Spectroscopy

Figure 1 shows the FT-IR spectra of PVA, graphite, GO, HDA-GS, and OMB. The spectrum of pure graphite does not exhibit any peaks while that of PVA and GO exhibit significant and broad absorption peaks characteristic for the OH and COOH functional groups [31]. Details of each substance are specified below.

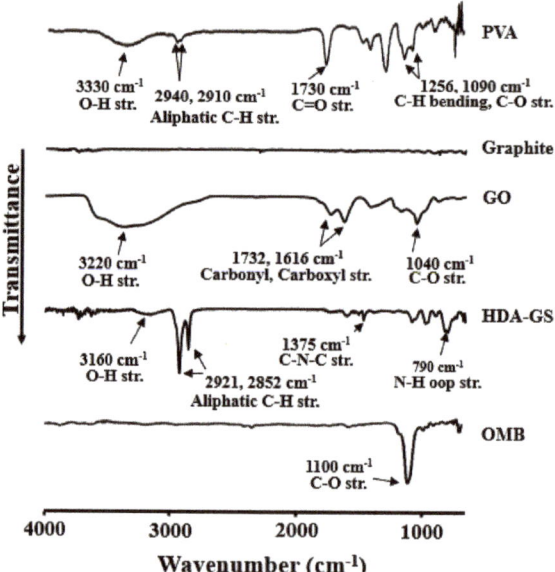

Figure 1. FT-IR spectra of PVA, graphite, and nanofillers.

The characteristic absorption peaks of the PVA are observed at 3330 cm^{-1} (O–H stretching), 2940 cm^{-1} and 2910 cm^{-1} (asymmetric stretching CH$_2$), 1730 cm^{-1} (due to water absorption), and 1256 and 1090 cm^{-1} (C-H bending and C-O stretching). In the case of the GO, the characteristic absorption peaks of the O-H are observed at 3220 cm^{-1} (stretching) and asymmetric epoxy appears in 3063 cm^{-1}, but does not appear to overlap with OH, 1732 and 1616 cm^{-1} (C=O stretching), and 1040 cm^{-1} (C–O stretching). The epoxide ring and C=C bond are much weaker compared to the others, whereas the O–H stretching is more intense. Figure 1 also shows the spectrum of HDA-GS: 3160 cm^{-1} (O–H stretching), 2921 and 2852 cm^{-1} (aliphatic C–H stretching), 1375 cm^{-1} (aromatic C–N–C symmetric stretching), and 790 cm^{-1} (N–H out-of-plane stretching). The OMB also exhibits a peak at 1100 cm^{-1} (C-O stretching).

3.2. Morphology of HDA-GS

The SEM images of unmodified natural graphite, GO, and HDA-GS are shown in Figure 2. The natural graphite exhibits a lamellar structure similar to that of graphene sheets with stacks denser than those observed for the other materials (Figure 2a). Figure 2b shows the translucent GO sheets that are wrinkled and folded in a manner resembling thin paper. Additionally, HDA-GS is prepared from chemically modified GO and exhibit an entangled morphology and a random distribution (Figure 2c).

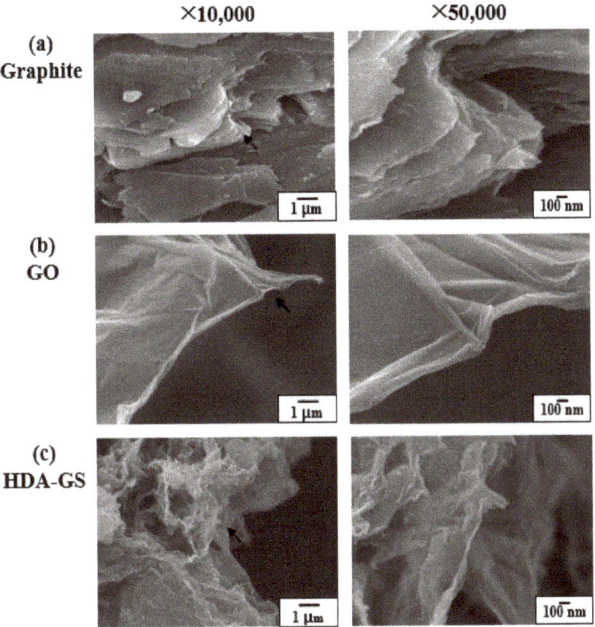

Figure 2. FE-SEM photographs of (**a**) graphite, (**b**) GO, and (**c**) HDA-GS at different magnifications (×10,000 and ×50,000).

Generally, a significant volume expansion and high porosity are observed in the FGSs, which results in low FGS bulk densities that can cause feed problems during melt compounding. Normally, master batches are required to solve the problem. However, in the study, solution blending of FGSs with polymers is successfully employed.

The AFM imaging provides more reliable information on the sheet dimensions and can also be used to probe the surface topology, defects, and bending properties. Additionally, stepped-height scans can also allow us to determine the lateral sizes and thicknesses of the particles lying on the substrates. Thus, AFM is used to indicate that the carbon sheets obtained in the present study are comprised of only a single atomic layer. Figure 3 shows the AFM image of HDA-treated GO sheets (HDA-GSs) on mica and a profile plot that reveals the average sheet thickness corresponding to 1.76 nm (Figure 3) when the thickness of the bare graphene sheet is approximately 1 nm [32], and the thickness of the layer of substituted HDA organic groups is approximately 0.76 nm. However, the layer of the substituted HDA organic groups exhibits a thickness of approximately 0.38 nm when the organic HDA groups on both sides of the graphene sheet in the synthesized FGS are significantly tilted. The change in thickness is associated with different orientations that are adopted by the long alkyl chains in the chemically modified HDA-GSs.

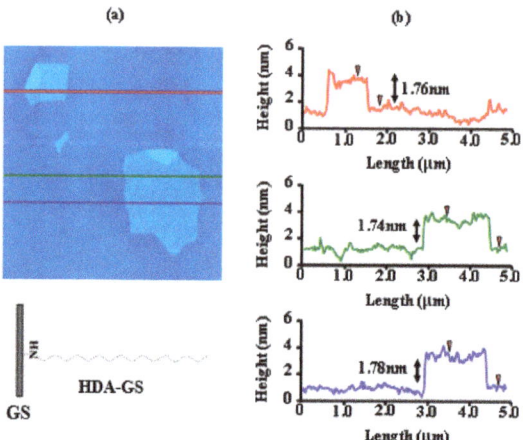

Figure 3. (a) Noncontact-mode AFM image of HDA-GS sheets with (b) three height profiles acquired in different locations.

3.3. Dispersion

The XRD traces of the pure nanofillers and their PVA hybrid films are shown in Figure 4. The d_{001} reflection of pristine SPT is present at $2\theta = 6.62°$, and this corresponds to an interlayer spacing d of 13.54 Å, as shown in Figure 4a. Specifically, Figure 4a also shows the XRD curves of the SPT/PVA hybrid films with clay contents in the range of 0–10 wt%. With respect to the PVA hybrids with a clay content ≤10 wt%, clay peaks did not appear in the XRD traces, which indicates that the clay particles are homogeneously dispersed in the hybrid polymer matrices. Figure 4b also shows the XRD curves of OMB in the region $2\theta = 2°-10°$. The d_{001} reflection of OMB is observed at $2\theta = 6.64°$, and this corresponds to an interlayer distance (d) of 13.30 Å. In the case of PVA hybrids containing 3 wt% of OMB, only a very slight peak appeared at $2\theta = 4.78°$, and this corresponds to an interlayer spacing of 18.46 Å (Figure 4b). The result indicates that a small amount of the clay is not aggregated in the PVA matrix. However, a significant increase in aggregation is observed for samples with OMB loadings reaching 10 wt%, as shown by the intensities of the XRD peaks. This suggests that perfect exfoliation of the layered structure of the clay did not occur. Further evidence of clay dispersion into PVA on a nanometer scale is obtained via TEM.

The XRD diffractograms of pure HDA-GS and HDA-GS/PVA hybrid films are shown in Figure 4c. The d_{001} reflection for HDA-GS is observed at $2\theta = 2.73°$ and corresponds to an interlayer spacing (d) of 32.32 Å. With respect to PVA hybrid films containing up to 7 wt% of HDA-GS, the peak observed at $2\theta = 2.73°$ for GS is observed to disappear from the diffraction patterns. The result indicates that the graphene layers are exfoliated and homogeneously dispersed throughout the PVA matrix and provide supporting evidence for the nanocomposite character of the HDA-GS/PVA hybrids. However, the intensities of the XRD peaks at $2\theta = 2.64$ ($d = 33.42$ Å) and $2\theta = 5.35$ ($d = 16.50$ Å) increase suddenly when the HDA-GS loading increases from 7 wt% to 10 wt%, which suggests that the dispersion is more effective at lower loadings as opposed to higher loadings of graphene.

Given the periodic arrangement of the graphite layers in the virgin GS and in intercalated hybrids, the XRD offers a convenient method to determine the interlayer spacing. However, although the XRD enables precise routine measurements of the GS layer spacings, it neither allows for the determination of the spatial distributions of GSs or the detection of any inhomogeneous sections of the hybrids. Initially, a few layered GSs do not exhibit well-defined basal reflections, and, thus, it is difficult to systematically follow peak broadening and reductions in intensity. Therefore, all conclusions on the mechanisms of hybrid formation and microstructure, based solely on XRD results, are only tentative. Therefore, further evidence of the GS dispersion in the PVA films on a nanometer scale is obtained via SEM and TEM as described in the next section.

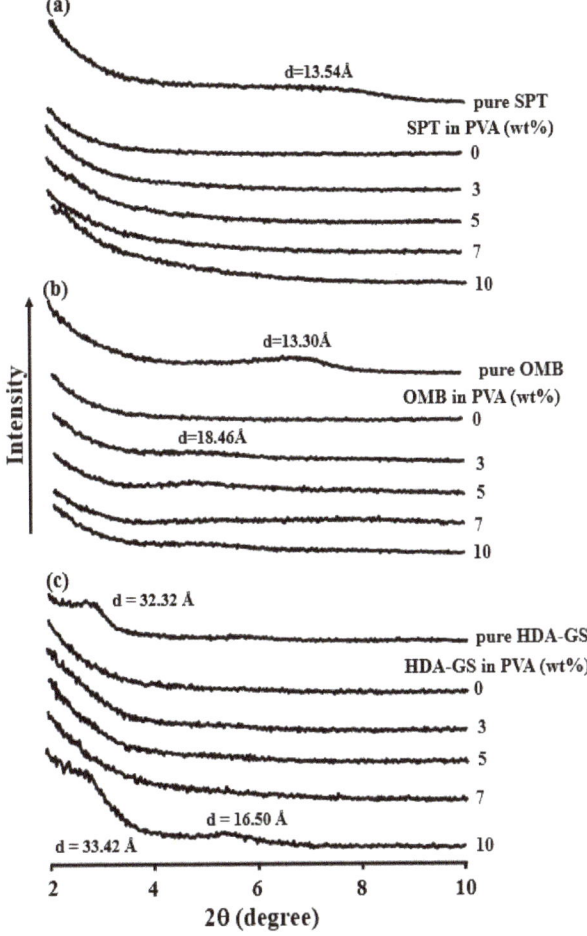

Figure 4. XRD patterns of pristine clay and PVA hybrid films with various nanofiller contents: (a) SPT, (b) OMB, and (c) HDA-GS in PVA.

3.4. Morphologies of PVA Hybrids

In addition to using XRD to measure the *d* spacings of the nanocomposites, SEM and TEM are used to evaluate the degree of intercalation and amount of aggregation in the nanofiller clusters. The morphologies of the aggregated fillers are characterized via SEM. Large filler aggregates can be easily imaged by SEM due to the difference between the scattering densities of the filler and the PVA matrix [33].

The morphologies of the hybrid films containing up to 10 wt% of SPT in the PVA matrix are examined by observing their fracture surfaces by SEM (Figure 5a). The PVA hybrid films containing 3 wt% and 5 wt% of SPT display uniform and dispersed phases. Conversely, the films containing 7 wt% and 10 wt% of SPT exhibit large particles and some deformed regions that can result from the coarseness of the fractured surface. A similar type of behavior was observed for PVA/OMB hybrid films. For example, the fractured surface of the 5-wt% OMB hybrid film (Figure 5b) exhibit uniform and dispersed phases. However, increased agglomeration is observed in the PVA matrix with clay content exceeding 7 wt% in the OMB system.

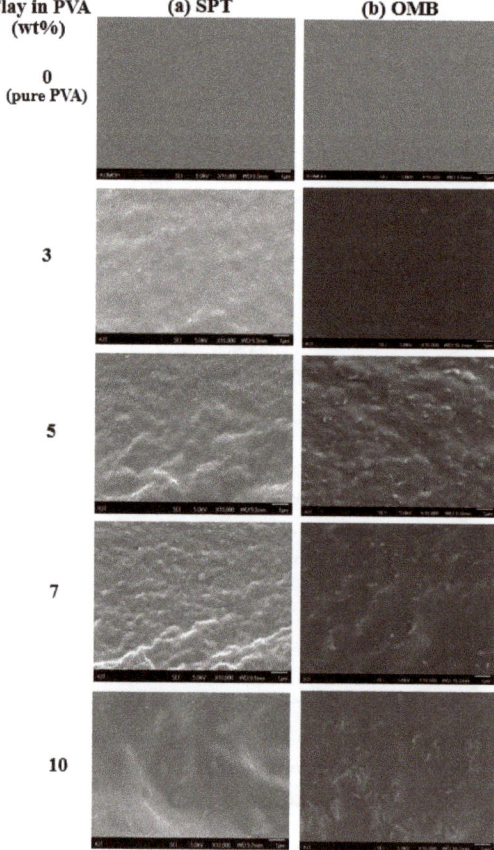

Figure 5. FE-SEM micrographs of PVA hybrid films with various clay contents: (a) SPT and (b) OMB in PVA (×10,000).

The fractured surfaces display increased levels of deformation for samples with higher clay contents. The trend is most likely linked to increases in the agglomeration of clay particles and indicate the lack of interfacial interactions between the clay and matrix polymers. Thus, several defects and significant agglomeration occur in interphase areas in high clay content nanocomposite PVA films. Figure 6 shows a comparative analysis of the SEM micrographs obtained for the PVA hybrids with different contents of HDA-GSs that exhibit platelet-orientation distribution morphology.

Graphene dispersions are readily observed in the SEM images due to the differences in the scattering densities of graphene and the matrix polymer. The SEM images of the fractured surfaces of PVA hybrid films containing 0–10 wt% of HDA-GS are compared in Figure 6. The hybrid film with 5 wt% of HDA-GS display a morphology consisting of graphene domains that are well dispersed throughout the continuous PVA phase. However, deformed surfaces and voids are observed in the case of the 7-wt% FGS hybrid film. Conversely, the micrographs of the 10-wt% HDA-GS/PVA hybrid films exhibit increased levels of voids and deformed regions when compared to the 3–7 wt% HDA-GS/PVA hybrid films due to the coarseness of the fractured surface. Overall, the comparison reveals that increases in the graphene content in the hybrid films increase the level of deformation of their fractured surfaces. The finding potentially results from the agglomeration of graphene particles. A comparison of the micrographs indicate that the fractured surfaces of the hybrid films with higher graphene content are more deformed than those of hybrid films with lower graphene content, and this is possibly due to the agglomeration of graphene particles [34,35]. It should be noted that Figure 6 also shows that most of the graphene remains in the form of straight and rigid platelets in the composite, which indicates that the graphene sheets are extremely stiff.

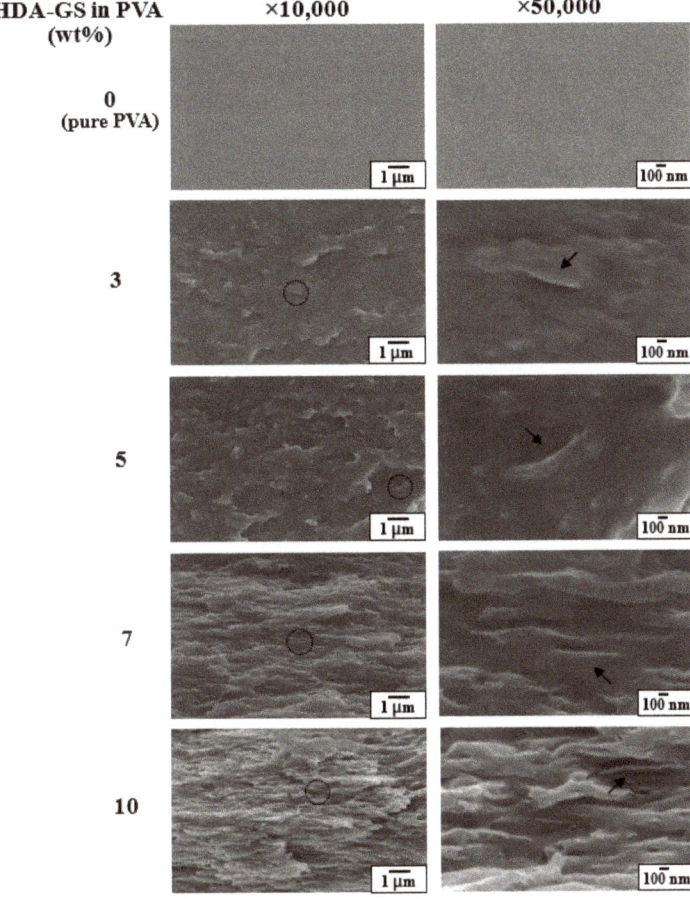

Figure 6. SEM micrographs of PVA hybrid films with various HDA-GS contents.

We extend the morphological analysis via TEM to evaluate the degree of intercalation and degree of aggregation of the nanofiller clusters. Additional direct evidence for the formation of a true nanocomposite is provided by the TEM analysis of ultra-microtomed sections. Figure 7 shows the micrographs of the PVA hybrid films with identical contents of the three different nanofillers. The dark lines in the photographs denote the intersections of the clays and GSs (1-nm thick) while the length between the dark lines denotes the interlayer distance. Figure 7a shows the morphologies of the PVA hybrids with 5 wt% and 10 wt% of SPT. With respect to the hybrid containing 5 wt% of SPT, each layer in the polymer matrix consists of well-dispersed individual clay layers (dark lines), and a few of the clays aggregate to a thickness of approximately 10 nm. In a manner similar to the case for the hybrid material containing 10 wt% of SPT, these clays are mainly aggregated in the polymer matrix. However, the average particle size is observed to be below 20 nm, as calculated from the TEM images. The presence of agglomerated particles in the SEM micrographs of hybrid materials with higher SPT contents (see Figure 5a) is attributed to the formation of aggregated layers.

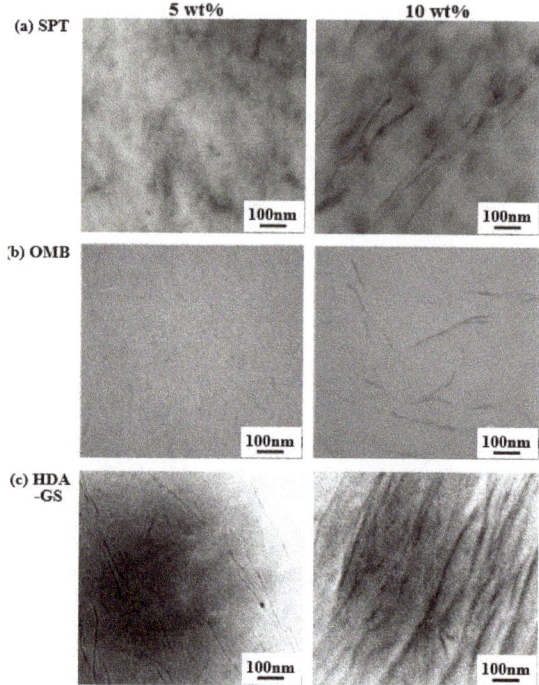

Figure 7. TEM micrographs of PVA hybrid films with different nanofiller contents: (**a**) SPT, (**b**) OMB, and (**c**) HDA-GS in PVA.

Typical TEM images of PVA hybrid films containing 5 and 10 wt% OMB are shown in Figure 7b. Evidently, the clays are well dispersed within the polymer matrix irrespective of the clay content. In contrast to the clay layers in the hybrids containing SPT, the clay layers in OMB hybrids are exfoliated within the matrix polymer.

The TEM micrographs of the 5 and 10 wt% HDA-GS hybrid films are shown in Figure 7c. The TEM micrographs indicate that the GS in the 5 wt% HDA-GS hybrid is dispersed in the polymer matrix, which indicates the formation of a nanocomposite. The findings suggest that GS breaks down into nanoscale building blocks during the intercalative polymerization process and is homogeneously dispersed in the polymer matrix to yield a polymer/GS nanocomposite. As in the case of the 10 wt% HDA-GS (see Figure 7c), the GSs are mostly agglomerated in the polymer matrix. In contrast to the hybrids containing 5 wt% of HDA-GS, the graphene layers of the 10 wt% hybrid exhibited agglomeration of the dispersed graphene phase and are not intercalated into the matrix polymer. The agglomeration of the dispersed graphene phase visibly increases with increases in graphene content, and the outcome is consistent with the XRD and SEM data shown in Figures 4c and 6.

The XRD, SEM, and TEM results indicate that the fillers are well dispersed throughout the PVA matrix at low filler contents while aggregated structures are present at higher filler contents. Additionally, the dispersions of SPT and OMB exceed that of HDA-GS in the PVA matrix (see Figures 4–7). The unusual thermal and gas barrier properties of these hybrid films are discussed in the following sections with respect to the dispersion of the nanofillers.

3.5. Thermal Properties

A comparison of the DSC results for pure PVA and PVA hybrids with approximately 3–10 wt% of clays (SPT and OMB) and graphene (HDA-GS) are listed in Table 1. The glass transition temperature

(T_g) of pure PVA corresponds to 69 °C. The T_g values of PVA hybrids containing various clay contents are virtually unchanged in the DSC results when compared to that of pure PVA irrespective of the filler loading, i.e., approximately 68–71 °C for the three different fillers. Generally, T_g increases when the filler content increases up to a critical concentration. The increase in T_g is potentially due to the confinement of the intercalated polymer chains within the filler galleries that prevents the segmental motion of the polymer chains [36,37]. However, the T_g values of the three types of PVA hybrid films remains constant irrespective of the filler content, which suggests that the variation in the filler content does not affect the confinement of the PVA chains.

Table 1. Thermal properties of PVA hybrid films with various nano-filler contents.

Nanofiller in PVA (wt%)	SPT				OMB				HDA-GS			
	T_g (°C)	T_m (°C)	$T_D{}^{i\,a}$ (°C)	$wt_R{}^{600\,b}$ (%)	T_g (°C)	T_m (°C)	$T_D{}^i$ (°C)	$wt_R{}^{600}$ (%)	T_g (°C)	T_m (°C)	$T_D{}^i$ (°C)	$wt_R{}^{600}$ (%)
0 (pure PVA)	69	165	227	3	69	165	227	3	69	165	227	3
3	70	172	248	6	70	170	240	6	69	166	236	7
5	70	176	249	12	69	184	252	12	70	168	237	11
7	69	168	240	12	70	180	251	12	71	173	245	12
10	70	166	238	15	68	178	246	14	70	156	238	15

[a] At a 2 % initial weight-loss temperature. [b] Weight percent of residue at 600 °C.

The endothermic peak of pure PVA appears at 165 °C and corresponds to its melting transition temperature (T_m) (see Table 1). The T_m values of the hybrid films are observed to increase from 165 to 176 °C when the SPT loading is increased from 0 to 5 wt% and, subsequently, decreases to 166 °C at an SPT content of 10 wt%. The increase in the T_m of the hybrid film potentially occurs as a result of the insulation effect of the clays and the interactions between the clay and PVA chains [38]. However, the decrease in the T_m value of the 10-wt% SPT hybrid suggests that its domains are more poorly dispersed in the PVA matrix than those in the 5-wt% SPT hybrid. Hence, increases in clay content led to the aggregation of clay particles, and this reduces the heat-insulation effect of the clay layers in the polymer matrix.

A similar trend is observed for both the OMB and HDA-GS hybrids. Specifically, the T_m of the PVA hybrids increases from 165 to 184 °C and from 165 to 173 °C when the filler loadings increase to 5 wt% for OMB and 7 wt% for HDA-GS. As observed for the SPT hybrid material, the maximum transition peaks of the PVA hybrids increase with the addition of the nanofiller only up to a certain content level and, subsequently, decreases when the content is increased above this point. For example, when the filler content of PVA reaches 10 wt%, the T_m decreases to 178 °C and 156 °C for materials employing OMB and HDA-GS nanofillers, respectively. The DSC thermograms of the PVA hybrids with various HDA-GS contents are shown in Figure 8. When the nanofiller is included, the peak is stronger than that of the pure PVA, and the more the amount of the filler increased up to 7 wt%, the greater the intensity of the peak becomes. Hence, the HDA-GS appears to act as a nucleating agent [39]. However, when the amount of HDA-GS reaches 10 wt%, the degree of dispersion decreases in conjunction with decreases in the peak intensity.

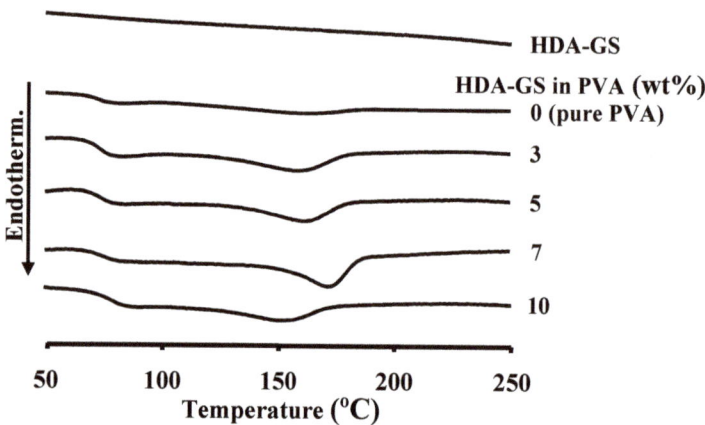

Figure 8. DSC thermograms of PVA hybrid films with various HDA-GS contents.

In a manner similar to the results for T_m, the initial thermal degradation temperatures ($T_D{}^i$) of the PVA hybrid films also increases linearly from 227 °C to 249 °C while increasing the SPT loading from 0 to 5 wt%, as shown in Table 1. With respect to the hybrids containing OMB, the $T_D{}^i$ value varies from 227 to 252 °C when the content of the OMB organoclay increases from 0 to 5 wt% in the PVA hybrids. The highest increase of 25 °C in the $T_D{}^i$ relative to that of pure PVA is observed for the 5 wt% OMB hybrid (252 °C). The clay enhances the $T_D{}^i$ by acting as an insulator and a mass-transport barrier to the volatile products generated during decomposition [40,41]. The increase in thermal stability is attributed to the high thermal stability of the clay and interactions between the clay particles and polymer matrix. In contrast to the behavior observed for clay contents ranging from 0 to 5 wt%, the $T_D{}^i$ values of the hybrids decreases when the clay content increases from 5 to 10 wt%. For example, the $T_D{}^i$ values of PVA hybrid film containing 10 wt% of clay loadings were 11 °C (238 °C) and 6 °C (246 °C) lower than those of PVA hybrids containing 5 wt% of SPT and OMB, respectively. The decrease in $T_D{}^i$ appeared to correspond to the result of clay aggregation that occurs when the clay content in the polymer matrix exceeds a critical value. Similar results are observed in samples containing HDA-GS, as shown in Table 1. The $T_D{}^i$ values of the PVA hybrid films also increases linearly from 227 to 245 °C when the HDA-GS loading is increased from 0 to 7 wt%. At 10 wt% content of HDA-GS in the PVA, the $T_D{}^i$ decreases again to 238 °C.

An analysis of the weight residue at 600 °C ($wt_R{}^{600}$) indicates that the weight increases with growth in the filler loading from 0 to 10 wt% and corresponds to a range from 3% to 15% for SPT, from 3% to 14% for OMB, and from 3% to 15% for HDA-GS (Table 1). The enhancement in char formation with increases in filler content is ascribed to the high heat resistance of the clays and graphene.

3.6. Gas Permeation

The mobility of the polymer chain segments in the polymer nanocomposite clearly differs from that of the pure polymer due to the confined environment, and this also affects the gas permeability. The following two main factors are responsible for permeability reduction [42–44]: (i) polymer chain-segment immobility and (ii) detour ratio, which is defined as the ratio of the film thickness in the nominal diffusion flow direction to the average length of the tortuous diffusion distance between nanolayers.

High aspect ratio nanolayers also lead to properties that are not possible for larger-scaled composites. The impermeable nano-sized layers mandate a tortuous pathway for a permeant to transverse the nanocomposite. Enhanced barrier characteristics, chemical resistance, reduced solvent uptake, and flame retardance of polymer hybrids benefit from the hindered diffusion pathways through the nanocomposite [43,44].

In order to further characterize the barrier properties of the PVA hybrids fabricated by the intercalation of polymer chains in the galleries of clays and graphene, the permeability of the resulting PVA hybrid films to O_2 is evaluated for various filler loadings in the range of 0–10 wt%. The results are summarized in Table 2. The thicknesses of all the films subjected to the gas permeation measurements are in the range of 20 to 24 µm. We discuss our results in terms of relative permeability, P_c/P_p, where P_p denotes the permeability of the pure polymer and P_c denotes the permeability of the composite. The results confirm that the mass-transfer process for O_2 as the penetrant is highly dependent on the level of filler loading. For example, the addition of only 7 wt% of SPT results in a 95% reduction in the permeability rate of O_2 (0.24 cc/m^2/day) relative to that of the pure PVA film (5.13 cc/m^2/day). With respect to the PVA hybrid films containing 3 to 7 wt% of OMB, the relative O_2 permeability rate is close to zero (see Table 2). The outcome is attributed to the increase in the length of the tortuous paths followed by the gas molecules and interactions between the O_2 and clay molecules. Furthermore, films containing higher amounts of clay appear as significantly more rigid, and this decreases their gas permeability. In a manner similar to the results observed for the clay hybrids, the addition of 5 wt% of HDA-GS results in an 81% reduction in the permeability rate of O_2 (0.98 cc/m^2/day) relative to that of the pure PVA film (5.13 cc/m^2/day). However, when the clay content increases from 7 to 10 wt%, the O_2 permeability rate slightly increases from 0.24 to 0.88 cc/m^2/day for SPT and more significantly from <10^{-2} to 3.18 cc/m^2/day for OMB. The increase in the HDA-GS loading from 5 to 10 wt% led to a similar increase in permeability from 0.98 to 4.27 cc/m^2/day. The increases in permeability values are primarily due to the aggregation of the filler particles in materials employing loadings that exceed the critical filler content levels. The present results are further corroborated via the electron micrographs shown in Figures 5–7.

Table 2. Oxygen permeations of PVA hybrid films with various nanofiller contents.

Nanofiller in PVA (wt%)	SPT			OMB			HDA-GS		
	Thickness (µm)	O_2TR [a] (cc/m^2/day)	P_c/P_p [b]	Thickness (µm)	O_2TR (cc/m^2/day)	P_c/P_p	Thickness (µm)	O_2TR (cc/m^2/day)	P_c/P_p
0 (pure PVA)	20	5.13	1.00	20	5.13	1.00	20	5.13	1.00
3	20	0.44	0.09	20	<10^{-2}	≈0	21	2.41	0.47
5	22	0.25	0.05	21	<10^{-2}	≈0	23	0.98	0.19
7	24	0.24	0.05	22	<10^{-2}	≈0	22	3.85	0.75
10	20	0.88	0.17	21	3.18	0.62	24	4.27	0.83

[a] Oxygen transmission rate. [b] Composite permeability/polymer permeability (i.e., relative permeability rate).

Collectively, the results of the gas permeation analysis reveal that the strongest gas barrier effect of OMB is observed with respect to the O_2TR among the three types of nanofillers that are studied. The enhanced gas barrier capacity of OMB stems from its hydrophilic character, and this allows the formation of hydrogen bonds between OMB molecules and the PVA polymer matrix as well as the reinforcement of chain packing, which significantly reduces the gas permeability of the material [45].

4. Conclusions

In the present study, we investigated the dispersibilities of three nanofillers including pristine clay SPT, hydrophilic organoclay OMB, and functionalized graphene sheets HDA-GSs, in PVA to improve the properties of the PVA hybrid films. Specifically, PVA hybrid films with varying filler contents ranging from 0 to 10 wt% were synthesized via the solution intercalation method. Their thermal properties, morphologies, and gas permeabilities were compared. The present results confirmed that the properties were dependent on the type and quantity of the nano-filler incorporated in the PVA polymer matrix.

The morphologies of the hybrid materials were examined via TEM, which confirmed that OMB exhibited better dispersion properties than SPT and HDA-GS with respect to the PVA matrix. This observation agreed with the thermal stabilities and gas barrier capabilities of these hybrid materials with the same filler loading levels. Furthermore, the results indicated that the addition of a small amount of nanofiller can sufficiently improve the properties of the PVA. Overall, the addition of OMB was more effective than the addition of SPT and HDA-GS to improve the thermal stability and O_2TR of the PVA hybrid composite due to the interactions that formed between the OMB and hydrophilic PVA.

In summary, we demonstrated a simple and effective method to fabricate PVA nanocomposites using the solution intercalation method. Improvements in the thermal property and gas barrier of the obtained composites were observed. It is expected that the use of the proposed methods will allow the widespread use of PVA hybrids in various applications such as permeation membranes, polymer electrolyte fuel cell, packaging films, and drug delivery. Application of the technique to the nano-sized fillers in other polymer composite materials can enhance the various advantages of polymeric materials. The technique can be utilized to further improve thermo-mechanical properties by using clay and other types of fillers based on carbon.

Funding: The study was supported by the Research Fund of Kumoh National Institute of Technology (2018-104-105).

Conflicts of Interest: The authors declare no conflict of interest.

Abbreviations

The following abbreviations are used in the manuscript:

AFM	Atomic force microscopy
DSC	Differential scanning calorimetry
FGSs	Functionalized graphene sheets
FT-IR	Fourier-transform infrared
GO	Graphene oxide
HDA-GSs	Hexadecylamine-functionalized graphene sheets
OMB	Organically modified bentonite
O_2TRs	O_2 transmission rates
PVA	Poly(vinyl alcohol)
PET	Poly(ethylene terephthalate)
SEM	Scanning electron microscope
SPT	Saponite
T_D^i	Initial thermal degradation temperatures
TEM	Transmission electron microscopy
TGA	Thermogravimetric analyzer
T_g	Glass transition temperature
T_m	Melting transition temperature
wt_R^{600}	Weight residue at 600 °C
XRD	Wide-angle X-ray diffraction

References

1. Zhou, K.; Gui, Z.; Hu, Y. Facile synthesis of LDH nanoplates as reinforcing agents in PVA nanocomposites. *Polym. Adv. Technol.* **2017**, *28*, 386–392. [CrossRef]
2. Kashyap, S.; Pratihar, S.K.; Behera, S.K. Strong and ductile graphene oxide reinforced PVA nanocomposites. *J. Alloys Compd.* **2016**, *684*, 254–260. [CrossRef]
3. Yahia, I.S.; Mohammed, M.I. Facile synthesis of graphene oxide/PVA nanocomposites for laser optical limiting: Band gap analysis and dielectric constants. *J. Mater. Sci. Mater. Electron.* **2018**, *29*, 8555–8563. [CrossRef]
4. Rathod, S.G.; Bhajantri, R.F.; Ravindrachary, V.; Naik, J.; Kumar, D.J.M. High mechanical and pressure sensitive dielectric properties of graphene oxide dope PVA nanocomposites. *RSC Adv.* **2016**, *6*, 77977–77986. [CrossRef]
5. Chang, J.-H.; Ham, M.; Kim, J.-C. Comparison of properties of poly(vinyl alcohol) nanocomposites containing two different clays. *J. Nanosci. Nanotechnol.* **2014**, *14*, 8783–8791. [CrossRef] [PubMed]

6. Heo, C.; Chang, J.-H. Syntheses of functionalized grapheme sheets and their polymer nanocomposites. In *Handbook of Functional Nanomaterials Volume 2—Characterization and Reliability*; Niknam, Z.A., Ed.; Nova Science Publishers, Inc.: Hauppauge, NY, USA, 2013; Chapter 4.
7. Jan, R.; Habib, A.; Akram, M.A.; Zia, T.; Khan, A.N. Uniaxial drawing of graphene-PVA nanocomposites: Improvement in mechanical characteristics via strain-induced exfoliation of graphene. *Nanoscale Res. Lett.* **2016**, *11*, 377–385. [CrossRef] [PubMed]
8. Lagaly, G. *Smectitc Calys as Ionic Macromolecules. Developments in Ionic Polymers*; Elsevier Science: London, UK, 1986; Volume 2, pp. 77–140.
9. Fukushima, Y.; Inagaki, S. Synthesis of an intercalated compound of montmorillonite and 6-polyamide. *J. Incl. Phenom. Macrocycl. Chem.* **1987**, *5*, 473–482. [CrossRef]
10. Giannelis, E.P. Polymer layered silicate nanocomposites. *Adv. Mater.* **1996**, *8*, 29–35. [CrossRef]
11. Manias, E.; Touny, A.; Wu, L.; Strawhecker, K.; Lu, B.; Chung, T.C. Polypropylene/ montmorillonite nanocomposites. Review of the synthetic routes and materials properties. *Chem. Mater.* **2001**, *13*, 3516–3523. [CrossRef]
12. Cendoya, I.; Lopez, D.; Alegria, A.; Mijangos, C. Dynamic mechanical and dielectrical properties of poly(vinyl alcohol) and poly(vinyl alcohol)-based nanocomposites. *J. Polym. Sci. Part B Polym. Phys.* **2001**, *39*, 1968–1971. [CrossRef]
13. Hernandez, M.C.; Suarez, N.; Martinez, L.A.; Feijoo, J.L.; Monaco, S.L.; Salazar, N. Effects of nanoscale dispersion in the dielectric properties of poly(vinyl alcohol)-bentonite nanocomposites. *Phys. Rev. E* **2008**, *77*, 1–10. [CrossRef] [PubMed]
14. Turhan, Y.; Alp, Z.G.; Alkan, M.; Dogan, M. Preparation and characterization of poly(vinyl alcohol)/modified bentonite nanocomposites. *Microporous Mesoporous Mater.* **2013**, *174*, 144–153. [CrossRef]
15. Chang, J.-H.; Jang, T.G.; Ihn, K.J.; Lee, W.K.; Sur, G.S. Poly(vinyl alcohol) nanocomposites with different clays: Pristine clays and organoclays. *J. Polym. Sci.* **2003**, *90*, 3208–3214. [CrossRef]
16. Gilman, J.W. Flammability and thermal stability studies of polymer layered-silicate (clay) nanocomposites. *Appl. Clay Sci.* **1999**, *15*, 31–49. [CrossRef]
17. Messersmith, P.B.; Giannelis, E.P. Polymer-layered silicate nanocomposites: In situ intercalative polymerization of e-caprolectone in layered silicates. *Chem. Mater.* **1993**, *5*, 1064–1066. [CrossRef]
18. Zhu, J.; Morgan, A.B.; Lamelas, F.J.; Wilkie, C.A. Fire properties of polystyrene−clay nanocomposites. *Chem. Mater.* **2001**, *13*, 3774–3780. [CrossRef]
19. Davis, C.H.; Mathias, L.J.; Gilman, J.W.; Schiraldi, D.A.; Shields, J.R.; Trulove, P.; Sutto, T.E.; Delong, H.C. Effects of melt-processing conditions on the quality of poly(ethylene terephthalate) montmorillonite clay nanocomposites. *J. Polym. Sci. Part B Polym. Phys.* **2002**, *40*, 2661–2666. [CrossRef]
20. Pinnavaia, T.J. Intercalated clay catalysts. *Science* **1983**, *220*, 365–371. [CrossRef] [PubMed]
21. Potts, J.R.; Dreyer, D.R.; Bielawski, C.W.; Ruoff, R.S. Graphene-based polymer nanocomposites. *Polymer* **2011**, *52*, 5–25. [CrossRef]
22. Pradhan, S.K.; Nayak, B.B.; Sahay, S.S.; Mishra, B.K. Mechanical properties of graphite flakes and spherulites measured by nanoindentation. *Carbon* **2009**, *47*, 2290–2292. [CrossRef]
23. Srinivas, G.; Zhu, Y.; Piner, R.; Skipper, N.; Ellerby, M.; Ruoff, R. Synthesis of graphene-like nanosheets and their hydrogen adsorption capacity. *Carbon* **2010**, *48*, 630–635. [CrossRef]
24. Ansari, S.; Giannelis, E.P. Functionalized graphene sheet-poly(vinylidene fluoride) conductive nanocomposites. *J. Polym. Sci. Part B Polym. Phys.* **2009**, *47*, 888–897. [CrossRef]
25. Raghu, A.V.; Lee, Y.R.; Jeong, H.M. Preparation and physical properties of waterborne polyurethane/ functionalized graphene sheet nanocomposites. *Macromol. Chem. Phys.* **2008**, *209*, 2487–2493. [CrossRef]
26. Pradhan, B.; Setyowati, K.; Liu, H.; Waldeck, D.H. Carbon nanotube−polymer nanocomposite infrared sensor. *Nano Lett.* **2008**, *8*, 1142–1146. [CrossRef] [PubMed]
27. Yang, X.; Li, L. Synthesis and characterization of layer-aligned poly(vinyl alcohol)/ graphene nanocomposites. *Polymer* **2010**, *51*, 3431–3435. [CrossRef]
28. Usuki, A.; Kawasumi, M.; Kojima, Y.; Okada, A.; Kurauchi, T.; Kamigato, O. Synthesis of nylon 6-clay hybrid. *J. Mater. Res.* **1993**, *8*, 1174–1184. [CrossRef]
29. Zhang, Y.; Tan, Y.W.; Stormer, H.L.; Kim, P.P. Experimental observation of the quantum hall effect and berry's phase in graphene. *Nature* **2005**, *438*, 201–204. [CrossRef] [PubMed]
30. Hummers, W.; Offman, R. Preparation of graphitic oxide. *J. Am. Chem. Soc.* **1958**, *80*, 1339. [CrossRef]

31. Pavia, D.L.; Lampman, G.M.; Kriz, G.S. *Introduction to Spectroscopy*; Harcourt: Washington, DC, USA, 2001.
32. Si, Y.; Samulski, E.T. Exfoliated graphene separated by platinum nanoparticles. *Chem. Mater.* **2008**, *20*, 6792–6797. [CrossRef]
33. Morgan, A.B.; Gilman, J.W. Characterization of polymer-layered silicate (clay) nanocomposites by transmission electron microscopy and X-ray diffraction: A comparative study. *J. Appl. Polym. Sci.* **2003**, *87*, 1329–1338. [CrossRef]
34. Galgali, G.; Ramesh, C.; Lele, A. A rheological study on the kinetics of hybrid formation in polypropylene nanocomposites. *Macromolecules* **2001**, *34*, 852–858. [CrossRef]
35. Heo, C.; Moon, H.G.; Yoon, C.S.; Chang, J.-H. ABS nanocomposite films based on functionalized-graphene sheets. *J. Appl. Polym. Sci.* **2012**, *124*, 4663–4670. [CrossRef]
36. Choi, I.H.; Chang, J.-H. Colorless polyimide nanocomposite films containing hexafluoro-isopropylidene group. *Polym. Adv. Technol.* **2011**, *22*, 682–689. [CrossRef]
37. Agag, T.; Takeichi, T. Polybenzoxazine–montmorillonite hybrid nanocomposites: Synthesis and characterization. *Polymer* **2000**, *41*, 7083–7090. [CrossRef]
38. Kumar, S.; Jog, J.P.; Natarajan, U. Preparation and characterization of poly(methyl methacrylate)–clay nanocomposites via melt intercalation: The effect of organoclay on the structure and thermal properties. *J. Appl. Polym. Sci.* **2003**, *89*, 1186–1194. [CrossRef]
39. LeBaron, P.C.; Wang, Z.; Pinnavaia, T.J. Polymer-layered silicate nanocomposites: An overview. *Appl. Clay Sci.* **1999**, *15*, 11–29. [CrossRef]
40. Fornes, T.D.; Yoon, P.J.; Hunter, D.L.; Keskkula, H.; Paul, D.R. Effect of organoclay structure on nylon 6 nanocomposite morphology and properties. *Polymer* **2002**, *43*, 5915–5933. [CrossRef]
41. Pertrvic, X.S.; Javni, L.; Waddong, A.; Banhegyi, G.L. Structure and properties of polyurethane–silica nanocomposites. *J. Appl. Polym. Sci.* **2000**, *76*, 133–151. [CrossRef]
42. Sakaya, T.; Osaki, N. The potential of nanocomposite barrier technology. *J. Photopolym. Sci. Technol.* **2006**, *19*, 197–202. [CrossRef]
43. Xu, B.; Zheng, Q.; Song, Y.; Shangguan, Y. Calculating barrier properties of polymer/clay nanocomposites: Effects of clay layers. *Polymer* **2006**, *47*, 2904–2910. [CrossRef]
44. Jarus, D.; Hiltner, A.; Baer, E. Barrier properties of polypropylene/polyamide blends produced by microlayer coextrusion. *Polymer* **2002**, *43*, 2401–2408. [CrossRef]
45. Ham, M.R.; Kim, J.-C.; Chang, J.-H. Characterization of poly (vinyl alcohol) nanocomposite films with various clays. *Polymer (Korea)* **2013**, *37*, 225–231. [CrossRef]

© 2019 by the author. Licensee MDPI, Basel, Switzerland. This article is an open access article distributed under the terms and conditions of the Creative Commons Attribution (CC BY) license (http://creativecommons.org/licenses/by/4.0/).

Article

Magnesium Oxide Nanoparticles: Dielectric Properties, Surface Functionalization and Improvement of Epoxy-Based Composites Insulating Properties

Jaroslav Hornak [1,*], Pavel Trnka [1], Petr Kadlec [1], Ondřej Michal [1], Václav Mentlík [1], Pavol Šutta [2], Gergely Márk Csányi [3] and Zoltán Ádám Tamus [3]

[1] Department of Technologies and Measurement, Faculty of Electrical Engineering, University of West Bohemia, Univerzitní 8, 306 14 Pilsen, Czech Republic; pavel@ket.zcu.cz (P.T.); kadlecp6@ket.zcu.cz (P.K.); mionge@ket.zcu.cz (O.M.); mentlik@ket.zcu.cz (V.M.)
[2] New Technologies-Research Centre, University of West Bohemia, Univerzitní 8, 306 14 Pilsen, Czech Republic; sutta@ntc.zcu.cz
[3] Department of Electric Power Engineering, Faculty of Electrical Engineering and Informatics, Budapest University of Technology of Economics, Egry J. Street 18., H-1111 Budapest, Hungary; csanyi.gergely@vet.bme.hu (G.M.C.); tamus.adam@vet.bme.hu (Z.A.T.)
* Correspondence: jhornak@ket.zcu.cz; Tel.: +420-37763-4530

Received: 9 May 2018; Accepted: 23 May 2018; Published: 30 May 2018

Abstract: Composite insulation materials are an inseparable part of numerous electrical devices because of synergy effect between their individual parts. One of the main aims of the presented study is an introduction of the dielectric properties of nanoscale magnesium oxide powder via Broadband Dielectric Spectroscopy (BDS). These unique results present the behavior of relative permittivity and loss factor in frequency and temperature range. Following the current trends in the application of inorganic nanofillers, this article is complemented by the study of dielectric properties (dielectric strength, volume resistivity, dissipation factor and relative permittivity) of epoxy-based composites depending on the filler amount (0, 0.5, 0.75, 1 and 1.25 weight percent). These parameters are the most important for the design and development of the insulation systems. The X-ray diffraction patterns are presented for pure resin and resin with optimal filler amount (1 wt %), which was estimated according to measurement results. Magnesium oxide nanoparticles were also treated by addition of silane coupling agent (γ-Glycidoxypropyltrimethoxysilane), in the case of optimal filler loading (1 wt %) as well. Besides previously mentioned parameters, the effects of surface functionalization have been observed by two unique measurement and evaluation techniques which have never been used for this evaluation, i.e., reduced resorption curves (RRCs) and voltage response method (VR). These methods (developed in our departments), extend the possibilities of measurement of composite dielectric responses related to DC voltage application, allow the facile comparability of different materials and could be used for dispersion level evaluation. This fact has been confirmed by X-ray diffraction analyses.

Keywords: broadband dielectric spectroscopy; dielectric strength; loss factor; magnesium oxide; nanocomposite; relative permittivity; surface functionalization; voltage response

1. Introduction

Magnesium oxide, often called periclase [1] (from Greek word periklao, peri—"around", klao—"to cut"), is white hygroscopic solid mineral. Its empirical formula is MgO and its lattice consist of Mg^{2+} ions and O^{2-} ions, together bonded by ionic bond (Figure 1). Magnesium oxide is generally produced by the calcination of magnesium hydroxide Mg(OH)$_2$ or magnesium carbonate MgCO$_3$. Thermal treatment, used when calcination process occurs, affects the surface area and pore size and also the final reactivity of formed magnesium oxide. Used temperature can be divided into three groups, 700 °C to 1000 °C, where caustic calcined magnesium oxide is formed, 1000 °C to 1500 °C, where lower chemical activity magnesium oxide is formed and calcination over 1500 °C, where reduced chemical activity type of refractory magnesium oxide is formed, that is mostly used for electrical and refractory applications [2].

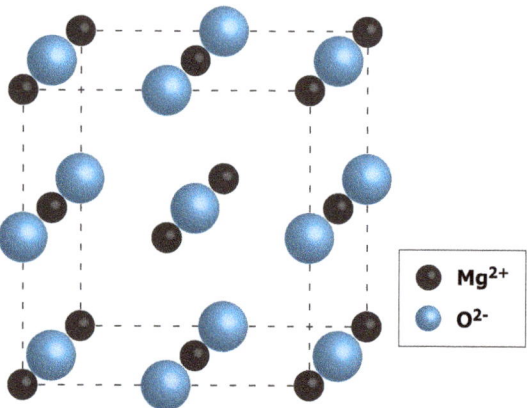

Figure 1. Structure of magnesium oxide crystal (Redrawn and adapted from [2]).

Physical properties (see [3]) make magnesium oxide a good candidate for various applications. It is colorless to brown or black (based on the presence of iron or other foreign element). Considering the surface structure, it is visible that MgO has simplest oxide structure, called Rock-Salt structure. Its density is around 3.579 g/cm^3 and hardness around 5 on Mohs scale. Thermal conductivity value of sintered magnesium oxide is defined at $T = 100$ °C as 36 W/(mK). Due to refractory properties, the melting and also boiling points of magnesium oxide are very high (melting point: 2800 °C, boiling point: 3600 °C). Value of electrical resistance is depended on the purity of magnesium oxide. For high purity magnesia, the values of electrical resistivity can reach 10^{16} Ω·m. Specific resistance is mostly depended on chemical purity, but for higher values of temperature, i.e., 2000 °C and more, the purity of magnesia does not have any influence on values of electrical resistivity. The dielectric constant of magnesium oxide is in the range from 3.2 to 9.8 at 25 °C and under frequency 1 MHz, also values of dielectric loss for same conditions are around 10^{-4}.

Chemical properties and surface composition of magnesium oxide are also influenced by the calcination procedure [4,5] (used temperature and used medium, i.e., air or vacuum) and also by the source of the precursor. Based on the various result, physical adsorption of water only occurs if MgO contains surface defects, such as high quantity of pores [6].

Applications of magnesium oxide includes various industry sectors. For their refractory properties, it is a valuable fireproofing ingredient in construction materials. Also in applications where corrosion [7] is not acceptable such as nuclear, chemical or superalloy industries. It has a usage in medical applications [8], where MgO is used for relief of heartburn and sour stomach, as an antacid, magnesium supplement, and as a short-term laxative. Other applications include insulators [9], fertilizers [10], water treatment [11], protective coating [12], etc.

Currently, there are trends to use nanoscale fillers [13]. In general, the nanotechnology is the production of functional structures in the range of 0.1–100 nm by various physical or chemical methods [14]. This fact also applies to magnesium oxide. The sol-gel technique [15] or hydrothermal technique [16] could be used for the production of nanoscale magnesium oxide. For an electrical application, e.g., in high voltage insulation, the MgO represents a prospective filler. Especially, due to wide band gap (7.8 eV) and high volume resistivity (10^{17} Ω·m). It is the highest value of volume resistivity from commonly used nanoscale oxides [17]. For this research, the MgO (supplied by NanoAmor [18]) with average diameter 20 nm, specific surface area more than 60 m^2/g and density 0.3 g/cm^3, was used.

A lot of studies have shown the effect of different types of nanofillers dispersed in epoxy-based composites on the mechanical [19,20], thermo-mechanical [21,22] or electrical [23] properties. The use of nanofillers has been also demonstrated to impove mechanical characteristics of many biomedical materials, mainly used in orthopedics [24] and dentistry [25]. However, this paper presents the unique results of nanoscale MgO dielectric properties itself, in the temperature and frequency range. These results are complemented by changes of dielectric properties of epoxy-based composites depending on the filler amount and by the effect of surface modification. These effects of surface functionalization were studied by two special measurement and evaluation techniques, i.e., reduced resorption curves (RRCs) [26] and voltage response method (VR) [27], which may be used as indirect method for evaluation of filler dispersion.

2. Dielectric Properties of MgO Nanoparticles via Broadband Dielectric Spectroscopy

Broadband Dielectric Spectroscopy (BDS) is a modern diagnostic method which allows interconnecting several measurement techniques to obtain a comprehensive view of the material behavior under an electric field with a frequency in very wide range. For this investigation, the main diagnostic unit of the Alpha-A measuring device (Novocontrol Technologies) has been used. It contains a frequency response analyzer with a sinusoidal signal generator and allows analyzing in the frequency range from 3×10^{-6} to 4×10^{7} Hz [28]. Used electrode system (ZGS type) is an active (incorporate a block for diagnostic signal processing) electrode system whose sample cell consists of two parallel cylindrical gold-plated electrodes and the tested flat sample is placed between these electrodes. The diameter of electrodes which are in direct contact with the sample (MgO pellet—see Section 2.1) is 10 mm and therefore the plate capacitor with a diameter of 10 mm is considered in calculations.

The aim of the BDS analysis is primarily to find out general trends of development (and some selected values) of dielectric constant and loss factor as results of measurement at variable frequency of the electric field and variable ambient temperature which is regulated by nitrogen vapor in the cryostat with inserted sample cell. Presented analysis was performed in the temperature range from 25 °C to 150 °C and in the frequency range from 0.5 Hz to 1 MHz. These ranges of set-up measurement parameters were chosen as sufficient with reference to the intended application of MgO as a filler for electrical insulating composites with a polymer matrix [29]. Entire measurement consists of two phases. The first phase called heating represents the period when the temperature in the cryostat was gradually increased from 150 °C to 25 °C with a step of 5 °C. Then, the temperature was decreased from 150 °C to 25 °C with same temperature step in the next phase called cooling. The frequency of the measuring voltage with amplitude of 1 V was gradually decreased in the chosen range for each selected temperature in both phases [30]. Several tens of pair of ϵ' and ϵ'' values for different frequencies were obtained for each temperature (for heating and cooling) after the processing of measured data as a final result of dielectric analysis via BDS.

2.1. Preparation of MgO Pellet

Used electrode system as a part of the Broadband dielectric spectroscope do not enable measurement of a powder filler in the delivered state. Measurement is possible only with a sample with defined shape and dimensions. In particular, it is necessary to prepare a pellet with a structure, which is as homogeneous as possible, and with a defined thickness. Optimally prepared pellet can be placed between measuring electrodes without pellet fragmentation. The preparation of the pellet represents a homogenization of the MgO powder (more than 200 mg) in a ShakIR sample grinder (PIKE Technologies, Fitchburg, WI, USA) in the first step. The second step is a pressing of the powder whereas the amount of 200 ±1 mg of powder was loaded into the evacuable pellet press (PIKE Technologies) with a pressing chambers diameter of 13 mm. Air is evacuated from the pressing chamber during a compression. The evacuable pellet press is inserted between parallel pressing plates of the hydraulic press H-62 (Trystom, Olomouc, Czech Republic). It was necessary to optimize the maximum applied pressure in order to avoid a excessive deformation and a fixation of filler particles in pellet volume. This pressure was set to a value of approximately 340 MPa (force of 45 kN applied to anvils) to minimize the adhesion of the pellet surface to the anvils surface that may cause a damage of the pellet during removal from the chamber [31]. This pressure was determined on the basis of self-optimization of pellet preparation from MgO.

2.2. Comprehensive Analysis of Relative Permittivity

The final result of dielectric analysis performed via BDS are interpreted primarily as complex 3D view (Origin®, OriginLab, Northampton, MA, USA) of frequency and temperature dependencies of dielectric constant and loss factor. These dependencies are shown in Figure 2, in which phases of the measurement under increasing and subsequently under decreasing temperature are distinguished in color.

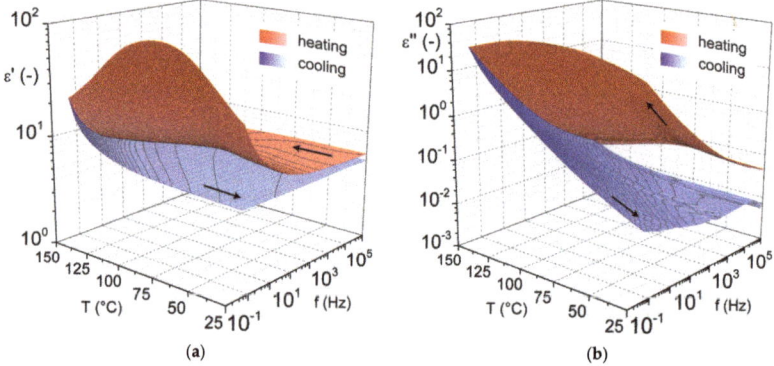

Figure 2. 3D interpretation of frequency-temperature dependencies of (**a**) dielectric constant and (**b**) loss factor for heating (red scale) and cooling (blue scale) of MgO pellet.

The results of the analysis show significant increase in ϵ' and ϵ'' values with the decreasing of measuring voltage frequency and also with the increasing of temperature when the material is heated in the delivered state. This trend of ϵ' and ϵ'' development is visible for cooling too, but with a different character. The results for the heating also denote the formation of noticeably visible peak in the temperature dependencies of dielectric constant and loss factor. The 3D interpretation of results of cooling shows a significantly smoother surface without visible peaks. In general, lower values of ϵ' and ϵ'' are recorded always for cooling, whereas primarily the loss factors decline by several orders of magnitude compared with heating is detected for lower temperatures.

The increase of ϵ' and ϵ'' caused by the frequency decreasing is primarily the effect of the electrical conductivity of the tested material, which is commonly visible in the low frequencies area of 3D interpretation of similar analysis results. The general cause of the most pronounced increase of dielectric constant and loss factor in the highest temperatures area is a disordered thermal movement of particles in the MgO pellet which is become more apparent under increasing temperature not only for this material. The peak occurrence in characteristics is a result of changes in material structure which are caused by the temperature rise during the measurement and which significantly influence dielectric properties of MgO. Specifically, changes in structure are related to a process of MgO dehydration. This material evidently contains a significant amount of water molecules in the delivered state. However, these water molecules are with high probability only absorbed or very weakly bonded in the volume of MgO powder if they are released at temperatures below 100 °C. The effect of dehydration of MgO (differences of ϵ' and ϵ'' between heating and cooling) is significant in the case of lower frequencies and mean temperatures which are the most important for the intended application in electrical engineering. Differences in values of dielectric constant and loss factor for selected temperatures and industrial frequency of 50 Hz are shown in Table 1. Results in this table prove the fact that the usage of MgO without dehydration in composites with thermoset matrix (with values of the loss factor lower by orders of magnitude than for MgO at evaluated temperatures) cured at room temperature can have a significantly negative effect on dielectric properties of the composite. On the other hand, after dehydration, MgO exhibits very similar or even lower values of the loss factor than the pure thermoset at evaluated temperatures.

Table 1. Selected values of dielectric constant and loss factor for the industrial frequency of 50 Hz.

T (°C)	ϵ' (-)	ϵ'' (-)
25	7.77 (5.80)	1.55 (0.205)
50	8.43 (5.67)	2.68 (0.0801)
75	9.56 (5.57)	4.90 (0.0390)
100	8.65 (5.51)	3.86 (0.0220)
125	7.25 (5.46)	1.90 (0.0145)

Numbers before brackets are values for heating and in brackets are values for cooling.

3. Improvement of Epoxy Based Composites Insulating Properties

The industrial epoxy resin (composition according to supplier safety sheet: Bisphenol-A and epichlorohydrin 50–70% (Figure 3); 1,4-Bis(2,3-epoxypropoxy)butane: 10–20%; Alkyl (C12–C14) glycidyl ether: 5–10%) with low processing viscosity and high bond strength was used for this experiment. This epoxy resin is curable at elevated temperatures (140 °C, 4–6 h or 160 °C 3–6 h) without additional hardener and it is commonly used in industry due to low viscosity and this is the main reason of our choice. Because lower viscosity of basic material ensures better dispersion of nanoparticles [32]. This resin is evaluated for thermal class H (IEC Standard 60085:2007 [33]), it is free of solvents and it is recommended for vacuum pressure impregnation (VPI) of rotating machines [34]. The density of selected epoxy resin is 1.12 g/cm^3.

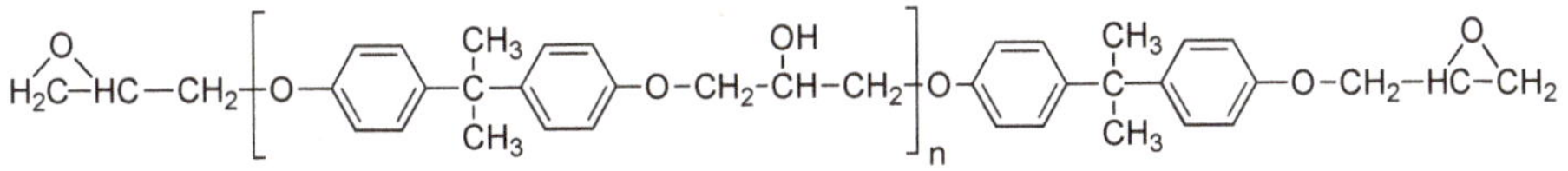

Figure 3. Structure of resin based on on Bisphenol-A diglycidyl ether.

3.1. Production of Epoxy-Based Nanocomposites Samples

In this part, the sample production of pure epoxy resin and resin with dispersed nanoparticles will be described. Amount of 60 g of epoxy resin was used for the creation of a collection of the samples (5 pcs). In the case of pure epoxy, only magnetic stirring of the epoxy resin was carried out together with the vacuum venting (8 mbar) for 3 h. After this time, the epoxy resin was placed in the preheated Teflon molds with a silicone frame determining the height of the sample. The resin was then cured in a hot-air oven (140 °C, 6 h). In case of matrices with dispersed particles, nanoparticles (0.5, 0.75, 1 and 1.25 wt %) were added to the already heated resin (75 °C, 600 rpm, 3 h). An ultrasonic thorn was then used to break agglomerates (30 min) with simultaneous magnetic stirring and heating (70 °C, 300 rpm). Further, the vacuuming (8 mbar) process was combined with magnetic stirring and simultaneous heating (90 °C, 300 rpm, 3 h). The epoxy resin mixture was further placed in dried Teflon molds and cured under the same conditions as in the previous case.

3.2. Dielectric Properties of Epoxy-Based MgO Nano-Composites

Four experimental measurements were performed for observation of MgO effect and finding the optimal filler ration to achieve the best dielectric properties in comparison with unfilled matrix (Figure 4). Namely dissipation factor and relative permittivity (IEC 60250:1969 [35]), dielectric strength (IEC 60243-1:2013 [36]) and volume resistivity (IEC 62631-3-2:2015 [37]) were measured according to mentioned standards. All measurements were performed according to Standard conditions given by IEC 60212:2010 [38]. Because of the basics of these parameters, no further information about step-by-step measurement procedure is provided in this article.

From presented results is visible that the addition of MgO nanofiller causes the changes of selected parameters. Possible reasons are discussed in following text. There is a slight increase of relative permittivity and dissipation factor, respectively. These parameters characterize the degree of polarizability of the matrix, filler and their interfaces and losses caused by their interactions with electric field. Different behavior can be explained by changing the curing reaction of the whole composite and by changing the degree of crosslinking [39].

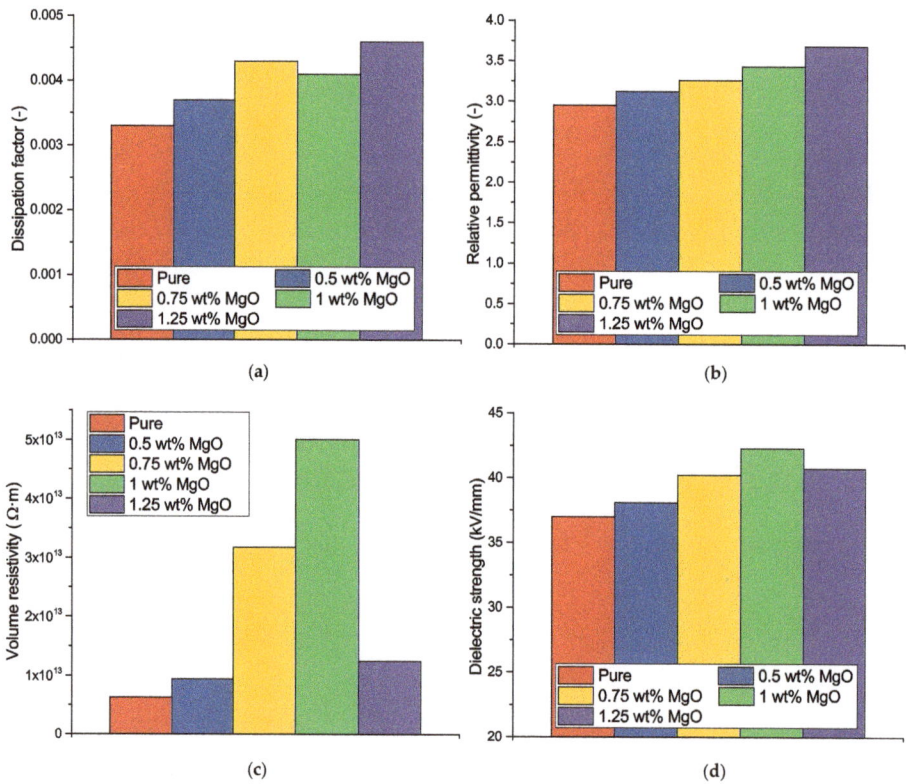

Figure 4. Changes of dielectric properties depending on the filler amount. (**a**) dissipation factor—500 V AC, 50 Hz; (**b**) relative permittivity—500 V AC, 50 Hz; (**c**) Volume resistivity—500 V DC; (**d**) Dielectric strength—increase 1.5 kV/s AC.

Study [40] highlighted the positive effect of magnesium oxide of nanometric dimensions on the reduction of trapped charge in the internal structure of the material at a fill volume in the range 0.5–2%. Due to the nature of the particles, when their volume resistivity is in the order of 10^{17} Ω·m, the resistivity of whole composite may increased at low filling rates. This fact was confirmed by performed measurements. An increase of volume resistivity of the composites can be attributed to an increase in resistance to injection of the charge carriers and their generation in the internal dielectric structure [41]. Dielectric breakdown phenomena of nanodielectrics is affected mainly due to low quantity of agglomeration at low filler loadings [42]. Some studies [43,44] also shown the changes of relation between the dielectric breakdown and free-volume in polymers. In these cases, also a percolation threshold could plays a role, but not more works have been presented for percolation threshold estimation in the case of dispersed nanoscale insulating particles, where the character of added filler causes improvement of the electro-insulation properties of whole composite. On the other hand, the behavior of dispersed conductive fillers is very-well known [45–48] and percolation threshold could be estimated based on the significant increase of conductivity. However, the dielectric parameters of basic material could be improved by addition of relatively low amount of nanofiller, as is evident from our previous studies [49–51]. In connection with these claims, the study [52] describes a theory of percolation and interfacial characterisation via breakdown voltage measurement. It may be also used in this case for the confirmation of presented results (Figure 4d). In the case of pure epoxy resin, deeper traps generally exist. It results in relatively easy charge capture. If the charge carriers are released, the breakdown occurs due to their energy. The increase of breakdown voltage can be attributed to increasing of shallow traps inside the material by addition of nanofiller up to percolation threshold. The expected decrease will occur if the double-layers [53] on the particles surface are overlapped. It leads to the easier movement of the charge carriers in double-layers. Due to this fact, the conductive path will be formed [54] and breakdown can occur more easily. These effects are better noticable on the values of volume resistivity (Figure 4c), which goes hand in hand with the breakdown voltage measurement. According to the presented results, the percolation threshold could be estimated greater than 1 wt %.

Taking into account the preliminary measurements of the basic electrical properties (Figure 4), especially the volume resistivity and breakdown voltage, together with the the above-presented data, the optimal weight ratio for further investigation was set to 1 wt %.

3.3. X-ray Diffraction of Epoxy Resin and Epoxy Resin with Dispersed MgO Nanoparticles

The X-ray difraction has been used for the characterisation of internal structure of the tested material. These measurements were performed on a Panalytical X'Pert Pro (Malvern Panalytical) automated powder X-ray diffractometer using an X-ray lamp (IKa1 = 0.154 nm, 40 kV, 30 mA) and a semiconductor ultra-rapid PIXcel detector in the geometric Bragg-Bretan arrangement. The results from the diffractometer were aligned with the Pearson VII curves. From the X-ray diffraction analysis is visible the character of the amorphous material. This symbolizes very wide diffraction (6–8 degrees in the diffraction angles 2-theta), which are shown in Figure 5.

These diffractions, resp. their positions on the x-axis, corresponding with the results of presented studies [55,56]. On this diffractogram, only the diffraction pattern of MgO (200), (220) and (222) are noticed. Other lines are weak, or we do not notice them at all. Analysis of the profile of diffraction lines (200) showed that the size of the coherent dispersion region of X-ray crystallization (crystallite) is in all cases about 23–25 nm and the micro deformation is relatively low (0.0022–0.0025).

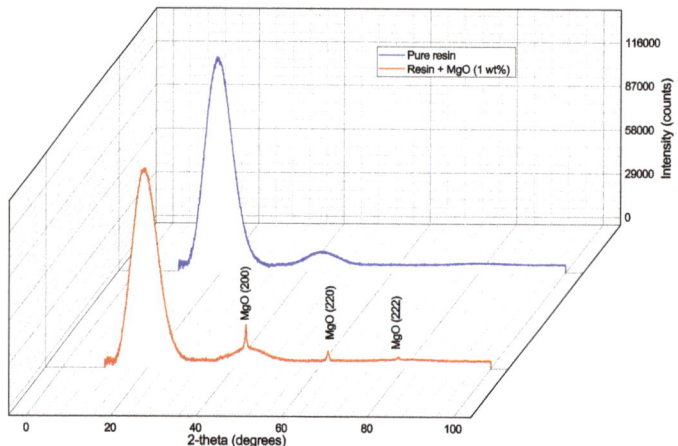

Figure 5. X-ray diffraction pattern of investigated materials.

4. Surface Functionalization and Effect on Dielectric Properties

With regard to the fact that the fillers used for electro-technical applications are in more cases of inorganic origin, it is very difficult to achieve sufficient dispersion in the organic matrix under normal conditions [57]. One possible solution is the use of silane-based coupling agents. The most popular ones are γ-Glycidyloxypropyltrimethoxysilane (GLYMO) and γ-Aminopropyltriethoxysilane (APTES) which provide covalent interface links to prevent a phase separation [58].

A lot of studies have shown the effect of functionalization on different material properties. The surface of inorganic particles [59], glass fibers [60] and also the natural fibers e.g., jute fibers [61] or hemp fibers [62] have been already modified by addition of GLYMO or APTES, respectively. In general, the formula of silane coupling agent can be written as $R(CH_2)nSiX_3$ where the silane molecule is silicon (Si) and two functional substituents (R, X) that provide a bonding effect between the inorganic filler and the organic matrix [63]. The substituent X represents hydrolyzable groups (e.g., methoxy, ethoxy, alkoxy), and R represents an organofunctional group attached to the silicon atom by a hydrolytically stable bond. Most of the coupling agents comprise three hydrolyzable groups X and one organofunctional group R [64]. Coupling agents and their linear formulas are shown in Table 2.

Table 2. Silane coupling agents characterizations [65–67].

Coupling Agent	Linear Formula
Trichlorovinylsilane	$H_2C=CHSiCl_3$
Triethoxyvinylsilane	$H_2C=CHSi(OC_2H_5)_3$
γ-Glycidoxypropyltrimethoxysilane	$C_9H_{20}O_5Si$
γ-Aminopropyltrimethoxysilane	$H_2N(CH_2)_3Si(OCH_3)_3$
[β-(3,4-Epoxycyclohexyl)-ethyl]trimethoxysilane	$C_{14}H_{28}O_4Si$
γ-Mercaptopropyltrimethoxysilane	$HS(CH_2)_3Si(OCH_3)_3$

The reaction of the γ-Glycidoxypropyltrimethoxysilane with the magnesium oxide filler can be explained as follow and is illustrated in Figure 6. The corresponding silanol molecules are formed after hydrolysis of the hydrolysable groups [68]. Furthermore, the process of chemisorption is going. The hydrogen bonds are formed between silanol and –OH groups on the surface of magnesium oxide.

A polysiloxane layer bonded with covalent bonds to the surface of the magnesium oxide is formed while water is released due to the condensation reaction [69,70].

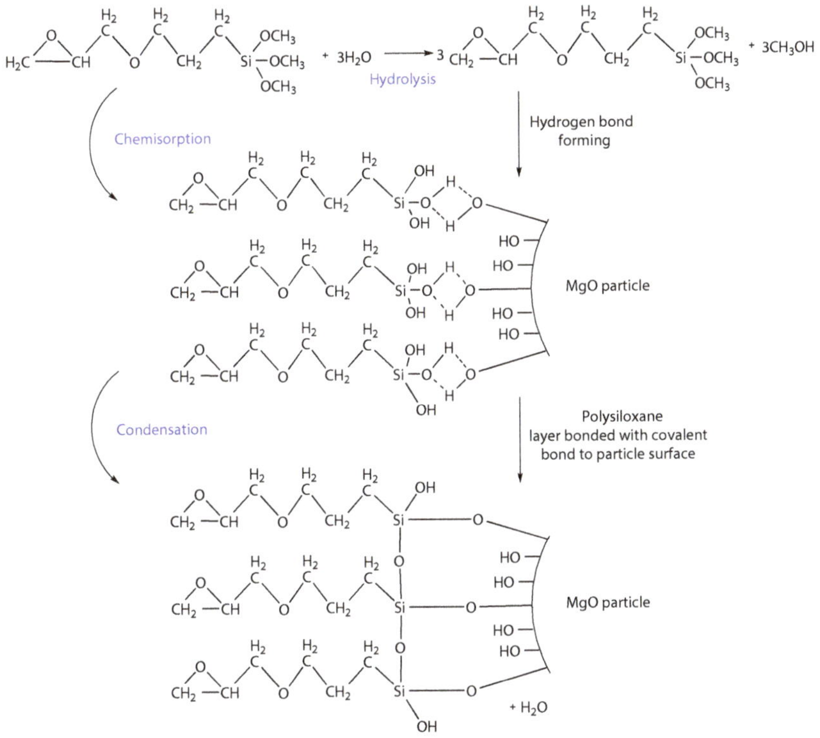

Figure 6. Simplified illustration of reaction of γ-Glycidoxypropyltrimethoxysilane with magnesium oxide surface (Redrawn and adepted from: [71–73]).

The determination of the correct ratio [67] of the coupling agent can be based on the relationship (1)

$$X = \frac{A}{w} \cdot f, \qquad (1)$$

where X (g) is the amount of coupling agent to form the minimum cover layer, A (m^2/g) is the specific surface area of the nanoparticle, w (m^2/g) is the wetting specific area of the coupling agent and f (g) is the weight of nanoparticles. Formic acid or hydrofluoric acid may be applied first to the nanoparticle to increase the electro-kinetic potential of its surface [74]. For the ability to react with different types of matrices, γ-Glycidoxypropyltrimethoxysilane was chosen for this experiment. The density of the selected coupling agent is 1.07 g/cm^3 and the wetting specific area is 331 m^2/g. It is an epoxysilane coupling agent, in particular, an organofunctional trialkoxysilane having a high reactivity between epoxide rings and amino groups [63]. The amount of GLYMO was determined as 18.12% from the total weight of nanoparticles according to Equation (1) and above-mentioned parameters.

4.1. Production of Epoxy-Based Nano-Composites Samples with Treated Surface of MgO Filler

In this case, the dried nanoparticles (1% of total weight of epoxy resin) were first added to a solution of 96% ethanol 4% H$_2$O (10 mL) and ultrasonic mixed (30 min). The coupling agent GLYMO (18.2% of total weight of nanoparticles) was then added to the mixture with re-application of ultrasonic mixing (2 h). Treated nanoparticles were added to already heated resin. The following procedure is the same as in the previous case.

4.2. Dielectric Properties of Epoxy-Based Nanocomposites with Treated Surface of MgO Filler

For comparison of the effect of the surface treatment, the measurement of dissipation factor (500 V AC, 50 Hz), relative permittivity (500 V AC, 50 Hz), volume resistivity (500 V DC) and dielectric strength (increase 1.5 kV/s AC) were repeated. The average values are shown in Table 3.

Table 3. Comparison of selected parameters after surface treatment.

Sample	Dissipation Factor	Relative Permittivity	Volume Resistivity	Dielectric Strength
Pure resin	0.0033	2.95	6.28×10^{12}	37 kV/mm
Resin + MgO	0.0041	3.43	5.01×10^{13}	42.3 kV/mm
Resin + MgO + GLYMO	0.0036	3.15	7.14×10^{14}	43.1 kV/mm

From the measurement results, it is clearly visible that the addition of coupling agent improved dielectric properties of the whole composite in comparison previous case. The relative permittivity was reduced approx. by 8%, due to the surface modification. The lower value of relative permittivity also guaranties a lower level of local stress inside the electrical insulation system. For example, in the case of an imperfect technological process during manufacturing of the insulating system of electrical machines and equipment. Mentioned decrease of relative permittivity could be caused by a changes of the degree of crosslinking due to the reaction of polymeric groups. They react with the coupling agent molecules on the nanoparticle surface and form a linear polymer chains in the interphase region [75]. The results further show that the dissipation factor of the composite is not negatively affected by the addition of the coupling agent, as confirmed other studies [39,75], as well. Addition of the coupling agent results in higher volume resistivity values, which can be attributed to a higher degree of the filler dispersion in the matrix and also to an increase of the energy levels of the electron traps [57], which results in a higher resistance to charge accumulation in the inner structure of the material.

4.3. Dielectric Response Measurement

Different optical methods and measurement techniques [76–79] are used for evaluation of the surface treatment effect or dispersion level, respectively. However, the idea of this paper is to evaluate the effect of surface treatment and particle dispersion by measurement of dielectric responses by special measurement techniques, i.e., reduced resorption curves (RRCs) and voltage response method (VR), which evaluate the conditions of the dielectric materials during charging and discharging process.

4.3.1. Reduced Resorption Curves Analyses

Dielectric absorption is a non-stationary phenomenon in dielectric materials after dc voltage application. Dielectric material is not able to follow the step change of the applied voltage. It means that the dielectric is charged for a certain time interval which is given by the relaxation time. This also applies to the discharging phenomenon. Both effects are caused by slow polarizations [80]. Here, the attention of our investigation was focused on the resorption characteristics. Resorption current can be used for reduced resorption curves (RRC) [26] determination. This method is based on the mathematical processing of time variable resorption current, which is transformed to relative resorption characteristic. The mathematical process of this methodology is expressed by Equations (2) and (3)

$$x = \ln(t) - \ln(15), \qquad (2)$$

$$y = \ln[ABS(i_t)] - \ln[ABS(i_{15})], \tag{3}$$

where x, y are transformed axes (-), t (s) is time, i_t (A) is current in time t, i_{15} (A) is current in 15th seconds. The main parameter for this investigation is the slope of the linear fit from transformed data. The higher slope of the curve generally means better resistance to charge trapping. The adequate interval length is important for appropriate linear fitting. In general, the interval between 15 and 300 s is usable for this determination. This procedure is illustrated in Figure 7.

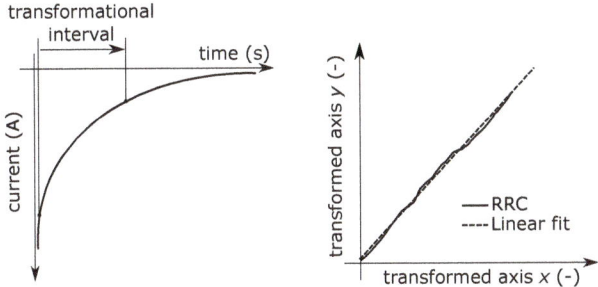

Figure 7. Transformation of resorption current to RRCs.

The electrometer KEITHLEY 6517A with suitable electrode system KEITHLEY 8009 Resistivity fixture was used for this measurement. The flat samples were conditioned (25 °C, 35% RH) and short-circuited in shielding room for 24 h before the measurement. After that, samples were charged by DC voltage 1000 V for 3600 s. After charging, the resorption current was measured up to 600 s and was recorded by the developed script in VEE Pro software. Transformation interval was set in the range 15...300 s. Average values of resorption currents were transformed to reduced resorption curves (RRCs) according to Equations (2) and (3). The RRCs are compared in Figure 8.

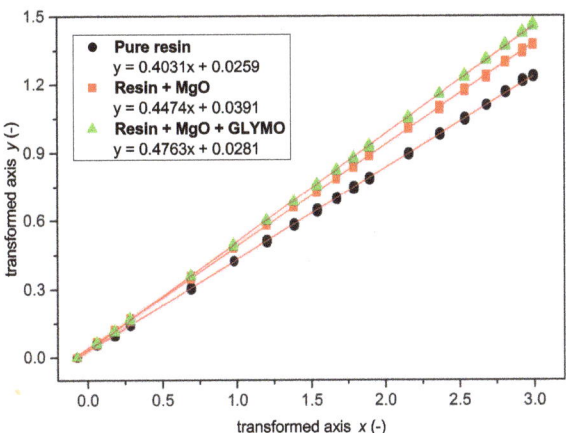

Figure 8. Reduced resorption curves for individual sample sets.

From the results is clearly visible the slope increase of the trendline due to addition of filler resp. filler and coupling agent. In the first case, the slope increase may be caused by the reduction of bulk charge accumulation due to better resistance to charge injection and ionic carriers generation in the bulk of dielectric [41]. The increase of the slope in the case of addition of silane coupling agent is caused by better dispersion and by a higher level of miscibility between organic matrix and an inorganic filler. The other reason is the increase of the trap depth which may contribute to charge recombination [81]. It means that the charge suppression is more effective. From this point of view, there is the higher ability of the material to discharge the charge accumulated in the inner structure after dc voltage charging, because a higher level of energy is needed for its trapping.

4.3.2. Voltage Response Analyses

Originally, the Voltage Response measurement method (VR) was developed for investigation of oil-paper insulated cables and measures the initial slopes of the decay and return voltages [27]. The timing diagram of the measurement can be seen in Figure 9.

After a long duration (100...1000 s) charging period (t_{ch}) the discharge voltage ($V_d(t)$) is measured on the insulation for t_{idp} time (<0.5 s). After a few seconds of short-circuiting (t_{dch}) return voltage ($V_r(t)$) is measured on charged insulation for t_{rvp} time (0.1...2 s).

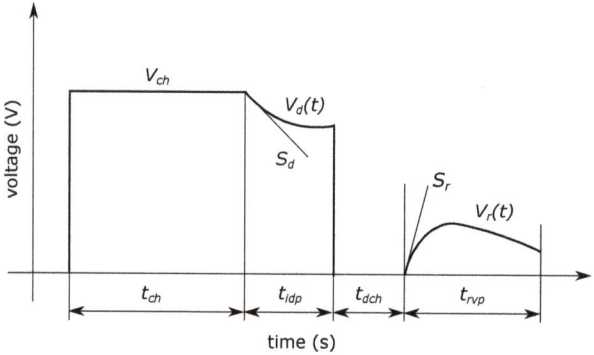

Figure 9. Timing diagram of Voltage Response measurement.

The initial slope of $V_d(t)$ (marked with S_d) is directly proportional to the conductivity of the insulation and the initial slope of $V_r(t)$ (marked with S_r) is directly proportional to the polarization conductivity, in other words to the intensity of the slow polarization processes. Therefore, the separate investigation of conductive and polarization processes is ensured by the measurement of S_d and S_r since they have the same information content as I_c conductive and I_p polarization component of the leakage current, respectively [82]. The measured values can be seen in Table 4.

Table 4. Results of voltage response measurement.

Sample	S_d(V/s)	S_r(V/s)
Pure resin	5.26	27.69
Resin + MgO	3.20	16.48
Resin + MgO + GLYMO	2.25	15.33

According to the measurement results, the pure resin had the highest conductivity and polarization conductivity from all the samples. By adding MgO filler, the conductive and polarization processes decreased significantly. However, the best results were measured after the addition of the silane coupling agent.

4.4. X-ray Diffraction of Epoxy-Based Nanocomposites with Treated Surface of MgO Filler

In this case, also the X-ray diffraction analyse has been performed for confirmation of our uttered assumptions. The X-ray diffraction signals of investigated materials have been deconvolute and diffraction pattern of epoxy matrix has been removed (Figure 10).

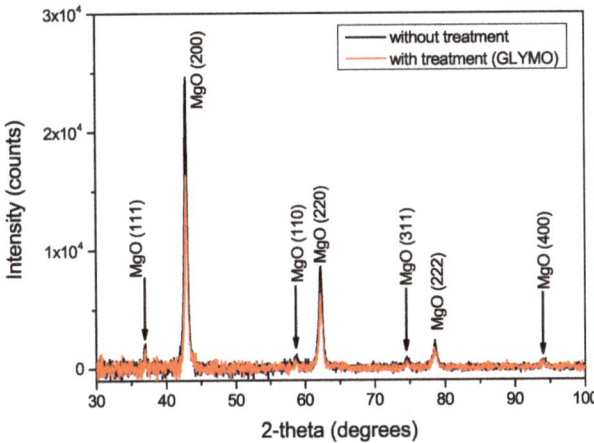

Figure 10. X-ray diagram of treated and untreated MgO nanoparticles.

There is clearly visible that addition of γ-Glycidoxypropyltrimethoxysilane do not contribute to additional chemical reactions inside the material and the phase structure has not been changed. Hovewer, on the first view, there are a differences between the heights of intensity peaks. The lower intensity peaks may be described by theory of presence of polysiloxane layer on the surface of the magnesium oxide powder [83]. It means that coupling agent occupies a part of volume. This measurement indirectly supports our previous results and statements on the suitability of dielectric response diagnostic methods for evaluation of nanofiller dispersion level.

5. Conclusions

This paper presents a few unique results. From the measurement results is visible the effect of dehydration on parameters of complex permittivity. This material evidently contains a significant amount of water molecules in the delivered state. However, these water molecules are with high probability only absorbed or very weakly bonded in the volume of MgO powder if they are released at temperatures below 100 °C. For this reason, it is necessary to dry nanoparticles to remove the surface moisture content before their applications.

Following on the first measurement, the effect of filler loading has been tested for improvement of epoxy matrix properties. The optimal filler loading was set to 1 wt % after dielectric parameters measurement. There are significant changes in investigated parameters, especially in the case of volume resistivity. The increase of volume resistivity is a consequence of the increase in resistance to injection of the charge carriers and their generation in the internal dielectric structure.

The effect of silane coupling agent was also investigated on the optimal filler loading. For the ability to react with different types of matrixes, γ-Glycidoxypropyltrimethoxysilane, has been used. The verification of improvement of dielectric parameters were also carried out. The main reason, why the parameters were improved, are due to a change in the degree of crosslinking due to the reaction of polymeric groups which react with the coupling agent molecules on the nanoparticle surface and form linear polymer chains in the interphase region. Also due to higher degree of dispersion of the filler in the matrix and also to an increase in the energy levels of the electron traps.

Two measurement and evaluation techniques (RRCs, VR) have been taken together in this paper. The possibility of interconnection between these two different techniques is clearly visible from the experimental results. In both cases, the effect of surface treatment of nanofiller was observed on the dielectric response. From this point of view, this methodology can be used not only for evaluation of dielectric parameters, such as conductivity or relaxation time of polarization mechanisms. According to presented results and previous presumptions, these methods may be used for observation of the proper dispersion of nanofiller in a polymer base.

Author Contributions: Conceptualization, J.H., P.T. and V.M.; Formal analysis, J.H., P.K. and G.M.C.; Investigation, J.H., P.K., P.Š. and G.M.C.; Methodology, J.H.; Supervision, P.T., V.M. and Z.Á.T.; Validation, P.T., V.M. and Z.Á.T.; Visualization, J.H., O.M., P.Š. and G.M.C.; Writing—original draft, J.H., P.T., P.K., O.M. and G.M.C.; Writing—review & editing, P.T., V.M. and Z.Á.T.

Acknowledgments: This work is supported by the Ministry of Education, Youth and Sports of the Czech Republic under the RICE-New Technologies and Concepts for Smart Industrial Systems, project No. LO1607 and by the Student Grant Agency of the West Bohemia University in Pilsen, grant No. SGS-2018-016 Diagnostics and materials in electrotechnics.

Conflicts of Interest: The authors declare no conflict of interest.

References

1. Marfunin, A.S. *Advanced Mineralogy: Volume 1 Composition, Structure, and Properties of Mineral Matter: Concepts, Results, and Problems*; Springer: Berlin, Germany, 1994; ISBN 978-3-642-78525-2.
2. Shand, M.A. *The Chemistry and Technology of Magnesia*; John Wiley & Sons: Hoboken, NJ, USA, 2006; ISBN 9780471980575.
3. Haynes, W.M.; Lide, D.R. *CRC Handbook of Chemistry and Physics: A Ready-Reference Book of Chemical and Physical Data*; CRC Press: Boca Raton, FL, USA, 2011; ISBN 978-1-4398-5511-9.
4. Zhang, B.; Peng, J.; Zhang, L.; Ju, S. Optimization of preparation for magnesium oxide by calcination from basic magnesium carbonate using response surface methodology. In *Magnesium Technology*; Mathaudhu, S.N., Sillekens, W.H., Neelameggham, N.R., Hort N., Eds.; Springer: Cham, Germany, 2012; pp. 75–79, ISBN 978-3-319-48203-3.
5. Vu, A.H.; Jiang, S.; Kim, Y.H.; Lee, C.H. Controlling the physical properties of magnesium oxide using a calcination method in aerogel synthesis: its application to enhanced sorption of a sulfur compound. *Ind. Eng. Chem. Res.* **2014**, *53*, 13228–13235. [CrossRef]
6. Costa, D.; Chizallet, C. Water on extended and point defects at MgO surfaces. *J. Chem. Phys.* **2006**, *125*, 054702. [CrossRef] [PubMed]
7. Li, Q.; Zhang, S.I.; Wang, J.P.; Gao, H. Process analysis of MgO film on NdFeB magnet by sol–gel method. *Surf. Eng.* **2009**, *25*, 589–593. [CrossRef]
8. Sharma, G.; Soni, R.; Jasuja, N.D. Phytoassisted synthesis of magnesium oxide nanoparticles with Swertia chirayaita. *J. Taibah Univ. Sci.* **2017**, *11*, 471–477. [CrossRef]
9. Wilson, L.O. Magnesium oxide as a high-temperature insulant. *IEE Proc. A Phys. Sci. Meas. Instrum. Manag. Educ. Rev.* **1981**, *128*, 159. [CrossRef]
10. Senbayram, M.; Gransee, A.; Wahle, V.; Thiel, H. Role of magnesium fertilisers in agriculture: Plant-soil continuum. *Crop. Pasture Sci.* **2015**, *66*, 1219–1229. [CrossRef]
11. Purwajanti, S.; Zhou, L.; Nor, Y.A.; Zhang, J.; Zhang, H.; Huang, X.; Yu, C. Synthesis of magnesium oxide hierarchical microspheres: A dual-functional material for water remediation. *ACS Appl. Mater. Interfaces* **2015**, *7*, 21278–21286. [CrossRef] [PubMed]
12. Gray, J.E.; Luan, B. Protective coatings on magnesium and its alloys—A critical review. *J. Alloys Compd.* **2002**, *336*, 88–113. [CrossRef]
13. Calebrese, C.; Hui, L.; Schadler, L.S.; Nelson, J.K. A review on the importance of nanocomposite processing to enhance electrical insulation. *IEEE Trans. Dielectr. Electr. Insul.* **2011**, *18*, 938–945. [CrossRef]
14. Dikshit, V.; Bhudolia, S.K.; Joshi, S.C. Multiscale polymer composites: A review of the interlaminar fracture toughness improvement. *Fibers* **2017**, *5*, 38. [CrossRef]

15. Wang, J.A.; Novaro, O.; Bokhimi, X.; Lopez, T.; Gomez, R.; Navarrete, J.; Llanos, M.E.; Lopez-Salinas, E. Structural defects and acidic and basic sites in sol-gel MgO. *J. Phys. Chem. B* **1997**, *101*, 7448–7451. [CrossRef]
16. Ding, Y.; Zhang, G.; Wu. H.; Hai, B.; Wang, L.; Qian, Y. Nanoscale magnesium hydroxide and magnesium oxide powders: Control over size, shape, and structure via hydrothermal synthesis. *Chem. Mater.* **2001**, *13*, 435–440. [CrossRef]
17. Andritsch, T. *Epoxy Based Nanodielectrics for High Voltage DC-Applications—Synthesis, Dielectric Properties and Space Charge Dynamics*; Delft University of Technology: Delft, The Netherlands, 2010; ISBN 9789053353318.
18. Magnesium Oxide (MgO, 99+%, 20 nm). Available online: www.nanoamor.com/inc/sdetail/11013/2543 (accessed on 20 April 2018).
19. Kim, M.I.; Kim, S.; Kim, T.; Lee, D.K.; Seo, B.; Lim, C.-S. Mechanical and thermal properties of epoxy composites containing zirconium oxide impregnated halloysite nanotubes. *Coatings* **2017**, *7*, 231. [CrossRef]
20. Domun, N.; Paton, K.R.; Hadavinia, H.; Sainsbury, T.; Zhang, T.; Mohamud, H. Enhancement of fracture toughness of epoxy nanocomposites by combining nanotubes and nanosheets as fillers. *Materials* **2017**, *10*, 1179. [CrossRef] [PubMed]
21. Zhang, X.; Wen, H.; Wu, Y. Computational thermomechanical properties of silica–epoxy nanocomposites by molecular dynamic simulation. *Polymers* **2017**, *9*, 430. [CrossRef]
22. Simcha, S.; Dotan, A.; Kenig, S.; Dodiuk, H. Characterization of hybrid epoxy nanocomposites. *Nanomaterials* **2012**, *2*, 348–365. [CrossRef] [PubMed]
23. Wang, Z.; Liu, J.; Cheng, Y.; Chen, S.; Yang, M.; Huang, J.; Wang, H.; Wu, G.; Wu, H. Alignment of boron nitride nanofibers in epoxy composite films for thermal conductivity and dielectric breakdown strength improvement. *Nanomaterials* **2018**, *8*, 242. [CrossRef] [PubMed]
24. Yamaguchi, S.; Inoue, S.; Sakai, T.; Abe, T.; Kitagawa, H.; Imazato, S. Multi-scale analysis of the effect of nano-filler particle diameter on the physical properties of CAD/CAM composite resin blocks. *Comput. Methods Biomech. Biomed. Eng.* **2017**, *20*, 714–719. [CrossRef] [PubMed]
25. Sfondrini, M.F.; Massironi, S.; Pieraccini, G.; Scribante, A.; Vallittu, P.K.; Lassila, L.V.; Gandini, P. Flexural strengths of conventional and nanofilled fiber-reinforced composites: A three-point bending test. *Dent. Traumatol.* **2014**, *30*, 32–35. [CrossRef] [PubMed]
26. Mentlík, V. *Dielektrické Prvky a Systémy*; BEN—Technická Literatura: Praha, Czech Republic, 2006; ISBN 9788073001896.
27. Németh, E. Measuring voltage response: A non-destructive diagnostic test method HV of insulation. *IEE Proc. Sci. Meas. Technol.* **1999**, *146*, 249–252.:19990651. [CrossRef]
28. Novocontrol Technologies. *User's Manual: Alpha-A High Resolution Dielectric, Conductivity, Impedance and Gain Phase Modular Measurement System*; Novocontrol Technologies: Hundsangen, Germany, 2012.
29. Čermák, M.; Kadlec, P.; Kruliš, Z.; Polanský, R. Dielectric analysis of halloysite nanotubes LLDPE nanocomposite compounds. *AIP Conf. Proc.* **2016**, *1713*, 090007. [CrossRef]
30. Polanský, R.; Kadlec, P.; Kolská, Z.; Švorčík, V. Influence of dehydration on the dielectric and structural properties of organically modified montmorillonite and halloysite nanotubes. *Appl. Clay Sci.* **2017**, *147*, 19–27. [CrossRef]
31. Evacuable Pellet Press for 13 Mm Pellets. Available online: www.piketech.com/pm-evacuable-pellet-press.html (accessed on 17 May 2018).
32. Harvánek, L. *Nanomaterials for Electrotechnic*; University of West Bohemia: Pilsen, Czech Republic, 2017.
33. International Electrotechnical Commission (IEC). *Electrical Insulation—Thermal Evaluation and Designation*; IEC 60085:2007; International Electrotechnical Commission: Geneva, Switzerland, 2007.
34. Product Information—Epoxylite®, 3750 LV, Elantas® Italia S.r.l, Collechion, Italy, 2009. Available online: http://www.elantas.com/europe/products/impregnating-materials.html (accessed on 17 May 2018).
35. International Electrotechnical Commission (IEC). *Recommended Methods for the Determination of the Permittivity and Dielectric Dissipation Factor of Electrical Insulating Materials at Power, Audio and Radio Frequencies Including Metre Wavelengths*; IEC 60250:1969; IEC: Geneva, Switzerland, 1969.
36. International Electrotechnical Commission (IEC). *Methods of Test for Electric Strength of Solid Insulating Materials*; IEC 60243-1:2013; IEC: Geneva, Switzerland, 2013.
37. International Electrotechnical Commission (IEC). *Dielectric and Resistive Properties of Solid Insulating Materials—Part 3-2: Determination of Resistive Properties (DC Methods)*; IEC 62631-3-2:2015; IEC: Geneva, Switzerland, 2015.

38. International Electrotechnical Commission (IEC). *Standard Conditions for Use Prior to and During the Testing of Solid Electrical Insulating Materials*; IEC 60212:2010; IEC: Geneva, Switzerland, 2010.
39. Li, H.; Wang, C.; Guo, Z.; Wang, H.; Zhang, Y.; Hong, R.; Peng, Z. Effects of silane coupling agents on the electrical properties of silica/epoxy nanocomposites. In Proceedings of the 2016 IEEE International Conference on Dielectrics, Montpeliere, France, 3–7 July 2016; pp. 1036–1039. [CrossRef]
40. Andritsch, T.; Kochetov, R.; Morshuis, P.H.F.; Smit, J.J. Dielectric properties and space charge behavior of MgO-epoxy nanocomposites. In Proceedings of the 2010 10th IEEE International Conference on Solid Dielectrics, Potsdam, Germany, 4–9 July 2010; pp. 1–4. [CrossRef]
41. Tanaka, T.; Imai, T. *Advanced Nanodielectrics: Fundamentals and Application*; Pan Stanford: Singapore, 2017; ISBN 9814745022.
42. Do Nascimento, E.; Ramos, A.; Windmoller, D.; Reig Rodrigo, P.; Teruel Juanes, R.; Ribe Greus, A; Amigo Borras, V.; Coelho, L.A.F. Breakdown, free-volume and dielectric behavior of the nanodielectric coatings based on epoxy/metal oxides. *J. Mater. Sci. Mater. Electron.* **2016**, *27*, 9240–9254. [CrossRef]
43. Artbauer, J.J. Electric strength of polymers. *J Phys. D Appl. Phys.* **1996**, *29*, 446–456. [CrossRef]
44. Nelson, J.K.; Utracki, L.A.; MacCrone, R.K.; Reed, C.W. Role of the interface in determining the dielectric properties of nanocomposites. In Proceedings of the 17th Annual Meeting of the IEEE Lasers and Electro-Optics Society, Boulder, CO, USA, 20 October 2004; pp. 314–317. [CrossRef]
45. Nan, C.W.; Shen, Y.; Ma, J. Physical properties of composites near percolation. *Annu. Rev. Mater. Res.* **2010**, *40*, 131–151. [CrossRef]
46. Zhang, L.; Bass, P.; Cheng. Z. Revisiting the percolation phenomena in dielectric composites with conducting fillers. *Appl. Phys. Lett.* **2014**, *105*, 042905. [CrossRef]
47. Pelíšková, M.; P. Sáha. The Effect of Expanded Structure on Electric Properties of Polymer Composites with Electroconductive Fillers. *Chem. Listy* **2012**, *106*, 1104–1109.
48. Motaghi, A.; Hrymak, A.; Motlagh, G.H.; Electrical conductivity and percolation threshold of hybrid carbon/polymer composites. *J. Appl. Polym. Sci.* **2015**, *132*, 41744. [CrossRef]
49. Mentlík, V.; Michal, O. Influence of SiO_2 nanoparticles and nanofibrous filler on the dielectric properties of epoxy-based composites. *Mater. Lett.* **2018**, *223*, 41–44. [CrossRef]
50. Mentlík, V.; Trnka, P.; Hornak, J.; Totzauer, P. Development of a Biodegradable Electro-Insulating Liquid and Its Subsequent Modification by Nanoparticles. *Energies* **2018**, *11*, 508. [CrossRef]
51. Boček, J.; Matějka, L.; Mentlík, V.; Trnka, P.; Šlouf, M. Electrical and thermomechanical properties of epoxy-POSS nanocomposites. *Eur. Polym. J.* **2011**, *47*, 861–872. [CrossRef]
52. Gao, M.; Zhang, P.; Wang, F. Effect of percolation and interfacial characteristics on breakdown behavior of nano-Silica/Epoxy composites. In Proceedings of the 2013 8th International Forum on Strategic Technology, Ulaanbaatar, Mongolia, 28 June–1 July 2013; pp. 120–123. [CrossRef]
53. Lewis, T.J. Interfaces: Nanometric dielectrics. *J. Phys. D* **2005**, *38*, 202–212. [CrossRef]
54. Zou, C.; Fothergill, J.C.; Rowe, S.W. The effect of water absorption on the dielectric properties of epoxy nanocomposites. *IEEE Trans. Dielectr. Electr. Insul.* **2008**, *15*, 106–117. [CrossRef]
55. Abdullah, S.I.; Ansari, M.N.M. Mechanical properties of grapheme oxide (GO)/epoxy composites. *HBRC J.* **2015**, *11*, 151–156. [CrossRef]
56. Kherzi, T.; Sharif, M.; Pourabas, B. Polythiophene–graphene oxide doped epoxy resin nanocomposites with enhanced electrical, mechanical and thermal properties. *RSC Adv.* **2016**, *6*, 93680–93693. [CrossRef]
57. Chen, G.; Li, S.; Zhong, L. Space charge in nanodielectrics and its impact on electrical performance. In Proceedings of the 2015 IEEE 11th International Conference on the Properties and Applications of Dielectric Materials, Sydney, Australia, 19–22 July 2015; pp. 36–39. [CrossRef]
58. Ponyrko, S.; Kobera, L; Brus, J.; Matějka, L. Epoxy-silica hybrids by nonaqueous sol-gel process. *Polymer* **2013**, *54*, 6271–6282. [CrossRef]
59. Lin, J.; Siddiqui, J.A.; Ottenbrite, R.M. Surface modification of inorganic oxide particles with silane coupling agent and organic dyes. *Polym. Adv. Technol.* **2001**, *12*, 285–292. [CrossRef]
60. Park, S.J.; Jin, J.S. Effect of silane coupling agent on tnterphase and performance of glass fibers/unsaturated polyester composites. *J. Colloid Interface Sci.* **2001**, *242*, 174–179. [CrossRef]
61. Zhou, Q.; Cho, D.; Song, B.K.; Kim, H.J. Novel jute/polycardanol biocomposites: Effect of fiber surface treatment on their properties. *Compos. Interfaces* **2009**, *16*, 781–795. [CrossRef]

62. Sawpan, M.A.; Pickering, K.L.; Fernyhough, A. Effect of fibre treatments on interfacial shear strength of hemp fibre reinforced polylactide and unsaturated polyester composites. *Compos. A Appl. Sci. Manuf.* **2011**, *42*, 1189–1196. [CrossRef]
63. Mittal, K.L. *Silanes and Other Coupling Agents*; CRC Press: London, UK, 2009; ISBN 9789004193321.
64. Rothon, R.N. *Particulate-Filled Polymer Composites*; Rapra Technology: Shrewsbury, UK, 2003; ISBN 9780582087828.
65. Han, G.; Zhang, C.; Zhang, D.; Umemura, K.; Kawai, S. Upgrading of urea formaldehyde-bonded reed and wheat straw particleboards using silane coupling agents. *J. Wood Sci.* **1998**, *44*, 282–286. [CrossRef]
66. Luštická, I.; Vyskočilová-Leitmannová, E.; Červený, L. Functionalization of Mesoporous Silicate Materials. *Chem. Listy* **2013**, *107*, 114–120.
67. Shokoohi, S.; Arefazar, A.; Khosrokhavar, R. Silane coupling agents in polymer-based reinforced composites: A review. *J. Reinf. Plast. Compos.* **2008**, *27*, 473–485. [CrossRef]
68. Xie, Y.; Hill, C.A.S.; Xiao, Z.; Militz, H.; Mai, C. Silane coupling agents used for natural fiber/polymer composites: A review. *Compos. A Appl. Sci. Manuf.* **2010**, *41*, 806–819. [CrossRef]
69. Plueddemann, E.P. Adhesion through silane coupling agents. *J. Adhes.* **2008**, *2*, 184–201. [CrossRef]
70. Merhari, L. *Hybrid Nanocomposites for Nanotechnology: Electronic, Optical, Magnetic and Biomedical Applications*; Springer: New York, NY, USA, 2009; ISBN 9780387723983.
71. Yu, Z.Q.; Wu, Y.; Wei, B.; Baier, H. Boride ceramics covalent functionalization and its effect on the thermal conductivity of epoxy composites. *Mater. Chem. Phys.* **2015**, *164*, 214–222. [CrossRef]
72. Mandhakini, M.; Lakshmikandhan, T.; Chandramohan, A.; Muthukaruppan, A. Effect of nanoalumina on the tribology performance of C_4-ether-linked bismaleimide-toughened epoxy nanocomposites. *Tribol. Lett.* **2014**, *54*, 67–79. [CrossRef]
73. Li, L; Li, B.; Dong, J.; Zhang, J. Roles of silanes and silicones in forming superhydrophobic and superoleophobic materials. *J. Mater. Chem. A* **2016**, *4*, 13677–13725. [CrossRef]
74. Foxton, R.M.; Nakajima, M.; Tagami, J.; Miura, H. Effect of acidic pretreatment combined with a silane coupling agent on bonding durability to silicon oxide ceramic. *J. Biomed. Mater. Res. B* **2005**, *73*, 97–103. [CrossRef] [PubMed]
75. Kochetov, R.; Andritsch, T.; Morshuis, P.H.F.; Smit, J.J. Anomalous behaviour of the dielectric spectroscopy response of nanocomposites. *IEEE Trans. Dielectr. Electr. Insul.* **2012**, *19*, 107–117. [CrossRef]
76. Issan, H.; Ristig, S.; Kaminski, H.; Asbach, C.; Epple, M. Comparison of different characterization methods for nanoparticle dispersions before and after aerosolization. *Anal. Methods* **2014**, *6*, 7324–7334. [CrossRef]
77. Li, X.; Zhang, H.; Jin, J.; Huang, D.; Qi, X.; Zhang, Z.; Yu, D. Quantifying dispersion of nanoparticles in polymer nanocomposites through transmission electron microscopy micrographs. *J. Micro Nano-Manuf.* **2014**, *2*, 021008. [CrossRef]
78. Bugnicourt, E.; Kehoe, T.; Latorre, M.; Serrano, C.; Philippe, S.; Schmid, M. Recent prospects in the inline monitoring of nanocomposites and nanocoatings by optical technologies. *Nanomaterials* **2016**, *6*, 150. [CrossRef] [PubMed]
79. Lively, B.; Bizga, J.; Zhong, W. Analysis tools for fibrous nanofiller polymer composites: Macro- and nanoscale dispersion assessments correlated with mechanical and electrical composite properties. *Polym. Compos.* **2013**, *35*, 10–18. [CrossRef]
80. Kao, K. *Dielectric Phenomena in Solids: With Emphasis on Physical Concepts of Electronic Processes*; Elsevier: San Diego, CA, USA, 2004; ISBN 9786610961399.
81. Lv, Z.; Wang. X.; Wu, K.; Chen, X.; Cheng, Y.; Dissado, L.A. Dependence of charge accumulation on sample thickness in Nano-SiO_2 doped LDPE. *IEEE Trans. Dielectr. Electr. Insul.* **2013**, *20*, 337–345. [CrossRef]
82. Csanyi, G.M.; Tamus, Z.A.; Ivancsy, T. Investigation of dielectric properties of cable insulation by the extended voltage response method. In Proceedings of the 2016 Conference on Diagnostics in Electrical Engineering, Pilsen, Czech Republic, 6–8 September 2016; pp. 1–4. [CrossRef]
83. Chen, H.; Zheng, J.; Qiao L.; Ying, Y.; Ji, L.; Che, S. Surface modification of $NdFe_{12}N_x$ magnetic powder using silane coupling agent KH550. *Adv. Powder Technol.* **2015**, *26*, 618–621. [CrossRef]

© 2018 by the authors. Licensee MDPI, Basel, Switzerland. This article is an open access article distributed under the terms and conditions of the Creative Commons Attribution (CC BY) license (http://creativecommons.org/licenses/by/4.0/).

Article

Preparation of High Refractive Index Composite Films Based on Titanium Oxide Nanoparticles Hybridized Hydrophilic Polymers

Makoto Takafuji [1,2,*], Maino Kajiwara [1], Nanami Hano [1], Yutaka Kuwahara [1] and Hirotaka Ihara [1,2,*]

[1] Department of Applied Chemistry and Biochemistry, Kumamoto University, 2-39-1 Kurokami, Chuo-ku, Kumamoto 860-8555, Japan; 181d8811@st.kumamoto-u.ac.jp (M.K.); Nanami_Hano@kumadai.jp (N.H.); kuwahara@kumamoto-u.ac.jp (Y.K.)
[2] Kumamoto Institute for Photo-Electro Organics (PHOENICS), 3-11-38 Higashimachi, Higashi-ku, Kumamoto 862-0901, Japan
* Correspondence: takafuji@kumamoto-u.ac.jp (M.T.); ihara@kumamoto-u.ac.jp (H.I.); Tel.: +81-96-342-3661 (H.I.)

Received: 21 February 2019; Accepted: 22 March 2019; Published: 2 April 2019

Abstract: Optical materials with high refractive index (n) have been rapidly improved because of urgent demands imposed by the development of advanced photonic and electronic devices such as solar cells, light emitting diodes (LED and Organic LED), optical lenses and filters, anti-reflection films, and optical adhesives. One successful method to obtain high refractive index materials is the blending of metal oxide nanoparticles such as TiO_2 and ZrO_2 with high n values of 2.1–2.7 into conventional polymers. However, these nanoparticles have a tendency to agglomerate by themselves in a conventional polymer matrix, due to the strong attractive forces between them. Therefore, there is a limitation in the blending amount of inorganic nanoparticles. In this paper, various hydrophilic polymers such as poly(N-hydroxyl acrylamide) (*p*HEAAm), poly(vinyl alcohol), poly(ethylene glycol), and poly(acrylic acid) were examined for preparation of high refractive index film based on titanium oxide nanoparticle (TiNP) dispersed polymer composite. The hydrogen bonding sites in these hydrophilic polymers would improve the dispersibility of inorganic nanoparticles in the polymer matrix. As a result, *p*HEAAm exhibited higher compatibility with titanium oxide nanoparticles (TiNPs) than other water-soluble polymers. Transparent hybrid films were prepared by mixing *p*HEAAm with TiNPs and drop casting the mixture onto a glass plate. The refractive indices of the films were in good agreement with calculated values. The compatibility of TiNPs with *p*HEAAm was dependent on the surface characteristics of TiNPs. TiNPs with the highest observed compatibility could be hybridized with *p*HEAAm at concentrations of up to 90 wt%, and the refractive index of the corresponding film reached 1.90. The high compatibility of TiNPs with *p*HEAAm may be related to the hydrophilicity and amide and hydroxyl moieties of *p*HEAAm, which cause hydrogen bond formation on the TiO_2 surface. The obtained thin film was slightly yellow due to the color of the original TiNP dispersion; however, the transmittance of the film was higher than 80% in the wavelength range from 480 to 900 nm.

Keywords: titanium oxide; hybrid material; anatase; organic-inorganic hybrid; nanocomposite; nanoparticles; high refractive index material

1. Introduction

High refractive index materials are required to improve the performance of eyeglass lenses, optical fibers, and optical devices, such as waveguide-based optical circuits, optical interference filters and mirrors, optical sensors, and solar cells [1,2]. High refractive index polymers and/or polymer composites have been attracting much attention as they are light-weight and have high flexibility and high formability compared to inorganic materials. The most useful applications for polymer-based high refractive index materials are as optical data storage [3], lenses [4], anti-reflective coatings [5] and immersion lithography [6]. The recent developments have been summarized in review articles [7,8] and the references therein. The refractive index (n) values of conventional polymers are lower (n = 1.3–1.7) [9] than those of inorganic materials [10,11]. Several methods to obtain high refractive index materials have been previously published. These methods can be categorized into two approaches; one is the incorporation of heavy atoms such as sulfur and/or a halogen into the polymer [12], and the other is the fabrication of polymer composites with high refractive index inorganic or metal nanoparticles (NPs) [13–22]. There are some challenges to increasing the refractive index in each approach. In the first case, it is technically difficult and costly to introduce heavy atoms into the polymer backbone. The maximum refractive index reported for a material prepared using this approach is 1.84, which was obtained for poly(sulfur-random-(1,3,5-triisopropenylbenzene)) [23]. In the second case, a variety of NPs such as those of zirconium oxide (ZrO_2) [13], titanium oxide (TiO_2) [14–18], alumina oxide [19], gold [20], and nanodiamond [21,22] have been used as nano-sized fillers. However, inorganic NPs agglomerate in the polymer matrix due to their high surface energies and low compatibilities with the polymer. Therefore, comparably higher amount of inorganic nanoparticles-involving polymer composites could be prepared by in situ polymerization of polymerizable monomers in the presence of inorganic nanoparticles [24,25], or in-site sol-gel reaction of inorganic precursors in the presence of a polymer matrix [26,27]. The polymer composites with higher amounts of titania NPs showed higher refractive indices of more than 1.7. For instance, poly(4-vinylbenzyl alcohol) was reported as a matrix polymer for titania NPs. The composites were prepared by in-site polymerization of 4-vinylbenzyl alcohol in the acid surface-modified TiO_2 nanoparticles dispersion. TiO_2 nanoparticles were dispersed in the polymer at up to 60 wt%, and the refractive index of the composite reached 1.77 [18]. The hydroxyl groups in the side chains of polymers may play an important role in the dispersion of the titania NPs into the polymer matrix. Recently we have reported a new method for fabrication of high refractive index materials based on simple blending of polymers and heteropoly acids [28,29]. In these cases, hydrophilic moieties of polymers such as carbonyl and hydroxyl groups exhibited high compatibility with heteropoly acids which were molecularly dispersed in the polymer matrix. The obtained hybrid showed a dramatic increase in the refractive index of the general polymer, while maintaining a high transparency. Based on these results, we focus on hydrophilic polymers as a matrix for polymer composites containing titania NPs. In this study, the hydrophilic polymers are examined as a matrix for titania NPs, and the optical properties including refractive index of the obtained composites are evaluated.

2. Materials and Methods

2.1. Materials

Poly(N-hydroxyethyl acrylamide) (pHEAAm) was synthesized by thermally initiated radical polymerization (see Supplementary Materials). Poly(acrylic acid) (pAA, Mw = 5000) and poly(vinyl pyrrolidone) (pVP, Mw = 36,000) were purchased from FUJIFILM Wako Pure Chemical Corporation (Osaka, Japan) and Honeywell Fluka (Morristown, NJ, USA), respectively, and poly(ethylene glycol) (pEG, Mw = 20,000) and poly(vinyl alcohol) (pVA, Mw = 500) were purchased from Nacalai Tesque (Kyoto, Japan). In this study, three different aqueous dispersions of titanium oxide nanoparticles (TiNP-x) were used for fabricating polymer composites. TiNP-1 (40 nm, 0.85 wt%, PTA) and TiNP-2 (20 nm, 0.85 wt%, TPX-HP) were purchased from Kon Corporation (Saga, Japan), and TiNP-3 (70 nm, 18.1 wt%, TisolA) was kindly provided by NYACOL Nano Technologies Inc. (Ashland, MA, USA). Deionized water was prepared with an Elix UV 3 (Merck KGaA, Darmstadt, Germany) and used as a solvent.

2.2. Preparation of Composites

The aqueous mixtures of polymers and TiNPs were prepared by mixing an aqueous solution of polymer and TiNP-x in predetermined ratios. Thin films of polymer/TiNP composites were prepared by drop casting 300 µL of the aqueous mixture onto glass and subsequently allowing it to dry either slowly, at 10 °C, for 24 h, or quickly, at 50 °C, for 1 h. The glass was treated with aqueous 1 mol L^{-1} sodium hydroxide solution before casting. The compatibilities of TiNPs with hydrophilic polymers were observed at 70 wt% TiNP.

2.3. Measurements

X-ray diffraction measurements (XRD) of TiNPs were performed on a SmartLab (Rigaku corporation, Tokyo, Japan) operating in reflection mode with CuKα radiation (λ = 1.54 Å, 45 kV, 200 mA) and a diffracted beam monochromator, using a step scan mode with a step of 0.02° (2θ) and 4 s per step. The morphologies of TiNPs were observed using transmittance electron microscopy (TEM) using a JEM-1400Plus (JEOL, Tokyo, Japan). Dynamic light scattering (DLS) measurements were performed with a Zetasizer Nano ZS (Malvern, Tokyo, Japan). Thermogravimetric (TG) analyses were carried out with TG/DTA-6200 (Hitachi-Hitech, Tokyo, Japan) under air at a flow rate of 200 mL/min and a heating rate of 10 °C/min.

The refractive index values and thicknesses of the films were evaluated using a prism coupler SPA-4000 (Sairon Technology, Inc., Gwangju, Korea) equipped with a He–Ne laser (λ = 632.8 nm) and a gadolinium gallium garnet (GGG) prism (n = 1.965). Transmittance spectra of the obtained films were measured using a V-560 spectrophotometer (JASCO, Tokyo, Japan).

3. Results and Discussion

3.1. Characterization of Titanium Oxide Nanoparticles

The TiNPs (TiNP-1, 2, and 3) used in this study were characterized by DLS, XRD, and TEM analyses (Figure 1). TEM images of TiNPs indicated that all TiNPs were nano-sized particles, however, the morphologies were different. The average sizes of TiNP-1, TiNP-2, and TiNP-3, determined by DLS measurements, were 40 nm, 20 nm, and 70 nm, respectively. These sizes were different from the sizes observed in the TEM images (average size of TiNPs were 36 nm, 27 nm (40 nm 12 nm for major and minor axes respectively), and 9 nm respectively), indicating that the TiNPs had coagulated in the aqueous dispersions. XRD patterns of all TiNPs used in this study exhibited high intensity peaks at 25°, 47°, and 55°, indicating that the TiNPs were predominantly in the anatase phase. All peaks were in good agreement with the standard spectrum (JCPDS no.: 88-1175 and 84-1286 (anatase)). All peaks were broad as compared with those of the micro-sized anatase phase titanium oxide crystal, and some small, unidentifiable shoulders were observed. These results suggest that the TiNPs were composed of irregular polycrystalline and amorphous phases.

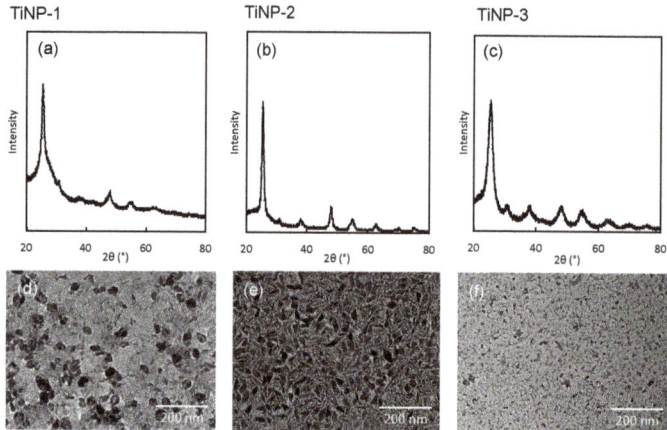

Figure 1. XRD patterns (a–c) and TEM images (d–f) of titanium oxide nanoparticles. TiNP-1 (a,d), TiNP-2 (b,e), and TiNP-3 (c,f).

3.2. Preparation of Hydrophilic Polymer Composite Films Containing TiNP

Conventional hydrophilic polymers such as poly(acrylic acid) (*p*AA), poly(ethylene glycol) (*p*EG), poly(vinyl pyrrolidone) (*p*VP), and poly(vinyl alcohol) (*p*VA) were tested as matrix polymers for comparison with *p*HEAAm. Aqueous mixtures of TiNP-1 with the hydrophilic polymers were prepared at 70 wt% TiNP-1. As shown in Figure 2, the aqueous mixtures with *p*AA and *p*EG formed hydrogels at this mixing ratio, and the aqueous mixture with *p*VP formed a viscous solution. No gelation was observed in TiNP-1 with *p*VA and *p*HEAAm, indicating that the hydroxyl groups in the polymer side chains increased the compatibility of the polymer with TiNP-1. Furthermore, the transparency of the aqueous mixture with *p*HEAAm was higher than that of the other mixtures. These results suggest that *p*HEAAm is a good dispersant for TiNP-1 in water. As *p*HEAAm possesses a relatively high number of hydrogen bonding sites, such as hydroxyl and amide groups, it is reasonable to postulate that they contribute to the dispersion of TiNP-1 via hydrogen bonding with the surface of titanium oxide.

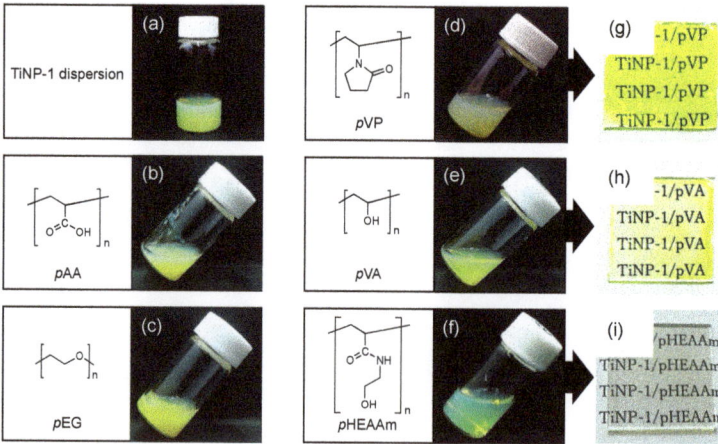

Figure 2. Photos of the aqueous dispersion of TiNP-1 (a), aqueous mixtures of TiNP-1 and the water-soluble polymers ((b): *p*AA, (c): *p*EG, (d): *p*VP, (e): *p*VA, (f): *p*HEAAm), and the composite films ((g): *p*VP, (h): *p*VA, (i): *p*HEAAm) prepared by casting of the aqueous mixture on a glass plate at 10 °C.

Aqueous mixtures of TiNP/polymer composites (300 µL) were dropped onto glass plates and dried at 10 °C to obtain thin films. It was difficult to obtain thin films of the aqueous mixtures with pAA, pEG and pVP, due to their high viscosity. Yellow, crack-free thin films were obtained from the aqueous mixtures with pVA and pHEAAm. The pHEAAm-based thin film prepared had higher transparency compared with pVA-based thin film. The TiNP-1/pHEAAm composite films were used for investigation of the refractive indices of the composite thin films.

3.3. Compatibility of TiNPs with the pHEAAm Polymer

The pHEAAm composite films containing different quantities of TiNP (50, 60, 70, 80, 90 and 95 wt%) were prepared by drop casting the aqueous mixtures onto a glass plate, followed by drying at ambient pressure and temperatures of 10, 25 or 50 °C. At the relatively low TiNP concentrations of 50 and 60 wt%, transparent yellow films were obtained from the aqueous mixtures of TiNP-1/pHEAAm and TiNP-2/pHEAAm, and transparent colorless films were obtained from mixtures of TiNP-3/pHEAAm. However, at higher TiNP concentrations, cracks and blurs were observed in the polymer composite films. TiNP-1 exhibited better compatibility with pHEAAm than TiNP-2 and TiNP-3. In the case of TiNP-1/pHEAAm, transparent and crack-free films were obtained at each temperature (10, 25 or 50 °C) for concentrations of up to 90 wt%, 80 wt% and 70 wt%, respectively. Figure 3 shows typical photos of TiNP-1/pHEAAm composite films with a variety of composition ratios, on glass plates, prepared at 10 or 50 °C. The lower drying temperature (10 °C) was suitable for the formation of TiNP/pHEAAm composite films with crack-free surfaces, on the glass plates. As a result, transparent and crack-free films were obtained with TiNP-1, TiNP-2, and TiNP-3 for concentrations of up to 90 wt%, 70 wt% and 60 wt%, respectively. At higher concentrations, cracks were observed on the surfaces of films prepared, even at a drying temperature of 10 °C (Figure 4).

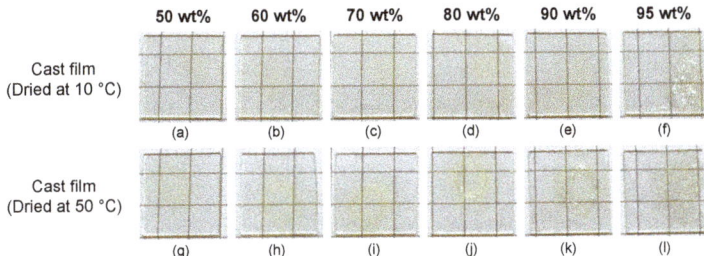

Figure 3. Photos of poly(N-hydroxyethyl acrylamide) (pHEAAm)-based thin films with different compositions of TiNP-1 on glass plates ((**a,g**): 50 wt%, (**b,h**): 60 wt%, (**c,i**): 70 wt%, (**d,j**): 80 wt%, (**e,k**): 90 wt%, (**f,l**): 95 wt%). The thin films were prepared by drop casting of the mixtures onto the glass plates and dried at 10 °C (**a–f**) or 50 °C (**g–l**).

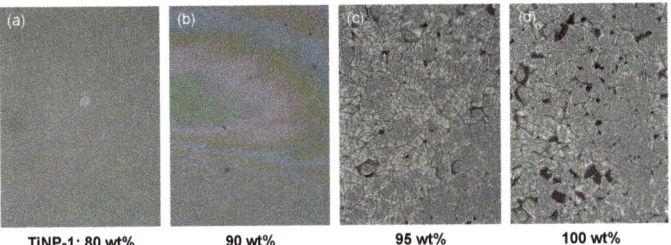

Figure 4. Laser microscopic images (120×) of the surfaces of TiNP-1/pHEAAm composite films with different compositions. (**a**) TiNP-1: 80 wt%, (**b**) TiNP-1: 90 wt%, (**c**) TiNP-1: 95 wt%, (**d**) TiNP-1: 100 wt%. The composite films were prepared by casting at 10 °C.

3.4. Refractive Indices of the TiNP/pHEAAm Composite Films

The refractive indices (n) of the polymer composite films were measured using a prism coupling refractometer, and the results are summarized in Table 1. The refractive index of the TiNP/pHEAAm composite films increased as the proportion of TiNP increased. The highest refractive index observed in this study was 1.90, which was obtained for a mixture of pHEAAm with TiNP-1 at 90 wt%. The n value was significantly improved compared with the n value of the pHEAAm film ($n = 1.53$) without TiNP.

Table 1. Refractive indices of TiNP/pHEAAm composite films under a variety of preparation conditions.

Composition TiNP:pHEAAm	TiNP-1			TiNP-2		TiNP-3	
	10 °C [a]	25 °C [a]	50 °C [a]	10 °C [a]	50 °C [a]	10 °C [a]	50 °C [a]
95:5	1.86	-	-	-	-	-	-
90:10	1.90	1.79	-	-	-	-	-
80:20	1.81	1.78	1.77	-	-	-	-
70:30	1.76	1.75	1.78	1.74	1.70	-	-
60:40	1.70	1.72	1.73	1.71	1.64	1.78	-
50:50	1.66	1.67	1.67	1.67	1.61	1.72	1.72
0:100	1.53	-	-	1.53	-	1.53	-

[a] Drying temperature of TiNP/pHEAAm composite film.

The observed n values were plotted against the concentration of TiNP (Figure 5) and compared with n values (dotted lines) calculated using Equation (1) and refractive indices of 2.50 for titania NPs and 1.53 for pHEAAm [30].

$$\frac{n^2-1}{n^2+2} = (1-c)\frac{\rho}{\rho_1}\frac{n_1^2-1}{n_1^2+2} + c\frac{\rho}{\rho_2}\frac{n_1^2-1}{n_1^2+2} \tag{1}$$

where n, n_1, and n_2 are the refractive indices of the composite, polymer (pHEAAm), and TiNP (TiNP-1), respectively, and ρ, ρ_1, and ρ_2 are the densities of the composite, polymer (pHEAAm), and TiNP (TiNP-1), respectively, and c is the concentration (wt%) of TiNP (TiNP-1).

The refractive indices of the thin films on the glass plates were measured using a prism coupler equipped with a HeNe laser and a GGG prism ($n = 1.965$). As shown in Figure 5, the observed n values of the TiNP-1/pHEAAm composites prepared at 50 °C were in good agreement with the theoretical values for TiNP-1, calculated with 0% impurity (Cal-A (1.52 (n_1), 4.29 (n_2), 1.11 g cm^{-3} (ρ_1), and 4.23 g cm^{-3} (ρ_2) were used for calculation), green line in Figure 5a), in the TiNP-1 composition range from 50 to 70 wt% TiNP-1, but lower at a TiNP-1 composition of 80 wt%. In contrast, the observed n values of the TiNP-1/pHEAAm composites prepared at 10 °C were lower than Cal-A in the TiNP-1 composition range from 50 to 90 wt%. As shown in Figure 5b, the thermogravimetric analysis indicated that the solid components in the aqueous TiNP-1 contained 9.6 wt% of evaporative and/or flammable impurities. The lower values for the experimentally obtained refractive indices compared with Cal-A is probably due to contamination by organic impurities. The Cal-B (1.52 (n_1), 4.29 (n_2), 1.47 (RI of unknown), 1.11 g cm^{-3} (ρ_1), 4.23 g cm^{-3} (ρ_2), 1.0 g cm^{-3} (density of unknown), and 9.6 wt% (composition of unknown) were used for calculation, blue broken line in Figure 5a) indicates the calculated n values with 9.6 wt% of impurities (for calculation, 1.0 g cm^{-1} and 1.47 were used for the density and the refractive index of impurity) added, and is in good agreement with the experimental n values of the composite films prepared at 10 °C. The experimental n value of the composite film at the higher TiNP-1 composition of 95 wt% was also significantly lower than the calculated value. As described in Section 3.2, cracks were observed on the surface of the composite films with higher TiNP-1 concentrations, whereas no cracks formed on films with lower TiNP-1 concentrations. The decreased n value at the higher concentration of TiNP-1 was probably due to the formation of cracks on the surface of the composite film.

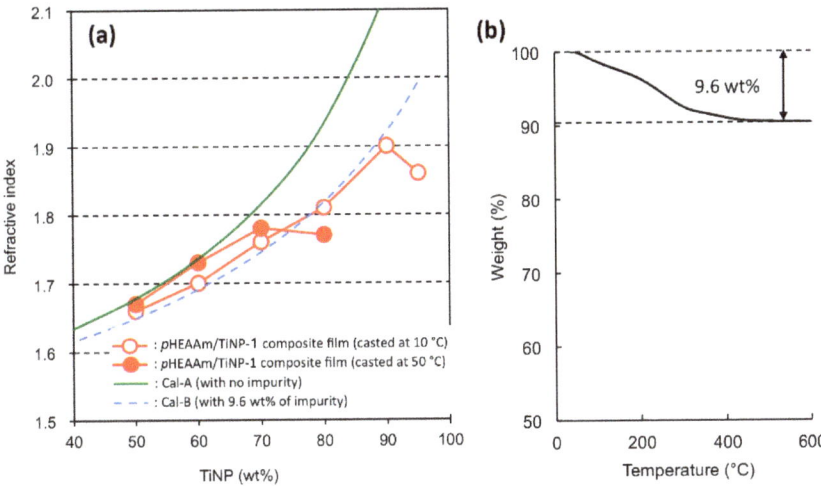

Figure 5. (a) Refractive indices of TiNP-1/pHEAAm composite films under a variety of preparation conditions and (b) thermogravimetric (TG) analysis of dried powder prepared from TiNP-1.

The transmittance UV-Vis spectra of composite films containing TiNP-1 at different concentrations were measured (Figure 6). Although the transmittance of the composite film gradually decreased with increasing TiNP-1 concentration, the composite film containing 90 wt% of TiNP-1 exhibited more than 80% transmittance in the wavelength range from 480 to 900 nm. Owing to absorption of the original TiNP-1 at wavelengths below 480 nm, the transmittances of the composite films decreased below 480 nm. It should be noted that the composite films were yellow but transparent. The yellow color was caused by the original TiNP-1 dispersion, which was probably due to the impurity; therefore, a colorless film may be obtained by removing the impurities from the TiNP dispersion.

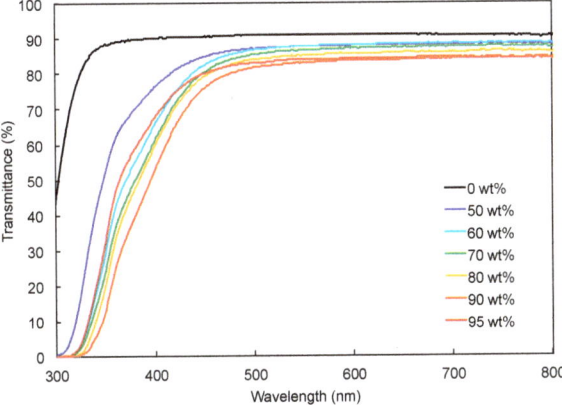

Figure 6. UV-Vis spectra of TiNP-1/pHEAAm composite films with different concentrations of TiNP-1. The composite films were prepared by drop casting the aqueous mixture of TiNP-1 and pHEAAm on a glass plate and drying at 10 °C.

4. Conclusions

In this study, various hydrophilic polymers with hydrogen bonding sites such as pHEAAm, poly(vinyl alcohol), poly(ethylene glycol), and poly(acrylic acid) were examined for the preparation of the polymer/titanium oxide nanoparticle composite films with high refractive indices. Among the hydrophilic polymers, *p*HEAAm exhibited the best compatibility with the titanium oxide nanoparticles. The compatibility was also affected by the nature of the titanium oxide. Further investigations into the hybridization behavior of *p*HEAAm with TiNP are needed, but the comparably higher compatibility of *p*HEAAm with titanium oxide nanoparticles is probably due to the formation of hydrogen bonding between the amide and hydroxide groups of *p*HEAAm, and the surface of titanium oxide nanoparticles. In the combination of *p*HEAAm and TiNP-1 (40 nm), the transparent composite thin films were obtained by casting the aqueous mixture of them at 10 °C even at the high concentration of 90 wt%. The experimental refractive indices were in good agreement with the calculated values, indicating that the TiNP-1 was homogeneously dispersed in the *p*HEAAm polymer matrix. The refractive index of the thin films was dropped at 95 wt% of TiNP-1. The composite thin films exhibited excellent refractive indices, with the highest refractive index reaching a value of 1.90. Laser microscopic observations indicated that the surface of the thin films was smooth and crackless at 90 wt% of TiNP-1 or less, but the cracks were observed on the surface of the thin films composed of TiNP-1 of more than 90 wt%. Thin films with high refractive indices are useful in a wide range of applications such as photo-devices, coatings, lighting equipment, and lenses.

Supplementary Materials: The following are available online at http://www.mdpi.com/2079-4991/9/4/514/s1.

Author Contributions: Conceptualization, M.T. and H.I.; methodology, M.T. and M.K.; validation, M.T., M.K., N.H., Y.K. and H.I.; formal analysis, M.K. and N.H.; investigation, M.T. and M.K.; data curation, M.T., M.K., N.H. and H.I.; writing—original draft preparation, M.T.; writing—review and editing, M.T., M.K., N.H. and H.I. project administration, M.T. and H.I.

Funding: This research received no external funding.

Acknowledgments: The authors express appreciation to Maki Horikawa and Shoji Nagaoka for providing the *p*HEAAm used in this study.

Conflicts of Interest: The authors declare no conflict of interest.

References

1. Lüab, C.; Yang, B. High Refractive Index Organic–inorganic Nanocomposites: Design, Synthesis and Application. *J. Mater. Chem.* **2009**, *19*, 2884–2901. [CrossRef]
2. Hasan, N.; Karkhanis, M.; Ghosh, C.; Khan, F.; Ghosh, T.; Kim, H.; Mastrangelo, C.H. Lightweight Smart Autofocusing Eyeglasses. *Proc. SPIE* **2018**, 1054507. [CrossRef]
3. Yetisen, A.K.; Montelongo, Y.; Butt, H. Rewritable three-dimensional holographic data storage via optical forces. *Appl. Phys. Lett.* **2016**, *109*, 061106. [CrossRef]
4. Kim, K.C. Effective graded refractive-index anti-reflection coating for high refractive-index polymer ophthalmic lenses. *Mater. Lett.* **2015**, *160*, 158–161. [CrossRef]
5. Li, X.; Yu, X.; Han, Y. Polymer thin films for antireflection coatings. *J. Mater. Chem. C* **2013**, *1*, 2266–2285. [CrossRef]
6. Sanders, D.P. Advances in Patterning Materials for 193 nm Immersion Lithography. *Chem. Rev.* **2010**, *110*, 321–360. [CrossRef] [PubMed]
7. Macdonald, E.K.; Shaver, M.P. Intrinsic high refractive index polymers. *Polym. Int.* **2014**, *64*, 6–14. [CrossRef]
8. Higashihara, T.; Ueda, M. Recent Progress in High Refractive Index Polymers. *Macromolecules* **2015**, *48*, 1915–1929. [CrossRef]
9. Fritsch, J.; Mansfeld, D.; Mehring, M.; Wursche, R.; Grothe, J.; Kaskel, S. Refractive Index Tuning of Highly Transparent Bismuth Containing Polymer Composites. *Polymer* **2011**, *52*, 3263–3268. [CrossRef]
10. Palik, E.D. *Handbook of Optical Constants of Solids*, 3rd ed.; Academic Press: Orlando, FL, USA, 1985; ISBN 978-0-08-054721-3.
11. Caseri, W. Nanocomposites of Polymers and Metals or Semiconductors: Historical Background and Optical Properties. *Macromol. Rapid Commun.* **2000**, *21*, 705–722. [CrossRef]
12. Olshavsky, M.A.; Allcock, H.R. Polyphosphazenes with High Refractive Indices: Synthesis, Characterization, and Optical Properties. *Macromolecules* **1995**, *28*, 6188–6197. [CrossRef]

13. Xia, Y.; Zhang, C.; Wang, J.X.; Wang, D.; Zeng, X.F.; Chen, J.F. Synthesis of Transparent Aqueous ZrO$_2$ Nanodispersion with a Controllable Crystalline Phase without Modification for a High-Refractive-Index Nanocomposite Film. *Langmuir* **2018**, *34*, 6806–6813. [CrossRef]
14. Tao, P.; Li, Y.; Rungta, A.; Viswanath, A.; Gao, J.; Benicewicz, B.C.; Siegel, R.W.; Schadler, L.S. TiO$_2$ Nanocomposites with High Refractive Index and Transparency. *J. Mater. Chem.* **2011**, *21*, 18623–18629. [CrossRef]
15. Dan, S.; Gu, H.; Tan, J.; Zhang, B.; Zhang, Q. Transparent Epoxy/TiO$_2$ Optical Hybrid Films with Tunable Refractive Index Prepared Via a Simple and Efficient Way. *Prog. Org. Coat.* **2018**, *120*, 252–259. [CrossRef]
16. Imai, Y.; Terahara, A.; Hakuta, Y.; Matsui, K.; Hayashi, H.; Ueno, N. Transparent Poly(bisphenol A Carbonate)-Based Nanocomposites with High Refractive Index Nanoparticles. *Eur. Polym. J.* **2009**, *5*, 630–638. [CrossRef]
17. Nakayama, N.; Hayashi, T. Preparation and Characterization of TiO$_2$ and Polymer Nanocomposite Films with High Refractive Index. *J. Appl. Polym. Sci.* **2007**, *6*, 3662–3672. [CrossRef]
18. Chieh, M.T.; Sheng, H.H.; Chun, C.H.; Yu, C.T.; Hsin, C.T.; Chung, A.W.; Wei, F.S. High Refractive Index Transparent Nanocomposites Prepared by In-situ Polymerization. *J. Mater. Chem. C* **2014**, *2*, 2251–2258. [CrossRef]
19. Cai, B.; Kaino, T.; Sugihara, O. Sulfonyl-Containing Polymer and Its Alumina Nanocomposite with High Abbe Number and High Refractive Index. *Opt. Mater. Exp.* **2015**, *5*, 1210–1216. [CrossRef]
20. Wang, C.; Cui, Q.; Wang, X.; Li, L. Preparation of Hybrid Gold/Polymer Nanocomposites and Their Application in a Controlled Antibacterial Assay. *ACS Appl. Mater. Interfaces* **2016**, *8*, 29101–29109. [CrossRef]
21. Voitylov, A.V.; Veso, O.S.; Petrov, M.P.; Rolich, V.I. Light Refraction in Aqueous Suspensions of Diamond Particles. *Colloids Surf. A Physicochem. Eng. Asp.* **2017**, *538*, 417–422. [CrossRef]
22. Morimune, S.; Kotera, M.; Nishino, T.; Goto, K.; Hata, K. Poly(vinyl alcohol) Nanocomposites with Nanodiamond. *Macromolecules* **2011**, *44*, 4415–4421. [CrossRef]
23. Tristan, S.K.; Ngoc, A.N.; Laura, E.A.; Soha, N.; Edward, A.L.; Sasaan, A.S.; Philip, T.D.; Clay, B.A.; Michael, S.M.; Jim, S.; et al. High Refractive Index Copolymers with Improved Thermomechanical Properties Via the Inverse Vulcanization of Sulfur and 1,3,5-Triisopropylbenzene. *ACS Macro Lett.* **2016**, *5*, 1152–1156. [CrossRef]
24. Liua, B.T.; Tanga, S.J.; Yub, Y.Y.; Linc, S.H. High-refractive-index Polymer/Inorganic Hybrid Films Containing High TiO$_2$ Contents. *Colloids Surf. A Physicochem. Eng. Asp.* **2011**, *377*, 138–143. [CrossRef]
25. Suac, H.W.; Chen, W.C. High Refractive Index Polyimide–nanocrystalline-titania Hybrid Optical Materials. *J. Mater. Chem.* **2008**, *18*, 1139–1145. [CrossRef]
26. Lu, C.; Cui, Z.; Guan, C.; Guan, J.; Yang, B.; Shen, J. Research on Preparation, Structure and Properties of TiO$_2$/Polythiourethane Hybrid Optical Films with High Refractive Index. *Macromol. Mater. Eng.* **2003**, *288*, 717–723. [CrossRef]
27. Lee, L.H.; Chen, W.C. High-Refractive-Index Thin Films Prepared from Trialkoxysilane-Capped Poly(methyl methacrylate)–Titania Materials. *Chem. Mater.* **2001**, *13*, 1137–1142. [CrossRef]
28. Jintoku, H.; Ihara, H. The Simplest Method for Fabrication of High Refractive Index Polymer–metal Oxide Hybrids Based on a Soap-free Process. *Chem. Commun.* **2014**, *50*, 10611–10614. [CrossRef]
29. Ishii, T.; Hoashi, Y.; Matsumoto, S.; Kuroki, M.; Jintoku, H.; Ogata, T.; Kuwahara, Y.; Takafuji, M.; Nagaoka, S.; Ihara, H. Facile Preparation of Transparent and High Refractive Index Polymer Composites by Polymerization of Monomer–Silicotungstic Acid Mixtures. *Chem. Lett.* **2017**, *46*, 489–491. [CrossRef]
30. Herráez, J.V.; Belda, R. Refractive Indices, Densities and Excess Molar Volumes of Monoalcohols + Water. *J. Solut. Chem.* **2006**, *35*, 1315–1328. [CrossRef]

© 2019 by the authors. Licensee MDPI, Basel, Switzerland. This article is an open access article distributed under the terms and conditions of the Creative Commons Attribution (CC BY) license (http://creativecommons.org/licenses/by/4.0/).

Article

Effect of Nano-SnS and Nano-MoS$_2$ on the Corrosion Protection Performance of the Polyvinylbutyral and Zinc-Rich Polyvinylbutyral Coatings

Zuopeng Qu [1,*,†], Lei Wang [1,†], Hongyu Tang [2,3,†], Huaiyu Ye [2] and Meicheng Li [4]

1. National Engineering Laboratory for Biomass Power Generation Equipment, School of Renewable Energy, North China Electric Power University, Beijing 102206, China
2. Delft Institute of Microsystems and Nanoelectronics, Delft University of Technology, 2628 CD Delft, The Netherlands
3. Changzhou Institute of Technology Research for Solid State Lighting, Changzhou 213161, China
4. State Key Laboratory of Alternate Electrical Power System with Renewable Energy Sources, School of Renewable Energy School, North China Electric Power University, Beijing 102206, China
* Correspondence: z.qu@ncepu.edu.cn
† These authors contributed equally to this work.

Received: 20 May 2019; Accepted: 17 June 2019; Published: 30 June 2019

Abstract: In this paper, four composite coatings of nano-SnS/polyvinylbutyral (PVB), nano-MoS$_2$/PVB, nano-SnS-Zn/PVB, and nano-MoS$_2$-Zn/PVB were prepared, and their anti-corrosion mechanism was analyzed by experimental and theoretical calculations. The results of the electrochemical experiments show that the effect of nano-MoS$_2$ on the corrosion protection performance of PVB coating is better than that of nano-SnS in 3% NaCl solution, and that the addition of Zn further enhances this effect, which is consistent with the results of weight loss measurements. Furthermore, the observation of the corrosion matrix by the field emission scanning electron microscope (FESEM) further confirmed the above conclusion. At last, the molecular dynamics (MD) simulation were carried out to investigate the anti-corrosion mechanism of the nanofillers/PVB composites for the copper surface. The results show that both nano-SnS and nano-MoS$_2$ are adsorbed strongly on the copper surface, and the binding energy of nano-MoS$_2$ is larger than that of nano-SnS.

Keywords: anti-corrosion; tin sulfide (SnS); molybdenum disulfide (MoS$_2$); electrochemical test; composite coating

---

1. Introduction

Copper has good thermal and electrical properties, and is a commonly used material in marine engineering. It is often used in many parts such as hulls, pipes, electronic devices, etc. [1–3]. However, the corrosive nature of seawater and gas above the ocean severely limits the service life of copper [4,5] and brings additional costs. Therefore, it is very meaningful to find a good way to delay the corrosion of copper. The protection of copper can be carried out in various ways such as inhibitor [6,7], film [8], coating [9,10]. The method of applying a composite coating to a metal substrate is convenient in application and low in cost, and is suitable for mass production in the industry. Thus many researchers have added various nanoparticles as fillers in polyvinylbutyral (PVB) as a physical barrier to enhance the barrier effect of the coating on water, oxygen, and other corrosive media [10].

Conventional graphene materials have been used as fillers in anti-corrosion coatings due to their high surface-to-volum ratio and excellent physical properties. However, they have a 'corrosion-promoting activity' when the coating is broken because of good electrical conductivity [9,11], which promotes the corrosion of the metal substrate. As a semiconductor material, tin sulfide (SnS) has good chemical stability and will not cause harm to human health and environment [12]. It is widely used in optoelectronics and sensors [13–16]. However, there are currently few reports on anti-corrosion applications for SnS. SnS has a very large resistivity and therefore may exhibit better corrosion resistance compared to graphene. MoS_2 is a transition metal disulfide with a variety of excellent physical properties. It has a layered structure and can be used to construct a variety of one-, two-, and three-dimensional materials [17–23], which has broad application prospects. In addition, MoS_2 has good hydrophobicity [24,25], which has been used by researchers to make composite coatings or to modify existing graphene products [17,26]. Therefore, we consider adding nano-MoS_2 as a filler to the coating to verify its protection against metallic copper. Moreover, nano-Zn, as an active metal, can be used as a cathodic protection in the coating to improve the protective ability of the coating. At present, many researchers have used Zn or zinc ions as additives to modify coatings [27,28] to enhance the protection of metals.

Our current work aims to study and compare the corrosion protection performance of nano-SnS and nano-MoS_2 on polyvinylbutyral (PVB) coatings and zinc-rich PVB coatings. Weight loss, potentiodynamic polarization, and electrochemical impedance spectroscopy (EIS) were used to evaluate the anti-corrosion performance of nano-SnS/PVB, nano-MoS_2/PVB, nano-SnS-Zn/PVB and nano-MoS_2-Zn/PVB coatings at first. In addition, field emission scanning electron microscope (FESEM) was used to characterize the morphology of the copper after corrosion to verify the test results. Furthermore, molecular dynamics (MD) simulations were used to study the adsorption properties of nano-SnS/PVB and nano-MoS_2/PVB on copper surfaces.

2. Experimental

2.1. Material and Sample Preparation

Some of the key materials used in this experiment are shown in Table 1, in which the average particle size of the nanoparticles is 50 nm.

Table 1. Materials used for preparing composite coatings.

Material	Manufacturer	Label
Polyvinylbutyral	MACKLIN	P815775
Tin sulfide	6Carbon Tech. Shenzhen	SC-CRYSTAL-SNS
Molybdenum disulfide	HANLANE	MoS_2-50
Zinc	HANLANE	Zn-50

Nano-SnS (0.1 g) was sonicated in 10 mL of methanol for more than 5 h to form a suspension. Then, 1.0 g of PVB powder was added to the suspension, and the mixture was thoroughly stirred for more than 36 h on a magnetic stirrer to prepare a uniform paint, and the resultant material was allowed to stand for use. The molecular structure of PVB is shown in Figure 1. Nano-MoS_2/PVB, nano-SnS-Zn/PVB, and nano-MoS_2-Zn/PVB coatings were prepared in the same manner, wherein the amount of Zn added was 0.05 g. The copper piece with a thickness of 0.05 mm was cut to a size of 1.0 cm × 1.0 cm, and it was carefully polished with 400, 800, and 2,000 mesh emerald paper. After the sanding was completed, the copper sheets were ultrasonically cleaned with ultrapure water for more than 5 min, then degreased with acetone and air dried naturally in a fume hood. The copper sheet was dipped into the prepared paint for 40 s and then taken out at a rate of about 0.5 mm s^{-1}. The samples were air dried naturally in a fume hood for 24 h. The thickness of the prepared coatings was essentially the same and they were controlled at 19.3 ± 1.2 µm, which was measured with a portable coating thickness gauge (EC-770, YOWEXA, Shenzhen, China). Some of the copper sheets were sealed with 705 silicone rubber on the back, leaving 1.0 cm^2 of working surface for electrochemical experiments, while the rest retained double working faces for weight loss measurements. A NaCl solution having a concentration of 3.0% was prepared for electrochemical experiments and weight loss measurements. All experiments were performed at room temperature (about 293 K).

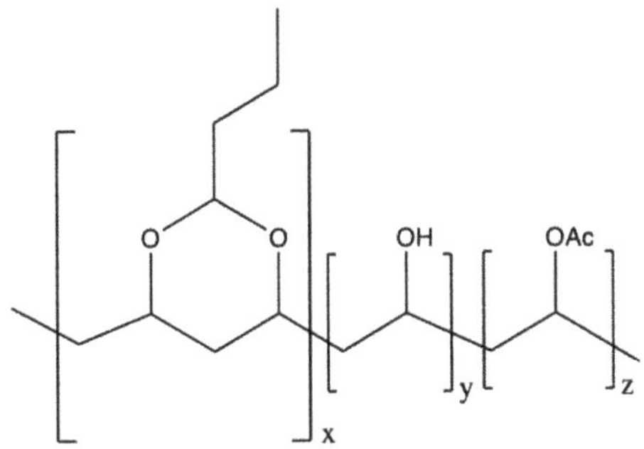

Figure 1. Molecular structure of PVB used in this work.

2.2. Weight Loss Measurements

Weight loss measurements were carried out in glass dishes at 293 K. Samples, without and with different coatings, prepared in triplicate were each immersed in a 3.0% NaCl solution. The samples were continuously immersed for 35 days, then soaked in methanol to dissolve the surface coating and thoroughly rinsed in 0.1 M HCl, water, and acetone in order to remove corrosion products and other impurities. The copper sheets were weighed with an analytical balance after drying. The average corrosion rate was finally calculated by the immersion time and the weight loss of each sample.

2.3. Electrochemical Experiments

The electrochemical experiments in this paper are performed using a conventional three-electrode system. The copper piece to be tested was the working electrode, the saturated calomel electrode (SCE) was used as the reference electrode, and the platinum mesh electrode (2.0 cm × 2.0 cm) was used as the counter electrode. The CHI660C electrochemical workstation was used for the electrochemical measurements. Prior to each test, the working electrode was immersed in an etching medium for about 30 min at an open circuit potential (OCP) until the OCP value reached an almost constant state. EIS measurements were made under open circuit potential with a sweep frequency range of 10 mHz–100 kHz and an amplitude of 5 mV AC sinusoidal disturbance. The EIS data were analyzed and fitted by Zsimpwin 3.60 software. Furthermore, a polarization experiment was performed at a potential range of ±250 mV relative to the OCP at a scan rate of 2 mV s^{-1}. Each sample was tested more than 4 times to ensure the reproducibility of the experiments.

2.4. Scanning Electron Microscope (SEM) Observations

Scanning electron microscopy (SEM, JSM6490LV, JEOL, Tokyo, Japan) was used to observe and analyze the copper matrix without and with different coatings at 293 K. Prior to observation, all samples were immersed in a 3.0% NaCl solution at 293 K for 10 days, and then the surface coated samples were immersed in anhydrous methanol to remove the surface coating.

2.5. Theoretical Study

In order to further discuss the interaction between the PVB/nanosheets composites and copper surface, molecular dynamics (MD) simulation were carried out to model the adsorption structure of SnS-PVB and MoS$_2$-PVB system on copper (111) surface. MD simulations were performed using commercially available software, Material Studio 8.0, purchased from Accelrys Inc (San Diego, CA, USA). The structure of PVB chain and molecules of SnS or MoS$_2$ nanosheet were generated and optimized through the Forcite module. The PVB model was prepared with the dimensions of

28 × 30 × 126 Å, with an initial density of 1.05 g cc^{-1}, which agrees well with experimental data of 1.07 g cc^{-1} [29]. The energy minimized structures of PVB (2 chains) and nanosheets were used for the construction of different amorphous cells. The optimization procedure follows convergence criteria: 2.0 × 10^{-5} kcal mol^{-1} for energy, 0.001 kcal mol^{-1} Å^{-1} for force, and 1.0 × 10^{-5} for displacement. After the geometry optimization, molecular dynamic calculation with constant number of particles, volume and temperature (NVT) and Universal force field (UFF) [30] was performed at time step of 1.0 fs up to total 3.0 ns, among which Andersen algorithm thermostat with 1.0 Collision ratio was used to maintain the temperature of the system at around 298 K. The interaction energy $E_{Interaction}$, post to equilibration is then calculated using Equation (1):

$$E_{Interaction} = E_{Total} - (E_{Cu} + E_{composite}), \quad (1)$$

where E_{Total} is the combined energy of the Cu surface and the PVB-nanosheets system, E_{Cu} is the energy of the solo Cu surface, and $E_{composite}$ is the energy of the PVB-nanosheets system taken independently. Based on Equation (1), it can be stated that the more negative the value, the better is the adhesion the composites coating applied on the surface.

3. Results and Discussion

3.1. Weight Loss Measurements

The corrosion resistance of various coatings to copper was investigated by weight loss measurements after immersion in a 3.0% NaCl solution at 293 K for 35 days. The corrosion rates (ω, mg m^{-2} h^{-1}) and protective efficiency (η_w) of these coatings were calculated as follows, and the results are shown in Table 2,

$$\omega = \frac{m_0 - m}{A\tau}, \quad (2)$$

$$\eta_w\% = \frac{\omega_0 - \omega}{\omega_0} \cdot 100, \quad (3)$$

where A is the total surface area of the sample; m_0, and m are the weights of the sample before and after immersion in the corrosive solution, respectively; τ is the soaking time; and ω_0 and ω are the corrosion rates of the copper samples containing and containing the coating, respectively.

Table 2. Corrosion rate and protection efficiency of copper sheets without and with different coatings in 3.0% NaCl solution at 293 K.

Coating	ω (mg m^{-2} h^{-1})	η_w (%)
Blank	59.52	-
PVB	41.67	30
nano-SnS	23.81	60
nano-SnS-Zn/PVB	15.48	74
nano-MoS$_2$/PVB	20.24	66
nano-MoS$_2$-Zn/PVB	14.29	76

As shown in Table 2, the corrosion rate of the coated copper sheet sample was smaller than that of the uncoated copper sheet sample. Compared to pure PVB coatings, copper samples with nanofiller coatings have a lower corrosion rate and higher protection efficiency. Furthermore, the addition of Zn can improve the protection efficiency of the coating by more than 10%. It is worth noting that for the nano-MoS$_2$-Zn/PVB coating, the η_w value is as high as 76%, which is 46% higher than the pure PVB coating.

3.2. Polarization Curve

The Tafel polarization curves of copper electrodes coated with different coatings measured in 3% NaCl solution are shown in Figure 2. We obtained the main electrochemical parameters by extrapolation, including corrosion potential (E_{corr}), corrosion current density (i_{corr}), anode and cathode Tafel slope (β_a, β_c), and protection efficiency (η_P). Their values were listed in Table 3. The value of the protection efficiency η is calculated as follows,

$$\eta_P(\%) = \frac{i_{corr,o} - i_{corr,k}}{i_{corr,o}} \cdot 100, \qquad (4)$$

where $i_{corr,o}$ (A cm^{-2}) is the corrosion current density of the uncoated copper sample, $i_{corr,k}$ (A cm^{-2}) is the corrosion current density of the sample containing the different composite coating.

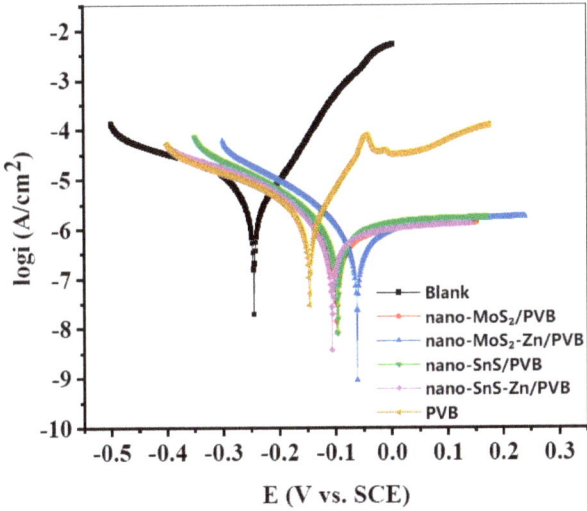

Figure 2. Tafel curve recorded for copper samples without and with different coatings in 3.0% NaCl solution at 293 K.

Table 3. Electrochemical parameters for copper sample without and with different coatings in 3.0% NaCl solution at 293 K.

Coating	E_{corr} (mV per SCE)	i_{corr} (µA cm^{-2})	β_c (mV dec^{-1})	β_a (mV dec^{-1})	Crate (mpy)	η_P (%)
Blank	−245	2.185	60	55	1.00	-
PVB	−147	1.839	179	75	0.85	15.8
nano-SnS/PVB	−97	0.921	105	280	0.42	57.9
nano-SnS-Zn/PVB	−107	0.757	99	317	0.35	65.4
nano-MoS$_2$/PVB	−98	0.710	102	297	0.33	67.5
nano-MoS$_2$-Zn/PVB	−62	0.591	84	190	0.27	72.9

Figure 2 shows that the current densities of the samples with coatings are significantly lower than that of the uncoated copper. More importantly, the coating with the nanofiller exhibited a significantly lower current density than the pure PVB coating. This indicates that after the surface of the copper sheet is coated, its corrosion strength and corrosion rate, in 3% NaCl becomes lower.

According to Table 3, the current density (i_{corr}) of MoS$_2$ is lower than that of SnS, and the i_{corr} higher compared with the corresponding Zn-containing composite coatings of SnS and MoS$_2$. This indicates that the corrosion protection ability of MoS$_2$ is better than that of SnS, and the addition of Zn has a certain improvement effect on the protective ability of the coating. The η_P value of the sample with the composite coating increased by more than 40% compared to the sample with a pure PVB coating, and the efficiency value of the nano-MoS$_2$-Zn/PVB coating was the largest, reaching 72.9%. The experimental results show that the composite coating has a significant protective effect on the metal matrix, which is in line with the result of weight loss measurements.

3.3. Electrochemical Impedance Spectroscopy (EIS)

In order to study the corrosion mechanism of the metal and the improvement of the corrosion resistance of the coating, we measured the electrochemical impedance spectroscopy of copper with pure PVB coatings and different nanofiller composite coatings. The experiments were carried out in 3.0% NaCl solution. As shown in Figure 3a, the Nyquist plot of the PVB coated and uncoated copper sheet show an incomplete semicircle in the high frequency region and an approximate straight line in the subsequent low frequency range. In general, the high frequency region semicircle is related to the charge transfer resistance (R_{ct}) and the double layer capacitance (C_{dl}). The low frequency impedance is the Warburg impedance (W), which means the diffusion of dissolved oxygen or soluble cuprous chloride complexes during the corrosion process.

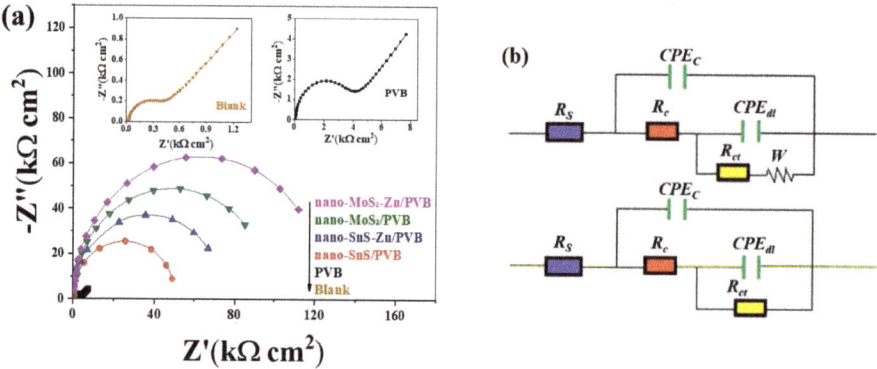

Figure 3. (a) Nyquist plots for copper samples in 3.0% NaCl solution without and with different coatings at 293 K and (b) equivalent circuit diagrams for fitting Electrochemical Impedance Spectroscopy (EIS) data.

Figure 3a shows that after the addition of the nanofiller, the shape of the curve appears to be approximately semicircular, and the Warburg impedance in the low frequency region disappears. This indicates that the addition of nanofiller inhibits the diffusion of dissolved oxygen and cuprous chloride complexes. At this time, the corrosion of copper depends on the charge transfer process. The order of the diameter of the semicircle obtained by different coating samples is: nano-MoS$_2$-Zn/PVB > nano-MoS$_2$/PVB > nano-SnS-Zn/PVB > nano-SnS/PVB, and the diameter of the semicircle of the sample added with Zn is obviously increased. According to this result, it can be judged that the nano-MoS$_2$ filler is better than the nano-SnS for the corrosion protection performance of the PVB coatings and the addition of Zn particles further improves the corrosion resistance of the composite coatings.

In order to quantitatively compare the corrosion inhibition properties of different coatings, we further fit the EIS data using the equivalent circuit diagram shown in the inset of Figure 3b, and the resulting electrochemical parameters are listed in Table 4, where R_s is the solution resistance, R_c is the resistance of the coating on the copper working surface, R_{ct} is the charge transfer resistance, and W is the Warburg impedance. CPE_c and CPE_{dl} are constant phase angle elements representing the coating capacitance (C_c) and the double layer capacitance (C_{dl}), respectively. The impedance of these circuits can be expressed as follows [31],

$$Z_W = R_S + \frac{1}{jwCPE_c + \frac{1}{R_c} + \frac{1}{jwCPE_{dl} + \frac{1}{R_{ct}+w}}} \tag{5}$$

$$Z = R_S + \frac{1}{jwCPE_c + \frac{1}{R_c} + \frac{1}{jwCPE_{dl} + \frac{1}{R_{ct}}}} \tag{6}$$

Table 4. Electrochemical parameters of EIS in copper samples without and with different coatings in 3.0% NaCl solution at 293 K.

Coating	R_s (Ω cm^2)	R_c (kΩ cm^2)	R_{ct} (kΩ cm^2)	C_c (μF cm^{-2})	C_{dl} (μF cm^{-2})	W	η_E (%)
Blank	7.94	0.04	0.32	13.08	128.30	0.002902	-
PVB	19.50	0.07	3.50	1.84	18.17	0.000620	90.86
nano-SnS/PVB	18.14	0.11	50.84	2.69	54.21	-	99.37
nano-SnS-Zn/PVB	22.54	0.30	74.36	2.49	69.49	-	99.57
nano-MoS$_2$/PVB	20.07	0.26	97.84	3.07	60.39	-	99.67
nano-MoS$_2$-Zn/PVB	25.09	0.30	126.20	1.92	43.34	-	99.75

The impedance of the CPE is defined as follows [31],

$$Z_{CPE} = \frac{w^{-n}}{Y\left(\cos\frac{n\pi}{2} + j\sin\frac{n\pi}{2}\right)^n} \tag{7}$$

where Y is the modulus of the CPE, w is the angular frequency, j is the imaginary number, and n is the deviation parameter. The η values of the copper electrode coatings in these 3% NaCl solutions are calculated as follows,

$$\eta_E(\%) = \frac{R_{ct} - R_{ct,0}}{R_{ct}} \cdot 100 \tag{8}$$

where R_{ct} and $R_{ct,0}$ are the charge transfer resistances of copper samples with and without various coatings in 3% NaCl solution, respectively.

As can be seen from Table 4, the R_c and R_{ct} values of the nano-MoS$_2$/PVB coated samples were larger than those of nano-SnS/PVB, and these values improved in the samples to which Zn was added. This again demonstrates that nano-MoS$_2$ enhances the corrosion resistance of the coating better than nano-SnS, and the addition of Zn further enhances the corrosion resistance of the coating. In addition, all copper samples with coatings incorporating nanofillers had η_E values above 99%, which is a significant increase compared to samples with pure PVB coatings. These demonstrate that the use of nanoparticles as a filler can effectively enhance the corrosion resistance of polymer coatings. The trend of these values is basically consistent with the weight loss measurements results. Several typical corrosion resistant materials and their values of η_E obtained in a NaCl solution are listed in Table 5. The higher η_E values further highlight the superior corrosion resistance of the composite coatings in this study compared to these materials [6,32,33].

Table 5. Anti-corrosion materials and their protection efficiency (η_E) by EIS test in 3.0–3.5% NaCl solution.

Classification	Samples	η_E (%)
Silicon carbide composite	POA-SiC/EP	87.54
Metal Organic Framework	ATT/ZIF-8	97.3
Organic inhibitor	polyaspartic acid	86.8

3.4. SEM Analyses

At 293 K, SEM high-resolution photographs of copper samples without and with different coatings immersed in 3% NaCl solution for 10 days are shown in Figure 4. After the copper with pure PVB coating (Figure 4b) was immersed in 3% NaCl solution, the substrate experienced severe corrosion with many obvious large-area corrosion marks. Although the nano-SnS/PVB sample (Figure 4c) also showed local corrosion, the degree of corrosion was lower than that of the pure PVB coating sample, and large-area local corrosion disappeared after the addition of Zn (Figure 4d). In addition, the nano-MoS_2/PVB sample (Figure 4e) exhibited slight signs of corrosion, while the substrate surface of the nano-MoS_2-Zn/PVB sample (Figure 4f) was well protected.

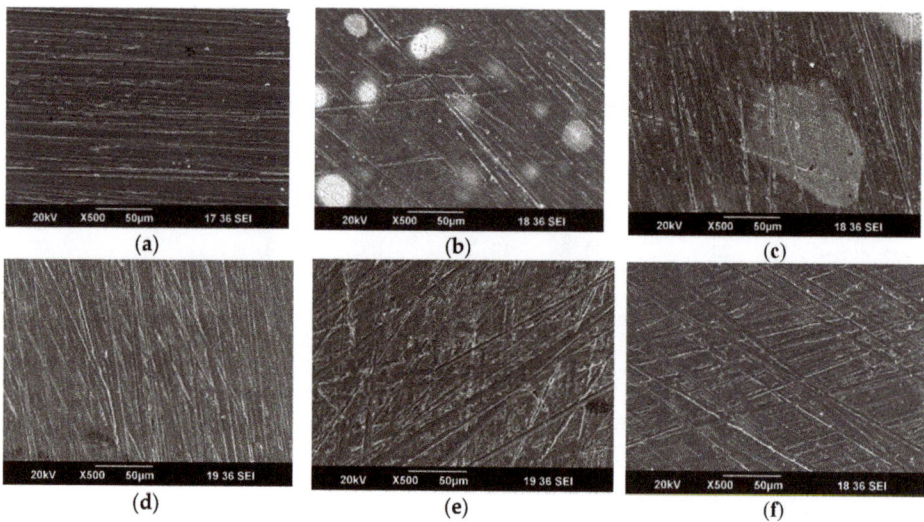

Figure 4. SEM images of (a) freshly polished copper specimen and the specimens immersed in 3% NaCl solution with (b) PVB, (c) nano-SnS/PVB, (d) nano-SnS-Zn/PVB, (e) nano-MoS_2/PVB, and (f) nano-MoS_2-Zn/PVB coating for 10 days at 293 K.

The observation of these high-definition pictures proves that MoS_2 can improve the anti-corrosion performance of PVB more than SnS, and the addition of Zn further enhances this performance. The SEM analysis further validated the results of electrochemical experiments and weight loss measurements.

3.5. Corrosion Mechanism Analysis

The corrosion process of Cu in NaCl solution has been described in numerous reports [34–37]. As we all know, in the corrosion process of copper with NaCl solution as the medium, the cathodic reaction is represented by the reduction of oxygen [1,38],

$$O_2 + 4e + 2H_2O \rightarrow 4OH^- \tag{9}$$

The anode undergoes the following series of complex reactions [39],

$$Cu \rightarrow Cu^+ + e, \tag{10}$$

$$Cu^+ + Cl^- \rightarrow CuCl, \tag{11}$$

$$CuCl + Cl^- \rightarrow CuCl_2^-, \tag{12}$$

$$CuCl_2^- \rightarrow Cu^{2+} + 2Cl^- + e \tag{13}$$

The insulation of the composite coating inhibits the transfer of current, hinders the formation of a closed loop between the substrate and the etching solution, and reduces the corrosion rate. Besides, the nanoparticle filler acts as a physical barrier in the polymer, preventing the penetration of O_2 and H_2O into the metal, and the diffusion of $CuCl_2^-$ into the 3% NaCl solution. Moreover, for the Zn-added coating, since Zn has a lower electronegativity than Cu, Zn is first corroded in the system, which delays the oxidation process of the anode Cu as shown in Equation (10). This further explains the test results of the EIS. The analysis of corrosion and protection mechanisms is consistent with Han et al. [9,40].

3.6. Molecular Dynamics Simulation

As shown in Figures 5 and 6, after sufficient relaxation and of equilibrium the PVB-nanosheets system, two types of composites slowly approached to Cu surface, indicating that the combined energy of the Cu surface and the PVB-nanosheets system is large. The interaction energy $E_{Interaction}$ of pure PVB, SnS/PVB, MoS$_2$/PVB with Cu surface are listed in Table 6. All the energies obtained have been found to be negative in sign, which means that all the formulated coatings show sufficient binding to the surface. The $E_{Interaction}$ of MoS$_2$/PVB composite ($-1,838.253$ kcal mol^{-1}) is larger than that of SnS/PVB composites ($-1,074.433$ kcal mol^{-1}) and PVB coatings (-852.33 kcal mol^{-1}), indicating that the anti-corrosion behavior of MoS$_2$/PVB composite coatings is better than that of SnS/PVB and pure PVB. The MD simulation results are in good agreement with the results obtained from potentiodynamic polarization measurements and EIS, which is further confirm that the excellent anti-corrosion performance of MoS$_2$/PVB is attributed to its high interfacial binding energy. In addition, the $E_{Interaction}$ is larger than that of previous work, such as -500 kcal mol^{-1} of Polyvinyl acetate (PVAc)-Perfluorooctane (PFO) systems and -27.3 kcal mol^{-1} graphene-based polymer coatings [41,42], which reveals the MoS$_2$/PVB and SnS/PVB composites coatings exhibit excellent corrosion protection performance than many nanosheets-polymer composites.

Table 6. The simulated surface energy for different systems at 298 K.

System	Interaction Energy (kcal mol^{-1})
Cu+PVB	−852.33
Cu+MoS$_2$/PVB	−1838.253
Cu+SnS/PVB	−1074.433

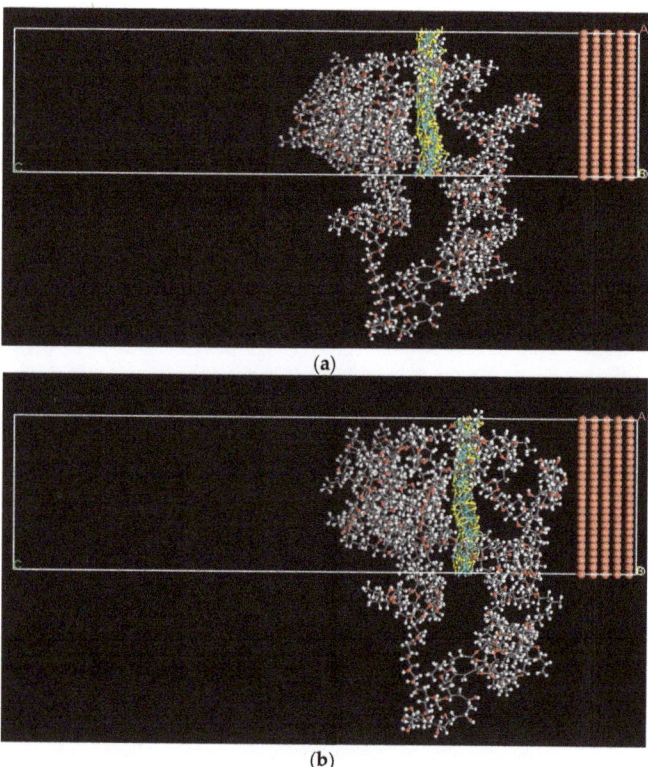

Figure 5. Illustration of (**a**) the MoS$_2$-PVB system consisting of copper (111) crystal as the substrate (shown in brown spheres) and PVB containing 10% MoS$_2$ at time 0 ps. (**b**) Final MoS$_2$-PVB system post-MD run of 500 ps.

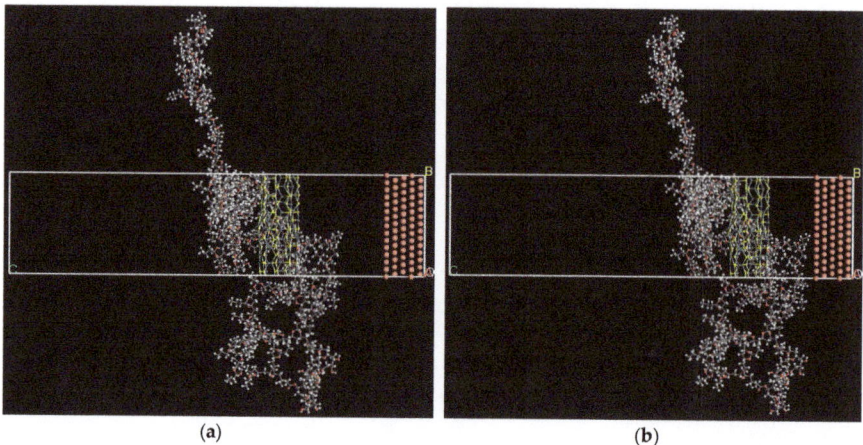

Figure 6. Illustration of (**a**) the SnS-PVB system consisting of copper (111) crystal as the substrate (shown in brown spheres) and PVB containing 10% SnS at time 0 ps. (**b**) Final SnS-PVB system post-MD run of 500 ps.

4. Conclusions

The anti-corrosion performance of nano-SnS/PVB, nano-MoS$_2$/PVB, nano-SnS-Zn/PVB, and nano-MoS$_2$-Zn/PVB was studied by experiments and theoretical calculations. All four coatings have good corrosion protection performance and their protective efficiencies calculated from the weight loss and polarization curves are consistent. Moreover, the Nyquist plot and fit to the EIS data indicate that nano-MoS$_2$/PVB has better anti-corrosion performance than nano-SnS/PVB. The addition of Zn further enhances this performance. These can be further confirmed by FESEM observation. Compared to graphene-based films or other composite coatings with nanomaterials as fillers, the composite coatings prepared herein have higher η_P and R_{ct} values [43,44], meaning that the coatings have better corrosion resistance. At last, the results of the molecular dynamics (MD) simulation show that both nano-SnS and nano-MoS$_2$ are adsorbed strongly on the copper surface, and the binding energy of nano-MoS$_2$ is larger than that of nano-SnS. Furthermore, compared to some other studies, these composite coatings have a larger interaction energy $E_{Interaction}$ with the copper surface. There are still many aspects of this research that need to be explored and improved in future work, such as sample preparation methods and coating modification.

Author Contributions: Conceptualization, Z.Q., L.W., and H.T.; methodology, L.W. and H.T.; software, H.T.; validation, Z.Q. and H.T.; formal analysis, L.W. and H.T.; investigation, Z.Q.; resources, Z.Q. and M.L.; data curation, L.W. and H.T.; writing—Original draft preparation, Z.Q., L.W., and H.T.; writing—Review and editing, Z.Q., H.T., and H.Y.; visualization, L.W.; supervision, M.L.; project administration, Z.Q.; funding acquisition, Z.Q.

Funding: This work was supported by the Fundamental Research Funds for the Central Universities 2014MS30.

Acknowledgments: The authors would like to thank Peng Peng for his technical support.

Conflicts of Interest: The authors declare no conflict of interest.

References

1. Qiang, Y.; Zhang, S.; Xu, S.; Yin, L. The effect of 5-nitroindazole as an inhibitor for the corrosion of copper in a 3.0% NaCl solution. *RSC Adv.* **2015**, *5*, 63866–63873. [CrossRef]
2. Al Kharafi, F.M.; Ghayad, I.M.; Abdallah, R.M. Corrosion inhibition of copper in seawater by 4-amino-4h-1,2,4-triazole-3-thiol. *Corrosion* **2013**, *69*, 58–66. [CrossRef]
3. Amin, M.A. Weight loss, polarization, electrochemical impedance spectroscopy, SEM and EDX studies of the corrosion inhibition of copper in aerated NaCl solutions. *J. Appl. Electrochem.* **2006**, *36*, 215–226. [CrossRef]
4. Ran, M.; Zheng, W.; Wang, H. Fabrication of superhydrophobic surfaces for corrosion protection: A review. *Mater. Sci. Technol.* **2019**, *35*, 313–326. [CrossRef]
5. Heakal, F.E.-T.; Fekry, A.M.; Fatayerji, M.Z. Influence of halides on the dissolution and passivation behavior of AZ91D magnesium alloy in aqueous solutions. *Electrochim. Acta* **2009**, *54*, 1545–1557. [CrossRef]
6. Zeino, A.; Abdulazeez, I.; Khaled, M.; Jawich, M.W.; Obot, I.B. Mechanistic study of polyaspartic acid (PASP) as eco-friendly corrosion inhibitor on mild steel in 3% NaCl aerated solution. *J. Mol. Liq.* **2018**, *250*, 50–62. [CrossRef]
7. Qiang, Y.; Zhang, S.; Xu, S.; Li, W. Experimental and theoretical studies on the corrosion inhibition of copper by two indazole derivatives in 3.0% NaCl solution. *J. Colloid Interface Sci.* **2016**, *472*, 52–59. [CrossRef]
8. Xu, S.; Gao, X.; Sun, J.; Hu, M.; Wang, D.; Jiang, D.; Zhou, F.; Weng, L.; Liu, W. Comparative study of moisture corrosion to WS$_2$ and WS$_2$/Cu multilayer films. *Surf. Coat. Technol.* **2014**, *247*, 30–38. [CrossRef]
9. Sun, W.; Wang, L.; Wu, T.; Pan, Y.; Liu, G. Inhibited corrosion-promotion activity of graphene encapsulated in nanosized silicon oxide. *J. Mater. Chem. A* **2015**, *3*, 16843–16848. [CrossRef]
10. Yu, Y.-H.; Lin, Y.-Y.; Lin, C.-H.; Chan, C.-C.; Huang, Y.-C. High-performance polystyrene/graphene-based nanocomposites with excellent anti-corrosion properties. *Polym. Chem.* **2014**, *5*, 535–550. [CrossRef]
11. Cui, C.; Lim, A.T.O.; Huang, J. A cautionary note on graphene anti-corrosion coatings. *Nat. Nanotechnol.* **2017**, *12*, 834–835. [CrossRef] [PubMed]
12. Ali, S.; Wang, F.; Zafar, S.; Iqbal, T. Hydrothermal synthesis, characterization and raman vibrations of chalcogenide SnS nanorods. In Proceedings of the 5th Annual International Conference on Material Science

and Engineering, Xiamen, China, 20–22 Ocotber 2017; Aleksandrova, M., Szewczyk, R., Eds.; IOP Publishing: Bristol, UK, 2018; Volume 275.
13. Gonzalez-Flores, V.E.; Mohan, R.N.; Ballinas-Morales, R.; Nair, M.T.S.; Nair, P.K. Thin film solar cells of chemically deposited SnS of cubic and orthorhombic structures. *Thin Solid Film.* **2019**, *672*, 62–65. [CrossRef]
14. Cabrera-German, D.; Garcia-Valenzuela, J.A.; Cota-Leal, M.; Martinez-Gil, M.; Aceves, R.; Sotelo-Lerma, M. Detailed characterization of good-quality SnS thin films obtained by chemical solution deposition at different reaction temperatures. *Mater. Sci. Semicond. Process.* **2019**, *89*, 131–142. [CrossRef]
15. Ye, J.; Qi, L.; Liu, B.; Xu, C. Facile preparation of hexagonal tin sulfide nanoplates anchored on graphene nanosheets for highly efficient sodium storage. *J. Colloid Interface Sci.* **2018**, *513*, 188–197. [CrossRef] [PubMed]
16. Reddy, N.K.; Devika, M.; Gopal, E.S.R. Review on tin (II) sulfide (SnS) material: Synthesis, properties, and applications. *Crit. Rev. Solid State Mater. Sci.* **2015**, *40*, 359–398. [CrossRef]
17. Li, Z.; Hui, S.; Yang, J.; Hua, Y. A facile strategy for the fabrication of superamphiphobic MoS_2 film on steel substrates with excellent anti-corrosion property. *Mater. Lett.* **2018**, *229*, 336–339. [CrossRef]
18. Thangasamy, P.; Partheeban, T.; Sudanthiramoorthy, S.; Sathish, M. Enhanced superhydrophobic performance of BN-MoS_2 heterostructure prepared via a rapid, one-pot supercritical fluid processing. *Langmuir* **2017**, *33*, 6159–6166. [CrossRef]
19. Rheem, Y.; Han, Y.; Lee, K.H.; Choi, S.-M.; Myung, N.V. Synthesis of hierarchical MoO_2/MoS_2 nanofibers for electrocatalytic hydrogen evolution. *Nanotechnology* **2017**, *28*, 105605. [CrossRef]
20. Zhou, Y.; Liu, Y.; Zhao, W.; Xie, F.; Xu, R.; Li, B.; Zhou, X.; Shen, H. Growth of vertically aligned MoS_2 nanosheets on a Ti substrate through a self-supported bonding interface for high-performance lithium-ion batteries: A general approach. *J. Mater. Chem. A* **2016**, *4*, 5932–5941. [CrossRef]
21. Gao, X.; Wang, X.; Ouyang, X.; Wen, C. Flexible superhydrophobic and superoleophilic MoS_2 sponge for highly efficient oil-water separation. *Sci. Rep.* **2016**, *6*, 27207. [CrossRef]
22. Shi, J.; Zhang, X.; Ma, D.; Zhu, J.; Zhang, Y.; Guo, Z.; Yao, Y.; Ji, Q.; Song, X.; Zhang, Y.; et al. Substrate facet effect on the growth of mono layer MoS_2 on Au foils. *ACS Nano* **2015**, *9*, 4017–4025. [CrossRef] [PubMed]
23. Hu, L.; Zhang, S.; Li, W.; Hou, B. Electrochemical and thermodynamic investigation of diniconazole and triadimefon as corrosion inhibitors for copper in synthetic seawater. *Corros. Sci.* **2010**, *52*, 2891–2896. [CrossRef]
24. Tang, Y.; Zhang, X.; Choi, P.; Xu, Z.; Liu, Q. Contributions of van der Waals interactions and hydrophobic attraction to molecular adhesions on a hydrophobic MoS_2 surface in water. *Langmuir* **2018**, *34*, 14196–14203. [CrossRef] [PubMed]
25. Kozbial, A.; Gong, X.; Liu, H.; Li, L. Understanding the intrinsic water wettability of molybdenum disulfide (MoS_2). *Langmuir* **2015**, *31*, 8429–8435. [CrossRef]
26. Chen, C.; He, Y.; Xiao, G.; Xia, Y.; Li, H.; He, Z. Two-dimensional hybrid materials: MoS_2-RGO nanocomposites enhanced the barrier properties of epoxy coating. *Appl. Surf. Sci.* **2018**, *444*, 511–521. [CrossRef]
27. Taheri, N.N.; Ramezanzadeh, B.; Mandavian, M.; Bahlakeh, G. In-situ synthesis of Zn doped polyaniline on graphene oxide for inhibition of mild steel corrosion in 3.5 wt.% chloride solution. *J. Ind. Eng. Chem.* **2018**, *63*, 322–339. [CrossRef]
28. Ding, R.; Zheng, Y.; Yu, H.; Li, W.; Wang, X.; Gui, T. Study of water permeation dynamics and anti-corrosion mechanism of graphene/zinc coatings. *J. Alloys Compd.* **2018**, *748*, 481–495. [CrossRef]
29. Zhang, X.; Hao, H.; Shi, Y.; Cui, J. The mechanical properties of Polyvinyl Butyral (PVB) at high strain rates. *Constr. Build. Mater.* **2015**, *93*, 404–415. [CrossRef]
30. Rappé, A.K.; Casewit, C.J.; Colwell, K.; Goddard, W.A., III; Skiff, W.M. UFF, a full periodic table force field for molecular mechanics and molecular dynamics simulations. *J. Am. Chem. Soc.* **1992**, *114*, 10024–10035. [CrossRef]
31. Tian, H.; Li, W.; Cao, K.; Hou, B. Potent inhibition of copper corrosion in neutral chloride media by novel non-toxic thiadiazole derivatives. *Corros. Sci.* **2013**, *73*, 281–291. [CrossRef]
32. Hu, C.; Li, Y.; Zhang, N.; Ding, Y. Synthesis and characterization of a poly(o-anisidine)-SiC composite and its application for corrosion protection of steel. *RSC Adv.* **2017**, *7*, 11732–11742. [CrossRef]
33. Tian, H.; Li, W.; Liu, A.; Gao, X.; Han, P.; Ding, R.; Yang, C.; Wang, D. Controlled delivery of multi-substituted triazole by metal-organic framework for efficient inhibition of mild steel corrosion in neutral chloride solution. *Corros. Sci.* **2018**, *131*, 1–16. [CrossRef]

34. Gonzalez-Rodriguez, J.G.; Porcayo-Calderon, I.; Vazquez-Velez, E.; de la Escalera, L.M.M.; Canto, I.; Martinez, L. Use of a palm oil-based imidazoline as corrsion inhibitor for copper in 3.5% NaCl solution. *Int. J. Electrochem. Sci.* **2016**, *11*, 8132–8144. [CrossRef]
35. Shi, J.M.; He, C.L.; Li, G.P.; Chen, H.Z.; Fu, X.Y.; Li, R.; Ma, G.F.; Wang, J.M. Corrosion behaviorsof pure copper and Cu-Ni-Zn alloy in NaCl solution and artificialsaltwater. In Proceedings of the 2nd International Conference on New Material and Chemical Industry, Sanya, China, 18–20 November 2017; Xin, S., Ed.; IOP Publishing: Bristol, UK, 2018; Volume 292.
36. Velazquez-Torres, N.; Martinez, H.; Porcayo-Calderon, I.; Vazquez-Velez, E.; Florez, O.; Campillo, B.; Gonzalez-Rodriguez, J.G.; Martinez-Gomez, L. Effect of plasma pre-oxidation on the Cu corrosion inhibition in 3.5% NaCl by an environmentally friendly amide. *Int. J. Electrochem. Sci.* **2018**, *13*, 8915–8930. [CrossRef]
37. Wang, D.; Xiang, B.; Liang, Y.; Song, S.; Liu, C. Corrosion control of copper in 3.5 wt.% NaCl Solution by domperidone: Experimental and theoretical study. *Corros. Sci.* **2014**, *85*, 77–86. [CrossRef]
38. Li, S.; Ma, Y.; Liu, Y.; Xin, G.; Wang, M.; Zhang, Z.; Liu, Z. Electrochemical sensor based on a three dimensional nanostructured MoS$_2$ nanosphere-PANI/reduced graphene oxide composite for simultaneous detection of ascorbic acid, dopamine, and uric acid. *RSC Adv.* **2019**, *9*, 2997–3003. [CrossRef]
39. Sherif, E.M.; Park, S.-M. Inhibition of copper corrosion in acidic pickling solutions by N-phenyl-1,4-phenylenediamine. *Electrochim. Acta* **2006**, *51*, 4665–4673. [CrossRef]
40. Kim, H.; Lee, H.; Lim, H.-R.; Cho, H.-B.; Choa, Y.-H. Electrically conductive and anti-corrosive coating on copper foil assisted by polymer-nanocomposites embedded with graphene. *Appl. Surf. Sci.* **2019**, *476*, 123–127. [CrossRef]
41. Ramezanzadeh, B.; Bahlakeh, G.; Moghadam, M.M.; Miraftab, R. Impact of size-controlled p-phenylenediamine (PPDA)-functionalized graphene oxide nanosheets on the GO-PPDA/Epoxy anti-corrosion, interfacial interactions and mechanical properties enhancement: Experimental and quantum mechanics investigations. *Chem. Eng. J.* **2018**, *335*, 737–755. [CrossRef]
42. Kumar, N.; Manik, G. Molecular dynamics simulations of polyvinyl acetate-perfluorooctane based anti-stain coatings. *Polymer* **2016**, *100*, 194–205. [CrossRef]
43. Hong, M.-S.; Park, Y.; Kim, J.G.; Kim, K. Effect of incorporating MoS$_2$ in organic coatings on the corrosion resistance of 316L stainless steel in a 3.5% NaCl solution. *Coatings* **2019**, *9*, 45. [CrossRef]
44. Kiran, N.U.; Dey, S.; Singh, B.P.; Besra, L. Graphene coating on copper by electrophoretic deposition for corrosion prevention. *Coatings* **2017**, *7*, 241. [CrossRef]

© 2019 by the authors. Licensee MDPI, Basel, Switzerland. This article is an open access article distributed under the terms and conditions of the Creative Commons Attribution (CC BY) license (http://creativecommons.org/licenses/by/4.0/).

Article

Fabrication of Flexible, Lightweight, Magnetic Mushroom Gills and Coral-Like MXene–Carbon Nanotube Nanocomposites for EMI Shielding Application

Kanthasamy Raagulan [1], Ramanaskanda Braveenth [1], Lee Ro Lee [1], Joonsik Lee [2], Bo Mi Kim [3], Jai Jung Moon [4], Sang Bok Lee [2,*] and Kyu Yun Chai [1,*]

1. Division of Bio-Nanochemistry, College of Natural Sciences, Wonkwang University, Iksan City 570-749, Korea; raagulan@live.com (K.R.); braveenth.czbt@gmail.com (R.B.); fjrtufl225@naver.com (L.R.L.)
2. Composite Research Division, Korea Institute of Materials Science, Changwon 51508, Korea; astro1228@kims.re.kr
3. Department of Chemical Engineering, Wonkwang University, Iksan 570-749, Korea; 123456@wku.ac.kr
4. Clean & Science Co., Ltd., Jeongeup 3 Industrial Complex 15BL, 67, 3sandan 3-gil, Buk-myeon 56136, Jeongeup-si 580-810, Korea; jjmoon@cands.kr
* Correspondence: leesb@kims.re.kr (S.B.L.); geuyoon@wonkwang.ac.kr (K.Y.C.);
  Tel.: +82-55-280-3318 (S.B.L.); +82-63-850-6230 (K.Y.C.);
  Fax: +82-55-280-3498 (S.B.L.); +82-63-841-4893 (K.Y.C.)

Received: 20 February 2019; Accepted: 26 March 2019; Published: 2 April 2019

Abstract: MXenes, carbon nanotubes, and nanoparticles are attractive candidates for electromagnetic interference (EMI) shielding. The composites were prepared through a filtration technique and spray coating process. The functionalization of non-woven carbon fabric is an attractive strategy. The prepared composite was characterized using X-ray photoelectron spectroscopy (XPS), X-ray diffraction (XRD), scanning electron microscope (SEM), energy-dispersive X-ray spectroscopy (EDX), and Raman spectroscopy. The MXene-oxidized carbon nanotube-sodium dodecyl sulfate composite (MXCS) exhibited 50.5 dB (99.999%), and the whole nanoparticle-based composite blocked 99.99% of the electromagnetic radiation. The functionalization increased the shielding by 15.4%. The composite possessed good thermal stability, and the maximum electric conductivity achieved was 12.5 S·cm^{-1}. Thus, the composite shows excellent potential applications towards the areas such as aeronautics, mobile phones, radars, and military.

Keywords: MXene; oxidized carbon nanotube (CNTO); nanoparticle decoration; functionalization; electromagnetic interference (EMI) shielding

1. Introduction

Electromagnetic interference (EMI) leads to inevitable interactions in electronic devices. Smaller and high-speed electronic systems are susceptible to issues related to EMI which can affect either adjacent electronic items or humans, thus potentially affecting the security of the nation [1,2]. In war, electromagnetic pulse weapons are being used to affect systems utilizing electromagnetic radiation (EMR) such as radar systems, high tech complex electronic devices, remote control armor, aircraft, and missiles. In addition, EMI affects the functions of sensors in modern electronic vehicles as they transmit signals using weak radiation and microcomputers. Hence, protecting electronic devices from malfunction and achieving electromagnetic compatibility is an essential requirement around the globe. Further, electromagnetic compatibility should be attained by diminishing incoming and outgoing electromagnetic radiation, ideally without affecting the function of the devices. This is because EMI not only affects electronic systems but also causes health issues in human beings [3–5].

Various substances have been used for EMI shielding, such as MXenes, graphene (GN), graphene oxide, carbon nanotubes (CNTs), nanoparticles, polymers, fabrics, textiles, composites, and metals in various frequency ranges [2,4–6]. The EMI shielding materials can be categorized in two types: Reflection and absorption domain materials. The reflection domain materials possess mobile charges while absorption domain materials contain magnetic and dielectric materials. The layered and implanted type structures influence the EMI shielding. EMI shielding can be accomplished by suppressing the incident wave, which has the three key mechanisms of absorption, reflection, and multiple reflection. The electric conductivity associated with primary shielding factor is reflection, where the mobile electron interacts with incident wave. The thickness, electric and magnetic dipole loss, magnetic permeability, defects, and structural features induce absorption. The ohmic loss can be achieved by conduction, electron hopping, and tunneling. The polarization loss occurs due to the rearrangement of the polarization while electromagnetic radiation (EMR) is passing through the shielding materials. Polarization can be induced by embedding functionalities, hybrid fillers, nanofillers, and defects in the matrix of the composite. The inhomogeneous scattering centers, layered structure, hollow structure, and interfaces generate multiple reflection, which finally leads to absorption. Further, the skin depth limits the EMI shielding effectiveness, which should be lower than that of the thickness of EMI shielding materials [2,6–11]. Certain properties, such as being lightweight, conductive, corrosion resistant, flexible, cost-effective, and high strength, are preferable for a modern EMI shielding material. Metals are conventionally exploited as shielding materials due to the fact that they possess excellent conductivity, but they are unable to fulfill the current needs of compact electronic systems. Hence, carbon-based substances are attracting attention, as their properties can be tuned by incorporating other materials like nanoparticles and polymers [1–5,8]. In addition, incorporating carbon nanotubes (CNTs), MXenes, polymers, and graphene with non-woven carbon fabric significantly increases EMI shielding. Thus, establishing a conductive network is an essential factor for good EMI shielding [5,8,12]. Different combinations of the constituents and amounts of filler loading have been used to fabricate multi-functional EMI shielding materials [13–17].

MXenes ($M_{(n+1)}X_nT_x$) are two dimensional (2D) material derived from a corresponding three dimensional (3D) MAX phase ($M_{(n+1)}AX_n$), where M is early transition elements (Ti, V, Cr, Nb, Ta, Zr, and Mo), A includes group 13/14 elements, X represents carbon or nitrogen, and T_x is surface functional groups (–OH, –F, and =O) [2]. A selective etching strategy is used in the production of MXenes. The minimally intensive layer delamination (MILD) method (LiF/HCl) has recently been endorsed, as it abridges the synthetic process, and HF, NH_4HF_2, and FeF_3/HCl have also been practiced. During the etching, the weaker M–A bonds are eradicated while the strong M–C bond remains with newly formed functionalities [2,18]. MXenes have a metal-like nature, and similar to graphene, have been used for various purposes such as in sensors, capacitors, storage, and EMI shielding materials. In general, MXenes are hydrophilic in nature, as –OH is one of the surface functional groups. Thus, MXenes can be incorporated with various materials like polymers in order to tune their properties [2]. There are various types of MXenes that have been studied such as Nb_2CT_x, Ti_3CNT_x, and Ti_2CT_x. In addition, MXenes are used to make hybrid composites such as TiO_2–$Ti_3C_2T_x$/graphene, $Ti_3C_2T_x$–sodium alginate, Ti_3C_2Tx/PVA, cellulose nanofibers–$Ti_3C_2T_x$ and $Ti_3C_2T_x$/paraffin. MXenes exhibit maximum EMI shielding of 92 dB with 45 μm thickness. Thus, the loading amount of MXenes in different polymers matrices, the morphology of composite, and the thickness influence the EMI shielding of the MXene [19–23].

In this study, we developed a layered pliable composite with different surface morphology and magnetic composite. Each layer of the composite consists of a layered and magnetic domain for which we employed spray coating and filtration technique under gravity. The functionalization of carbon fabric intercalation of the MXene, CNTs, and magnetic nanoparticle dramatically changed the EMI shielding. The spray coating samples were denoted as MXCNTCx, where x (x = 10, 25, or 30) was the number of coating cycles, and Ni coated fabric was expressed as MXCNTNi25. The composite fabricated using the filtration method was denoted as MXCBCM, where M is the type of metal nanoparticle such as Ni, Co, Fe, Cu, and Fe_3O_4. Its corresponding composites were MXCBCNi, MXCBCCo, MXCBCFe, MXCBCCu, and MXCBCFeO, respectively. Further, MXCB and MXCS were labelled based on the surfactant used, like cetyltrimethylammonium bromide (CTAB) and sodium dodecyl sulfate (SDS), respectively. MC and FC indicated uncoated carbon fabric and functionalized carbon fabric, respectively. The parameters of EMI shielding, elementals analysis, morphology, structural analysis, electric conductivity, surface property, magnetic property, and thermal stability were investigated in detail.

2. Materials and Methods

2.1. Materials

Multiwall carbon nanotube (MWCNTs) (CM-90, 90 wt %, diameter of 20 nm, and length of 100 μm) were purchased from Applied Carbon Technology Co. Ltd. (Pohang, Korea). Carbon fiber (fiber diameter 7-micron, 6 mm length) and polyethylene terephthalate (PET) binder (fiber diameter 2.2 dtex, 5 mm) were collected from TORAY product, (Tokyo, Japan). Sodium dodecyl sulfate (SDS) (98%), lithium fluoride (LiF) (98%, 300 mesh), sodium borohydride ($NaBH_4$), polyacrylamide (PAM), anhydrous $FeCl_2$, $FeCl_3$, $NiCl_2$, $CuCl_2$, $CoCl_2$, and cetyltrimethylammonium bromide (CTAB) were obtained from Sigma Aldrich (Seoul, Korea). Wet laid nickel coated non-woven carbon fabric (basic density of 19.2 g/m^2, thickness of 150 μm) was acquired from Clean & Science Co. Ltd., (Seoul, Korea). Nitric acid (HNO_3-70%) and hydrochloric acid (HCl-35%) were obtained from Samchun Chemical Co., Ltd. (Seoul, Korea). Chitooligosaccharide (Mw 5000) was issued by biomedical polymer lab, Sunchon National University (Suncheon, Korea).

2.2. Synthesis of Ti_3AlC_2

We reported that TiC, Ti, and Al powders were taken with molar ration of 2:1:1 ball milled by using Pulverisette 6 Planetary Mono Mill (Fritsch, Germany) in ethanol medium at 200 rpm for 1 h in a nitrogen environment. The resultant mixture was dried at 80 °C for 12 h. Then, 3 g of 12 mm diameter disc was prepared by applying 27.6 MPa pressure for 5 min in a laboratory press. The resultant disc was treated at 1350 °C with a heating rate of 20 °C/minute in argon gas for 2 h, then cooled down to room temperature. The treated disc was again ball milled in ethanol medium at 300 rpm for 3 h in a nitrogen environment. The powder yield was then dried at 80 °C for 12 h, and the obtained product was directly used for MXene synthesis [24].

2.3. Preparation of Oxidized Carbon Nanotube (CNTO)

1 g of MWCNT (CNT) and 90 mL of HNO_3 were sonicated by using a mini ultrasonic cleaner (Uil ultrasonic Co., Ltd., Gyeonggi-do, Republic of Korea) for 5 h at room temperature. The volume of the reaction mixture was doubled by adding deionized (DI) water and filtered. The acid solution and reacted CNT were separated by filtration. The resultant black color product known as CNTO was washed until reaching a neutral pH, then dried at 80 °C for 24 h. This CNTO was used for the coating process and synthesis decorated CNTO.

2.4. Preparation of Fe_3O_4 Decorated CNTO

100 mL of 0.1 M of Fe^{3+}, 100 ml of 0.1 M of Fe^{2+}, and 0.5 $g·L^{-1}$ of CNTO were stirred together at 30 °C for 30 min. A total of 10% of NH_4OH was added dropwise until the pH of the solution reached 11–12, along with 2 mL of 0.2 M SDS. Then, the temperature of the reaction mixture was raised up to 80 °C and stirred for 2 h. The volume of the solution was reduced by half via evaporation and then cooled down to room temperature. The resultant product was washed until reaching a neutral pH in a vacuum filter, then dried at 105 °C for 12 h.

2.5. Preparation of Nanoparticle Decorated CNTO

100 mL of 0.1 M^{2+} of metal ion solution (Fe^{2+}, Ni^{2+}, Cu^{2+}, and Co^{2+}) and 0.5 g·L^{-1} of CNTO were stirred for 30 min. 100 mL of 0.2 M cold $NaBH_4$ was added drop wisely in to the mixture, along with 0.2 g·L^{-1} of SDS. Fe, Ni, Co, and Cu decorated CNTO were successfully synthesized in a nitrogen environment. The product was filtered and washed by an ample amount of DI water. Then, the product was dried at 80 °C and 0.8 atm in a vacuum oven for 12 h. The obtained product was denoted as CM, where C is CNTO and M is nanoparticles. The corresponding decorated CNTOs were denoted as CFe, CNi, CCo, and CCu.

2.6. Preparation of Dispersed Solutions

0.1 g of CNTO and 0.1 g of SDS were mixed in 100 mL of deionized water and sonicated for 3 h. Then, the dispersed mixture was refluxed at 120 °C for 12 h. The obtained well dispersed solution was then used for the spray coating process. In addition, 0.4 g of CNTO and 0.4 g of CTAB were added together in 100 mL of deionized water and sonicated for 5 h (CTAB–CNTO). This procedure was repeated using SDS and CNTO (SDS–CNTO). The decorated nanoparticle dispersed solution was prepared accordingly; equal amounts of CTAB and CM (0.1 g) were sonicated in 25 mL of deionized water for 30 min. The obtained dispersed solutions were directly used for the filtration process.

2.7. Preparation of MXene and MXene Colloidal Solution

Equal amounts (1 g) of Ti_3AlC_2 and LiF were mixed together in 20 mL of 9 M HCl solution. This mixture was stirred at 35 °C for 24 h. The etched product was washed with DI water up to approximately pH 6 by centrifuging at 3500 rpm for 5 min, then $Ti_3C_2T_x$ was dried in a vacuum oven for 12 h. Then, 0.1 g of MXene was sonicated in 10 mL of DI water in a nitrogen environment at about 15–17 °C for 2 h. The sonicated MXene solution was centrifuged at 3500 rpm for 30 min, and the supernatant was collected in a Teflon container stored at 5 °C for the coating process. The concentration of the MXene colloidal solution was 0.175 g·L^{-1}. 100 mL of colloidal solution was filtered by using 0.45 μm of 47 mm of Nylon supported filter paper and dried at 80 °C (0.8 atm) until obtained constant weight. The concentration was calculated based on the weight differences.

2.8. Synthesis of Carbon Fabric by Wet Laid Method (MC)

We reported that 0.6 kg of carbon fiber, 0.15 kg of PET binder and 0.3 weight percent of dispersant (PAM) were dispersed in a sufficient amount of deionized (DI) water at 500 rpm for 10 min. The general wet laid method was used to produce the web. During the process, the drum dryer was used with a 140 °C surface temperature and 7 m·min^{-} speed. The areal density of obtained fabric was 30 g·m^{-2} [24].

2.9. Fabrication of Fabric Composite

2.9.1. Preparation of Functionalized Carbon Fabric (FC)

0.2 g of chitooligosaccharide was dissolved in 100 mL of deionized water and stirred for 15 min. Then, carbon fabric with a dimension of 29 × 21 cm^2 and a basic density of 30 g·m^{-2} was dipped in the chitooligosaccharide solution for 2 min and dried at 100 °C for 12 h. This fabric was directly used for the preparation of the filtration-based composite.

2.9.2. Fabrication of MXCNTCx Composite by Spray Coating

A series of MXene–CNTO composites were prepared by the spray coating process, for which 30 g·m^{-2} of MC with a dimension of 29 × 21 cm^2 was used, where one CNTO layer was sandwiched between two MXene layers. The thickness of the fabric was adjusted by a spraying and drying process. The drying was done using an air-drying gun while spraying was done by using an air compressor (Keyang compressors (KAC-25), Sichuan, China). This process was repeated for Ni coated fabric in order to compare the EMI shielding of carbon fabric. The coating on carbon fabric was denoted as MXCNTCx, while the Ni coated carbon fabric was denoted as MXCNTNiCx, where x was the number of coating cycles. MXCNTC30, MXCNTC25, MXCNTC10 and MXCNTNiC25 were successfully manufactured.

2.9.3. Fabrication of MXene–CNTO Composite by Filtration

100 mL of MXene colloidal solution, 100 mL of CM dispersed solution and 70 mL of dispersed CNTO were alternatively filtered through FC under gravity and dried using an air gun. The resultant composite was denoted as MXCBCM. Further, MXCB was prepared by using 100 mL of MXene colloidal solution and 70 mL of CTAB–CNTO dispersed while 100 mL of MXene colloidal solution and 70 mL of SDS–CNTO dispersed mixture were used to prepare MXCS. Finally, MXCBCFeO, MXCBCFe, MXCBCNi, MXCBCCo, and MXCBCCu were successfully prepared.

2.10. Characterization

The structural features of the composites were investigated using a high-resolution Raman spectrophotometer (Jobin Yvon, LabRam HR Evolution (Horiba, Tokyo, Japan). A Laser Flash Apparatus LFA457 (NETZSCH, Wittelsbacherstrabe, Germany) was used to measure the density of the composites. A field emission scanning electron microscope (SEM, S-4800 (Hitachi, Tokyo, Japan) was used to examine the surface morphology of the composites. XPS with a 30–400 µm spot size at 100 W of Emax (Al anode) (K-Alpha, Thermo Fisher, East Grinstead, UK) was used to analyze the chemical environment and elemental percentage of the composites. A High-power X-ray Diffractometer D/max-2500V/PC, (Ragaku, Tokyo, Japan) with Cu(Kα) was used to record the X-ray diffraction patterns of the composites. The EMI shielding effectiveness (SE) of the composite in S-band (1–3 GHz) was recorded using an EMI shielding tent ASTM-D4935-10, ASTM International (West Kentucky, PA, USA) at room temperature while X-band (8.2–12.4 GHz) EMI shielding was measured using a vector network analyzer (VNA, Agilent N5230A, Agilent Technologies, Santa Clara, CA, USA) with a sample size of 22.16 mm × 10.16 mm. The four-probe method FPP-RS8, DASOL ENG (Seoul, Korea) was used to measure the electric conductivity of the composites. The thermal stability of the composites was tested using a Thermal Analyzer DSC TMA Q400 (TA Instruments Ltd., New Castle, DE, USA). A Mitutoyo thickness 2046S dial gage (Mitutoyo, Kanagawa, Japan) was used to measure the thickness of the composites. The surface property was measured using a Phoenix-300A contact angle meter (S.E.O.Co., Ltd., Suwon, Korea). The magnetic property was measured using SQUID—VSM (Quantum Design, Inc., San Diego, CA, USA). The graphs were plotted using a Savitzky–Golay function (Origin 2017 graphing and analysis, Origin Lab (Boston, MA, USA).

3. Results and Discussion

3.1. Structural Analysis

3.1.1. Scanning Electron Microscopic Analysis of Morphology

SEM was used to characterize the morphology of nanoparticles and composites, the arrangement of CNTO, nanoparticles, and MXene flakes, the structural feature of fiber, and the topography of the composites. Figure 1 illustrates the differently decorated CNTOs by different types of nanoparticles. The oxidation of the carbon nanotube mainly occurred in the tip of the carbon nanotube, as confirmed by the SEM images (Figure 1 and Figure S1), and the decoration of carbon nanotube generated a cauliflower-like structure (Figure 1a–c,e and Figure S1a,b,d). The oxidized carbon nanotube consists of a carboxylic acid functional group which can act as anchoring side of the nanoparticles. According to the Figure 1 and Figure S1, the deposition of the nanoparticle grafting occurred in the terminal of the CNTO, indicating that the oxidation predominantly happed in the tip. The oxidation of multiwall carbon nanotube produced the terminal carboxylic group which helps terminal grafting, inhibits the aggregation and increases the solubility in water. This phenomenon leads the various morphologies in decorated CNTOs [25,26]. Fe$_3$O$_4$, Fe, and Cu nanoparticles behaved in a similar manner, whereas the Ni nanoparticle encircled all of the carbon nanotube and was densely packed like a cauliflower. The self-assembling of the Co nanoparticles was completely different from that of another nanoparticle used. This is because it consumed CNTO as a template and formed a structure-like bacterial chain (Figure 1d and Figure S1c) [27]. The precursor of MXene, which is Ti$_3$AlC$_2$, exhibited a layered structure (Figure S3h) [6]. The MILD etching created cleaves, due to the eradication of the Al layer and the evolution of the hydrogen gas (Figure 1f and Figure S3i,j). This gap is more prominent in the clay etching method (50% of HF) [18]. The etched MXene showed a layered structure with fewer gaps, and the folding of the single flake confirmed that the etching occurred. As it consisted of small gaps, it appeared like a MAX phase (Figure 1f and Figure S3i,j). Further, the presence of a single MXene flake in the composites affirmed the occurrence of effective exfoliation during the process (Figures 1f and 2b–i) [27].

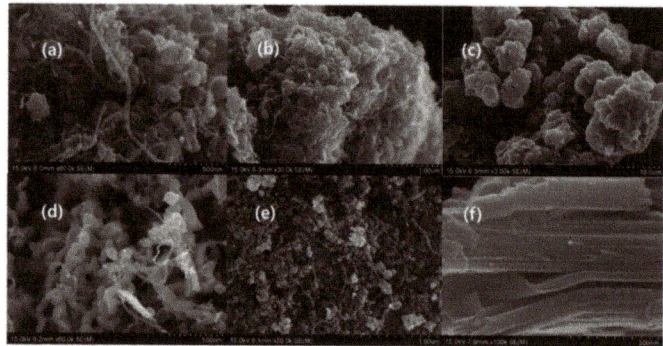

Figure 1. SEM image of oxidized carbon nanotubes (CNTOs) decorated by (**a**) Fe$_3$O$_4$ ($\times$60,000) (**b**) Fe ($\times$30,000) (**c**) Ni ($\times$3000) (**d**) Co ($\times$60,000) (**e**) Cu ($\times$50,000) and (**f**) Ti$_3$C$_2$T$_x$ ($\times$100,000).

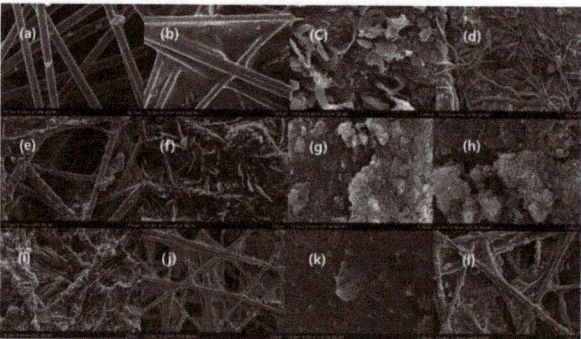

Figure 2. SEM image of carbon fabric composite of (**a**) Functionalized carbon fabric (FC) ($\times$1000) (**b**) MXCB ($\times$500) (**c**) surface of MXCS ($\times$10,000) (**d**) MXCBFeO ($\times$700) (**e**) MXCBFe ($\times$500) (**f**) MXCBNi ($\times$500) (**g**) MXCBCCo ($\times$10,000) (**h**) MXCBCCo ($\times$25,000) (**i**) MXCBCCu ($\times$200) (**j**) MXCNTC25 ($\times$300) (**k**) MXCNTC25 ($\times$10,000) and (**l**) MXCNTNi25 ($\times$300).

Figure 2 illustrates the morphology of the composites. The functionalized nonwoven carbon fabric exhibited a similar morphology of nonwoven carbon fabric, where the fibers were arranged capriciously (Figure 2b,j and Figure S3d). The fibers of the carbon fabric possessed annular gaps and cracks which were occupied by CNTO, MXene, and nanoparticle [5]. This functionalization and the nanomaterials altered the property of MC and dramatically changed the structural feature of the carbon fabric (Figure 2a–l). Filtration was an effective strategy over spray coating, because filtration closed most of the gaps between fibers and interconnected the fibers with fewer defects, while MXCNTNiC25 had prominent defects (Figure 2l and Figure S3b–j) [6]. MXCB formed like a film with well interconnected fibers, whereas few pore structures remained (Figure 2b and Figure S2b). MXene, CNTO, and Fe_3O_4 decorated CNTOs formed a structure like roots of a tree fixed on the soil surface, and some points of the MXene flake formed an unexfoliated MXene structure. Further, a root-like nature was given by the MXene flakes (Figure 2d). MXCBFe and MXCNTC25 formed similar structures and MXCBFe generated a highly interconnected network. In addition, the MXene flakes self-assembled in a random manner and showed a similar pattern of graphene–Polyvinylidene fluoride (PVDF) coated fabric (Figure 2c,f) [6]. Furthermore, MXCBCCu exhibited a mushroom gills-like structure which was generated by MXene flakes arranged in parallel among nanoparticles and CNTO [28]. The MXCBCCo and CNi exhibited a coral like morphology (Figure 2g,h) [29]. The coral structure was formed by cubic Co nanoparticles. In order to achieve the coral structure, CCo was used as mediator (Figure S2h,i). The surface of the MXCNTC25 displayed a network of CNTO encircling the MXene flake. Furthermore, the fibers in MXCNTNiC25 were interconnected with many defects and cracks which dramatically affected the EMI shielding and conductivity (Figure 2k,l). The etching caused the introduction of new elements such as F, Cl, and O while it eradicated most of the Al from the MAX phase, leaving Ti and C (Table S1). In addition, Cl, F, and O were derived from etchant. According to the EDX analysis, the O and C were major elements present in all composites, while other metals like Co, Ni, Fe, and Cu were also present based on the precursor used to manufacture composites. As the MXene was a structural unit of the composites, the Ti and F prevailed in most of the composites, and Al and Br were also found in some composites, which are derived from MXene colloidal solution and CTAB surfactant, respectively. In addition, exfoliated MXene had more than a single layer, which consisted of the little amount of remaining Al. Further, spray coated fabric consisted of S, which was derived from the SDS surfactant used (Table S1 and Figure S2c).

3.1.2. Raman Spectroscopic Analysis for Structure of Carbon-Based Material

The structural and crystalline nature of materials like MXene, graphene, and CNT can be investigated using Raman spectroscopy [2,5]. Figure 3a,b illustrate the Raman spectrum of the decorated carbon nanotube and composites as plotted between 250 and 3500 cm^{-1} Raman shifts. The peaks at 624, 394, and 263 cm^{-1} were attributed to the in-plane vibrational mode of surface functionalities, C, and Ti, respectively. In addition, $Ti_3C_2T_x$ engendered feeble wide D and G bands at 1353 and 1568 cm^{-1}, respectively, and the peaks at 624, 510, and 398 cm^{-1} exhibited the presence of TiO_2 anatase (Figure S4k) [30–33]. In addition, the missing peak at 263 cm^{-1} revealed the absence of the Al layer and the fixing of new surface functionalities in the eradicated Ti–C–Al bond. The G band and G' band of the CNT and CNTO were located at 1570.9 and 2675.9 cm^{-1}, respectively. The position of the D band slightly differed from that of the CNT peak located at 1336.3 cm^{-1} while CNTO generated a peak at 1341.5 cm^{-1}. In addition, CNTO gave rise to new weak peaks at 2435.5, 2916.2, and 3320.4 cm^{-1} while CNT formed weak peaks at 2420, 2371, and 3226.2 cm^{-1}. These differences were created due to the oxidation that was considered as oxidational effect of the carbon nanotube. Furthermore, the presence of defects and the amorphous nature of CNT generated the D band while the graphite structure induced the G band. The characteristic G' band at 2672 cm^{-1} was formed by an overtone of the D band. Further, the level of defect present in the carbon nanotube can be explained using the ratio between I_D/I_G and $I_D/I_{G'}$ [34–45]. I_D/I_G and $I_D/I_{G'}$ of CNT were 0.79 and 1.39, respectively, while those of CNTO were 0.96 and 1.76, respectively. The CNTO possessing higher values of I_D/I_G and $I_D/I_{G'}$ revealed that CNTO had more defect density than CNT. Hence, chemical oxidation created disorder in the carbon nanotube. The D band of the CNTO and nanoparticle decorated CNTO was located between 1340–1355 cm^{-1} while the G and G' bands were placed between 1570–1585 cm^{-1} and 2675–2700 cm^{-1}, respectively. All of the carbon nanotubes and decorated CNTOs exhibited weak peaks between 2910–2943 cm^{-1} and 3220–3240 cm^{-1}, respectively. Further, there were extra peaks at lower Raman shift, which were due to the carbon–metal and oxygen–metal vibration modes

(Figure 3a and Table S2) [46,47]. The G band intensity of CNTO was lower than that of CNT when compared with its corresponding D band. A similar pattern was shown by CFe whereas the other decorated CNTOs exhibited a higher D band intensity, implying that the introduction of nanoparticle generates the defect. The I_D/I_G values of CCu, CCo, CNi, CFe, CFeO, and FC were 1.01, 1.15, 1.02, 0.92, 1.4, and 0.94, respectively, whereas the corresponding $I_D/I_{G'}$ values were 1.84, 1.96, 1.68, 1.74, 2.71, and 4.18, respectively [47]. All of the fabric showed a similar Raman spectra pattern and a peak originated between 1339–1350 cm^{-1} which was responsible for D band of the composites, whereas the corresponding G and G' bands laid between 1567–1586 cm^{-1} and 2678–2688 cm^{-1}, respectively. The non-woven carbon fabric gave rise to D and G bands at 1363 and 1592.1 cm^{-1}, respectively, which is due to the graphite (HOPG), indicating the presence of the graphite-like structure and the generation of a feeble G' band at 2908.2 cm^{-1}. The I_D/I_G values of the MC, MXCS, MXCBCCu, MXCBCCo, MXCBCNi, MXCBCFe, MXCBCFeO, and MXCB were 0.91, 1.03, 0.81, 0.87, 1.01, 0.86, 0.81, and 1.01, respectively, and the corresponding $I_D/I_{G'}$ values were 4.13, 2.33, 1.48, 2.12, 1.88, 1.96, 2.09, and 2.04, respectively (Figure 3b and Table S2). All of the composite I_D/I_G values were relatively similar to MC, while the decreasing $I_D/I_{G'}$ value of composite confirmed that the defects of the fabric were diminished significantly. The disappearing of the lower Raman shift of MXene and decorated carbon nanotube and the formation of the new peaks confirmed that the proper link occurred between the fabric, MXene, carbon nanotube, and nanoparticles.

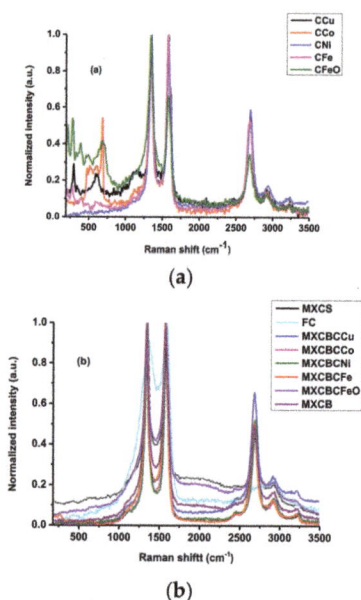

Figure 3. Raman spectra of (a) decorated carbon nanotube and (b) composites.

3.1.3. X-ray Diffraction (XRD) Analysis

The crystalline or amorphous nature of the material can be predicted based on XRD profile. The XRD profiles of the Ti$_3$AlC$_2$, Ti$_3$C$_2$T$_x$, CNT, CNTO, decorated CNTO, and composites are shown in Figure 4a,b, and drew a 2θ range between 5 to 90°. The crystalline MAX phase generated sharp peaks at 9.52° (002), 19.53° (004), 34° (101), 35.1° (102), 36.8° (103), 38.99° (008), 41.76° (104), 42.54° (105), 48.48° (107), 52.36° (108), 56.5° (109), 60.16° (110), 52.36° (1011), 64.98° (1011), 70.34° (1012), 74.02° (118), and other miscellaneous small peaks [30–36]. Following the etching process, the corresponding MAX phase peaks vanished or shifted, and the sequence of new diffraction peaks was formed. The formed MXene held a crystalline nature and the peak at 7.14° (002) was a characteristic peak of MXene interplanar crystal space.

In addition, the peaks originating at 14.36°, 19.12°, 28.98°, 38.86°, and 40.9° confirmed the crystalline nature of the MXene and attested to the occurrence of etching [32–35]. The peak shift of Ti$_3$AlC$_2$ from 9.59° to 6.96° and the formation of the MXene new peak at 21.57° indicated that the effective eradication of Al layers occurred (Figure 4a). The peak at 38.86° implied the remains of the layered MAX phase structure without an Al layer which confirmed the formation of MXene. Further, the separation of the layers after the etching was low; thus, the crystalline nature of the MXene remained the same as the structure of Ti$_3$AlC$_2$ (Figures 1f and 4a). The carbon nanotubes exhibited a crystalline nature, as confirmed by the 25.88° (002), 42.84° (100), 43.69° (101) 48.94° (102), and 54.07° (004) reflection peaks and implied the presence of the concentric cylindrical MWCNT [48,49]. In addition, the shifting of the position of 2θ of the corresponding MWCNT attested to the oxidation of MWCNT and increased the percentage of the sp2 hybridized carbons (Figure 4b) [36–40]. The CFeO generated peaks at 18.36°, 30.21°, 35.66°, 43.33°, 53.76°, 57.27°, 62.88°, 71.37°, and 74.42°, and its corresponding reflection plans were (111), (220), (311), (222), (422), (511), (440), and (533), respectively [50,51]. The CFe generated peaks of zero valent iron nanoparticle at 44.73° (110), 64.53° (200), and 82.39° (211), and the other peaks were corresponding CNTO signals [52]. The (200) reflection peaks of CNi, CCo, and CCu were located at 51.68°, 51.68°, and 50.31°, respectively. The (111) peak of CCo and CCu originated at 44.87° and 43.2°, respectively [53–55]. The new peaks were raised due to the CNTOs and the aggregation of nanoparticles in the CNTOs (Figure 4b).

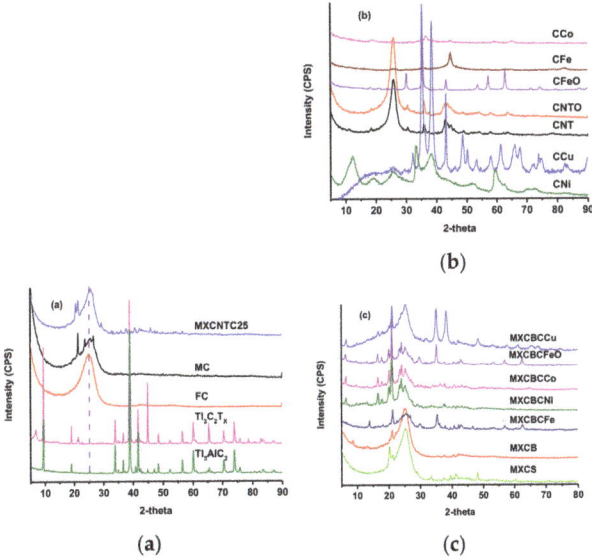

Figure 4. XRD of (a) MXene, MAX phase, and fabric (b) decorated carbon nanotubes and (c) composites.

The broad peaks of MC exhibited the amorphous nature of the fabric along with the 2θ peak at 25.52°, which is similar to the peak of carbon nanotubes and confirmed the presence of a graphite structure [5,40]. Further, MC consists of various small peaks which were not above 30°, and all small peaks disappeared with functionalization (Figure 4a). The MXene–fabric composite showed various small peaks, which were absent in FC. Hence, the introduction of the MXene, CNTOs, and nanoparticles generated many small peaks. All of the carbon fabric composites showed distinctive peaks between 20.3–21.24°, and all of the nanocomposites showed characteristic peaks at 6.69° and 16.82°, except for MXCBCFe, which showed a typical signal at 14°. MXCBCCu, MXCBCFe, and MXCBCFeO exhibited a peak at 35.25° among the MXCBCCu generated shoulder peak at 38.46°. Thus, these composites can easily be distinguished from these characteristic peaks and formed from the constitutional elements of the composites.

3.1.4. X-ray Photoelectron Spectroscopy Analysis

The functionalities, surface elemental composition, structure, and bonding nature of the composites can be explained using XPS. The fitting curve of MXene was plotted using overlapping curves of the Gaussian–Lorentzian function and the overlapping curve of the composites was plotted using origin pro. The fitting curves of Ti2p, O1s, and C1s, the F1s of $Ti_3C_2T_X$, and the C1s of CNTO displayed different peak positions and corresponding functional groups and bonds (Figure 5 and Table S2) [56–58]. The Ti2p fitting curve disclosed five different chemical environments and the corresponding binding energies were 454.5, 456.4, 458.5, 461.3, and 464.5 eV. These binding energies indicated the presence of functional groups such as Ti–C $(2p_{3/2})$, $Ti^{2+}(2p_{3/2})$, TiO_2 $(2p_{3/2})$, Ti^{2+} $(2p_{1/2})$, and $TiO_2(2p_{1/2})$, respectively (Figure 5a). The binding energies positions of the O1s fitting curves such as 529.6, 531.1, 532.3, and 533.8 eV showed corresponding functional groups like TiO_2, C–Ti–Ox, Al_2O_3, and H_2O, respectively, (Figure 5b–d). The presence of C–Ti–Fx generated a single peak at 685.5 eV in the F1s fitting curve (Figure 5c). The C1s fitting curves at 281.1, 283.2, 284.5, and 286.1 eV confirmed the functionalities such as C–Ti–Tx, C–C, and CHx–CO. (Figure 5d) [33,35,43,59–62]. Thus, MXene formed with the formula of $Ti_3C_2(OH, F)$. In the fitting curve of MWCNT, the intense peak at 284.13 eV was raised due to the C–C bond of graphite, while the presence of the oxygen generated weaker peaks between 287–291 eV, and its corresponding oxygenic species such as C=O, C–O, and carbonate were triggered. The fitting peak at 285.86 eV confirmed the presence of the defects, which was further backed by the I_D/I_G ratio and amount of oxygen [63,64]. The defects were comparatively low in CNT and increased by oxidation in CNTO (Figure 5e, Figure S4 and Table S3) [5]. The overlapping curve of XPS exhibited the constitutional elements and its corresponding peak positions. The constitutional elements of MXene were Cl, C, F, Ti, and O, and the corresponding binding energies were 198.7, 285.22, 685.78, 459.62, and 532.65 eV, respectively, while those given by N, Co, Ni, Fe, and Cu were 400.99, 780.1, 854.81, 710.7, and 933.65 eV, respectively (Figure 5f). In addition, the overlapping curve of the composite lying between 284.3–284.5 eV confirmed that the proper bonding occurred among the constitutional components (Figure S5) [56].

3.2. Electrical Conductivity (EC) and Surface Properties

CNT and MXene possess good EC, which is one of the factors influencing the EMI shielding effectiveness [2,57]. The electric conductivity and sheet resistance (R_s) of the fabric significantly changed due to the introduction of the MXene, CNTOs, and nanoparticles in the non-woven fabric matrix (Figure 6). The fabricated composite exhibited EC ranging from 12.5 to 2.65 S·cm^{-1}, and R_s lay between 13.98 and 2.08 Ω·sq^{-1}. The MXCNTNiC25 hit a maximum R_s of 13.98 Ω·sq^{-1} and a minimum conductivity of 2.65 S·cm^{-1}, which was due to the high surface defect (Figure 2l). The functionalization process increased EC by 10.1% and changed R_s by 20.9%. Thus, the functionalization process minimized the defect and increased the electron mobility. The filtration-based composite showed EC above 10 S·cm^{-1} while MXCBCFeO and MXCBCFe exhibited a maximum and minimum of 12.5 and 8.81 S·cm^{-1}, respectively. The EC of MXCB, MXCBCNi, MXCBCCo, MXCS, and MXCBCCu were 12.1, 11.3, 11.65, 11.8, and 11.22 S·cm^{-1}, respectively, whereas the corresponding R_s were 2.08, 2.38, 2.11, 2.2, and 2.56 Ω·sq^{-1}, respectively. Spray coated MXene–CNTO composites displayed EC below 10 S·cm^{-1}, among which MXCNTC30 exhibited 9.55 S·cm^{-1}. Hence, 14.62% of electric conductivity was increased by introducing MXene, CNTOs, and decorated carbon nanotubes into the MC network (Figure 6, Figure S6 and Table S6). The MXene has a high instinct electric conductivity due to its metal like nature [5]. In addition, the doping of CNT by nitrogen expands the EC while the degree of oxidation reduces the conductivity. However, the MXene–carbon fabric composite showed a maximum of 8.84 S·cm^{-1} EC while the CNTO–carbon fabric composite displayed 16.32 S·cm^{-1} of EC. Thus, MXene–CNTO-nanocomposite showed conductivity between the MXene and CNTO composites [2,5,24,65]. These defects limit the electron mobility and electron hopping along the fiber [66,67]. We attempted to explain this using the hydrophobic nature of the composites. The wetting ability of the surface can be explained based on the contact angle, which above 90° is called a water-repellent surface, while below 90° is considered a water loving surface. Further, the contact angle is dependent on the surface roughness and energy. Roughness increases the defect, thus increasing the surface energy and surface roughness raise the hydrophobic nature [5,16,56]. The contact angle, wetting energy, spreading coefficient, and work of adhesion of MXCNTNiC25 were 131.3°, −48.09 mN·m^{-1}, −120.89 mN·m^{-1}, and, 24.71 mN·m^{-1}, respectively, while FC exhibited 134.18 mN·m^{-1}, −50.7 mN·m^{-1}, −123.54 mN·m^{-1}, and 22.06 mN·m^{-1}, respectively. In addition, the other composites showed no coated angle that was due to the absorption of the water or well spread on the surface of the composites. The conductivities of most of the composites were high considering the fewer surface defects while the conductivity of the MXCNTNiC25 showed the lowest conductivity, confirming that MXCNTNiC25 possessed high defect and surface roughness (Figures 2l and 6, and Figure S7).

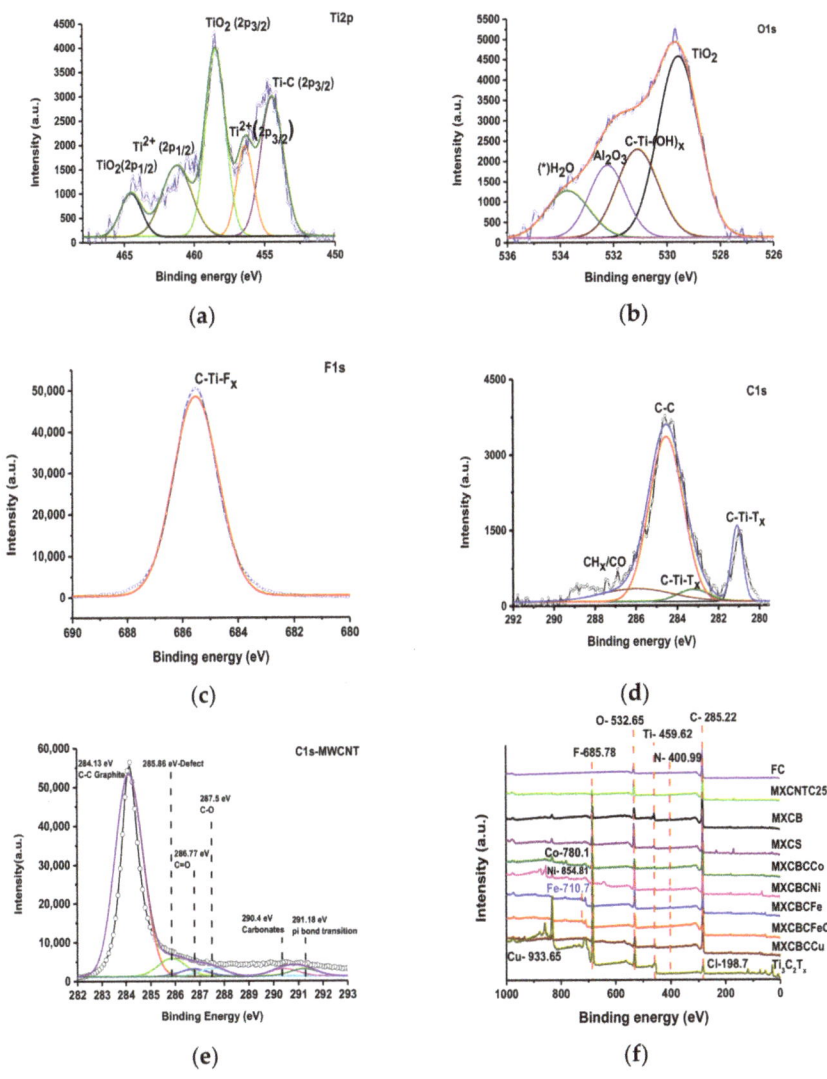

Figure 5. XPS fitting curves of (**a**) Ti2p, (**b**) O1s, (**c**) C1s, (**d**) F1s, (**e**) C1s of MWCNT, and (**f**) survey of the composites.

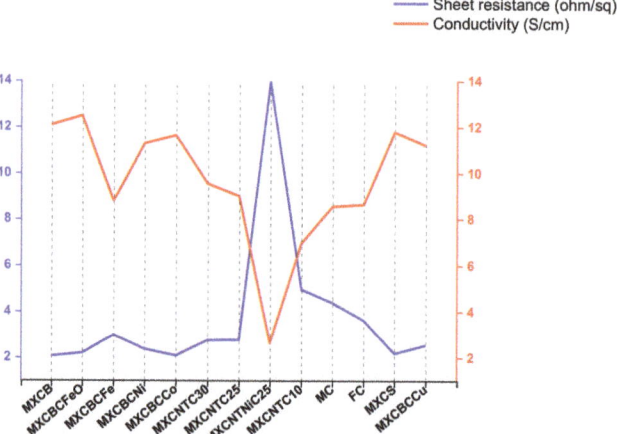

Figure 6. Electric conductivities and sheet resistances of the composites.

3.3. Magnetic Properties of the Composites

The magnetic properties of the composites at 300 K were studied using hysteresis loop measured at 300 K (Figure 7 and Figure S8). The saturation magnetization (Ms), remanence (Mr), coercivity (Hc), and coefficient of squareness of hysteresis loops (Kp) can be determined using hysteresis loop [68–71]. In addition, the Kp can be calculated using the Mr/Ms ratio. Alborzi et al. and co-workers showed that decreasing the Kp value enhanced the super magnetic property, whereas diminishing the particle size decreased the Hc and increased the Ms. In addition, super magnetic material can be created by minimizing the Hc [69]. The formation of a cluster structure improves the Ms and dropping the cluster size leads to lower magnetic energy and super magnetic behavior [71]. Thus, the magnetic properties of the materials were influenced by various factors such as the geometry, size, functional groups, morphology, and crystallinity [72]. Further, precursor salt also affects the magnetic properties of the material and the synthesis of Fe_3O_4 by using ferrous and ferric sulfate generates 46.7 emu·g^{-1} of Ms while ferrous and ferric chloride produce 55.4 emu·g^{-1} of Ms and the magnetization of bulk Fe_3O_4 is 93 emu·g^{-1} [69]. All the composites possessed nonlinear behavior against applied field and showed the hysteresis loop (Figure 7 and Figure S8). The Ms of the composites ranged from 0.45–0.009 emu·g^{-1} and the saturated magnetic strength was placed between 9.95–3284.4 Oe. The Kp ranging from 0.022 to 1.128 and Mr fluctuated between 2.2×10^{-4}–0.187 emu·g^{-1}, whereas a 1.27–232.3 Oe range of Hc was given by the composites (Figure 7, Figure S8 and Table 1). According to Figure 7b, the magnetization of the FC approached almost zero, compared with other composites, and the hysteresis loop was not smooth, like other composites. It was considered to be due to the irregular arrangement of the fibers (Figure S8c and Figure 2a), and a study by Lu et al. showed that carbon fibers are non-magnetic materials. Thus, functionalization induced the magnetic behavior of the MC [7]. Furthermore, the Ms, Mr, and Kp values were lower than those of the others. This confirmed that interconnecting fibers using nanomaterials alters the magnetic property of the nonwoven fabric (Table 1 and Figure 2b–l). MXCBCNi and MXCBCCo behaved differently when applied magnetic field increased the magnetization and also increased while others exhibited constant magnetization (Figure 7b). A part of the loop of MXCBCFe pass through the origin, did not have negative Mr and positive Hc value, and had more than one loop, whereas MXCBCFeO did not have a positive Hc value, and instead the loop was located near the Mr point. Further, MXCBCFeO and MXCBCFe possessed high Kp values of 1.128 and 0.095, where 0.095 was lowest among all composites. Additionally, MXCBCFeO and MXCBCFe contained the lowest Ms, Hc, and saturated magnetic strength, which affect the EMI shielding of the nanoparticle-based composites. According to this study, increasing Kp increased the EMI shielding.

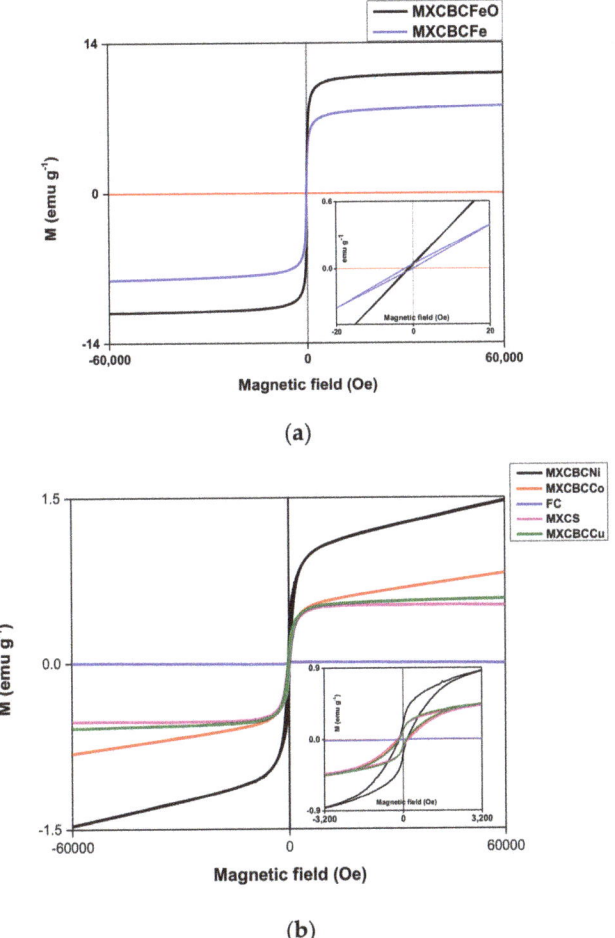

Figure 7. Magnetization against applied field at 300 K: (**a**) Fe- and Fe$_3$O$_4$-based composites and (**b**) Ni, Co, Cu, and non-nanoparticle-based composites.

Table 1. Comparison of saturation magnetization, retentivity, and coercivity of the composites.

Composites	Saturated Magnetic Strength (Oe)	Saturation Magnetization (Ms) (emu·g−1)	Remanence (Mr) (emu·g−1)	Coefficient of Squareness of Hysteresis Loops (Kp) = Mr/Ms (Dimensionless)	Coercivity (Hc) (Oe)
MXCBCCu	3284.4	0.45	0.08	0.178	144.59
MXCS	3200	0.43	0.1	0.233	232.3
FC	2587.7	0.0099	2.2 × 10−4	0.022	22.34
MXCBCCo	3195.9	0.45	0.096	0.213	185.57
MXCBCNi	3184.9	0.86	0.187	0.217	152.58
MXCBCFe	21.4	0.41	0.039	0.095	2.02
MXCBCFeO	9.95	0.39	0.44	1.128	1.27

3.4. Electromagnetic Shielding Effectiveness (EMI-SE) of Composites

The study employed filtration and spray coating techniques, for which 30 g·m^{-2} of areal density fabrics were used, while Ni coated fabric was 20 g·m^{-2} of areal density. For the spray coating, MXene colloidal solution and dispersed CNTO and SDS solution were used, while each coating used 40 mL of solution. A maximum of 100 mL of colloidal solution was used in the filtration process. In spray coating composites, one MXene layer was sandwiched between two layers of CNTO, while one MXene layer placed between CNTO and nanoparticles decorated CNTO in filtrated composites. For the spray coating process, SDS was used as surfactant and CTAB was used to disperse nanoparticles decorated carbon nanotube. The spraying and filtration process changed the thickness and pore size of the fabric. The EMI shielding was performed in the region of the X band and S band. X band (8–12.4 GHz) EMI-SE was performed for all composite, while S band (1–3 GHz) EMI-SE was only measured for spray coated composites (Figure 7a). The intercalation of nanomaterials with FC and MC significantly increased the EMI shielding. The spray coated composite showed a higher EMI-SE between 1.5 to 2.6 GHz and non-woven carbon fabric preferable over Ni coated carbon fabric as it showed a lower EMI-SE than the other composites (Figure 8a,b). The non-fabrics were flexible and contained physically interconnected fibers with a three-dimensional reticular structure. When the incident wave hit the surface of the shielding materials, the reflection, multiple reflection, absorption, and transmission occurred. The strength of this mechanism varies based on the materials used for EMI-SE. According to the Simon formalism, EMI-SE depends on the electric conductivity and thickness of the material, and the length of the carbon fiber has no influence on EMI shielding. Further, hiking areal and volume density increase the electric conductivity and EMI shielding. A study by Lu et al. showed that the EMI-SE of 50 g·m^{-2} of carbon fabric is 30.2 dB, while 30 g·m^{-2} produces 23.1 dB, and the EMI shielding of carbon fabric is independent of the frequency range [7,73,74]. Hence, increasing the areal density of the carbon fabric increases the EMI-SE. The 30 g·m^{-2} of MC gave rise to a maximum of 28.5 dB of EMI-SE in the S-band whereas 31.7 dB of EMI-SE was generated in the X-band region, and this was further enhanced by functionalization up to 43.9 dB of maximum (Table S5 and Figure 8a,b). This result was almost consistent with that of the Lu et al. study that EMI-SE was independent of frequency. This is despite the fact that the areal density and electric conductivity is low while functionalization had a greater effect on EMI-SE. After the functionalization, the MC possessing the magnetic property was advanced criteria for the high EMI-SE.

All the FC-based nanocomposites' minimum, maximum, and averaged shielding were above 99.99% of incident wave in the X-band region while spray coated composites' maximum shielding was just below 99.99% and the minimum and average laid between 99–99.9%. For the FC-based composite of SE, reflection (SE$_R$) and absorption (SE$_A$) prevented 90% and 99.9%, respectively. MXCNTC30 and MXCNTC25 displayed 99.99% of maximum, minimum, and average shielding whereas the others showed 99.9% of shielding in the X-band region. In addition, spray coated composites SE$_A$ were in the range of 99–99.9% and 90% of reflection. The spray coated composite showed maximum shielding of 99.9% (Figure 8a–d and Table S5). The maximum EMI-SE shown by MXCNTC30, MXCNTC25, MXCNTNiC25, MXCNTC10, and MC in the S band were 39.6, 39.9, 34.1, 33.2, and 28.5 dB, respectively, whereas 47.3, 47.1, 34.9, 39.9, and 39.6 dB were given in the X-band region. Thus, the shielding of composites comparatively increased when the measurement frequency reached from the S band to the X band. Further, it was obvious that above 25 coating cycles, the shielding of composite was significantly reduced in the S-band region. It was considered that increasing the amount of MXene and CNTO reduced the conductivity and shielding ability of composite in the S band. When MXCNTC30 went from the S band to the X band, the EMI-SE increased by 19.4%, while it increased by 18% for MXCNTC25. (Figure 7a and Table S5). Thus, EMI-SE for nanocomposites were studied in the X-band region. By contrast, MXCNTNiC25 showed the lowest shielding in the S and X bands, due to surface defects and low electric conductivity. The defect was due to the reduced adhesive ability of MXene–CNTO composites with Ni coated carbon fabric, and the carbon fabric showed low defects (Figures 2l and 8 and Figure S3f). The MXCNTNiC25 exhibited higher EMI shielding than MC as the porosity was diminished. Nonetheless, all the coated carbon fabric showed higher EMI shielding than MXCNTNiC25 [1–3]. The maximum EMI-SE of MXCB, MXCBCFeO, MXCBCFe, MXCBCNi, MXCBCCo, MXCBCCu, and MXCS were 47.6, 45.9, 46.7, 45, 46, 43.6, and 50.5 dB, respectively. The nanoparticle-free composites displayed higher blocking ability than others and the mushroom gills like structure significantly reduced the EMI-SE as compared with FC, whereas the coral like structure showed a higher EMI-SE than FC (Figures 2g–i and 8b and Table S5). Despite this, all showed above 45 dB, except for MXCBCCu (Figure 7 and Table S5). The absorption was dominant over reflection, and reflection stayed nearly constant for all composites. Thus, the changing of the shielding effectiveness was due to the absorption of the composites. The composite showed electric conductivity in the range of 2–13 $S \cdot cm^{-1}$, which was not sufficient to produce good SE_R. The effective percolation of the conductive network facilitates the electron mobility and reflection and enhances the ohmic loss [1–3,7]. The intercalation of the MXene, CNTO, and nanoparticle decorated CNTO increase the electric conductivity, though the lack of an effective conductive network minimizes the SE_R [7]. In this case, part of the incident wave was reflected, while most of the remaining part underwent absorption and multiple reflection. The maximum SE_A of MXCB, MXCBCFeO, MXCBCFe, MXCBCCo, MXCS, MXCNTC30, and MXCNTC25, were all above 33 dB, while the SE_R of all the composites were above 10 dB (Figure 8c,d and Table S5). In addition, the Kp value above 0.09 led to shielding above 99.99 % (Figure 7a,b and Table 1). Thus, absorption can be achieved by different internal geometry, functionalization, ohmic loss, defects, and magnetic property of materials. In addition, the CNTO contributed inner tube scattering leads to a higher SE_A because the blend of different components establishes diverse phases which enhance multiple reflection and absorption [1]. The functionalization considerably increased the specific shielding effectiveness (SSE) and absolute effectiveness (SSE/t), and the SSE and SSE/t of the FC were 401.93 dB $cm^{-3} \cdot g^{-1}$ and 12639.95 dB $cm^{-2} \cdot g^{-1}$, respectively. The intercalations of zero, one-, and two-dimensional materials in the three-dimensional fiber network significantly reduced the SSE and SSE/t. Further, iron-based composited displayed lower SSE and comparatively lower SSE/t (Figure 8e and Table S6).

Figure 8f and Table S4 compared the EMI-SE with previous corresponding work. The dip coated CNTO non-woven fabric (basic weight 20 $g \cdot m^{-2}$) showed a maximum of 33 dB, which is substantially lower than that of the MXene–CNTO–nanoparticle composite [5]. The polystyrene–MWCNT composite showed a maximum EMI-SE of about 22 dB. The Polyvinylidene fluoride (PVDF)–MWCNT composite exhibited 28.5 dB of EMI-SE with a thickness of 0.2 cm while the segregated carbon nanotube–polypropylene composite showed an EMI-SE of 48.3 dB with 0.22 cm of thickness and interconnected MWCNT polymeric matrix exhibited 27 dB at 18 GHz [1,12,13,74,75]. In addition, the spongy carbon nanotube composite disclosed 54.8 with thickness of 0.18 cm in the X–band region [16]. Most of the study showed the EMI shielding range of the CNT composite was 20–30 dB with higher thickness. The different types of fillers and geometry altered the shielding ability of the MWCNT. Thus, we analyzed various combinations of materials with different geometries in order to increase the shielding effectiveness of the non-woven fabric and MWCNT. MXene film crammed between Polyethylene terephthalate polymer film showed an excellent EMI shielding of 92 dB with 4600 $S \cdot cm^{-1}$ electrical conductivity [2]. Cao et al. and coworkers highlighted that nacre-like MXene–cellulose nanofiber composite showed an EMI shielding range of 5.3–25.8 dB based on the percentage of MXene added to one dimensional cellulose fiber [76]. The MXene–carbon fabric composite exhibited an EMI-SE of 43.2 dB [24]. Hence, a combination of the MXene–CNTO-based composite dramatically increased the EMI-SE. Further, the carbon fabric-based composite showed lower shielding; therefore, the manufactured composite was considered to have excellent EMI-SE ability (Figure 8f).

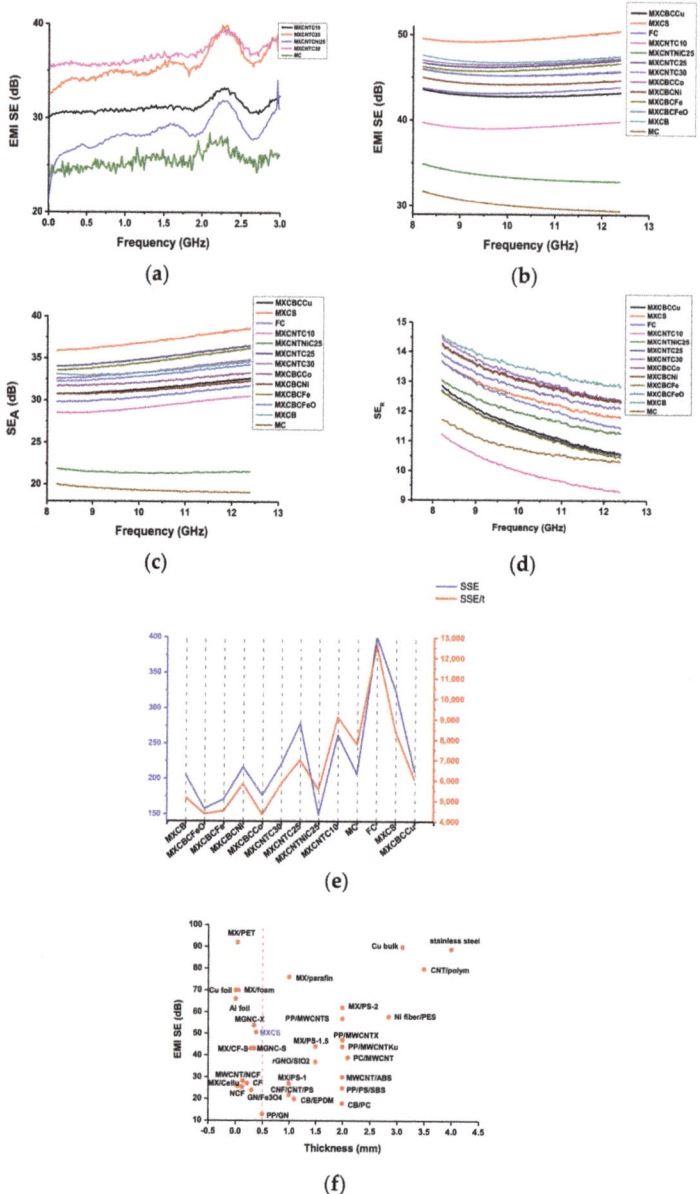

Figure 8. Electromagnetic interference (EMI) shielding of the composites: (**a**) Total EMI shielding (S band), (**b**) total EMI shielding (X band), (**c**) absorption (X band), (**d**) reflection (X band), (**e**) specific shielding effectiveness (SSE) and absolute effectiveness (SSE/t) of the composites, and (**f**) EMI shielding comparison of different composites.

3.5. Thermal Stability and Thermo Gravimetric Analysis of Composites.

Thermogravimetric analysis (TGA) and differential thermal analysis (DTA) were used to investigate the thermal stabilities of the composites. TGA graphs were plotted in the temperature range of 30–1000 °C while DTA graphs were plotted in the range of 30–900 °C (Figure 9a–d). The TGA and DTA analysis were carried out using Al$_2$O$_3$ crucible in a nitrogen environment with a heating rate of 10 °C·min^{-1}. The composites exhibited an outstanding thermal stability over 100 °C. Most of the composites had a starting point of degradation above 140 °C, and the degradation beginning lay between 120–255 °C, where MXCNTC30 showed 120 °C. In addition, MXCB, MXCBCFeO, MXCBCFe, MXCBCNi, MXCBCCo, and MXCNTNiC25 prohibited degradation above 150 °C among mushroom gill like MXCBCCu decomposed at lower temperature of 127.5 °C, whereas the degradation temperatures of the other composites were below 149 °C and above 127.5 °C (except MXCNTC30 at 120 °C). The rapid mass changing percentage of composites occurred between 120–897.5 °C where spray coated composite hit the higher value (above 700 °C), while filtered gave rise below 653 °C (Figures 2i and 9a,c and Table S7). The temperature between the 30–1000 °C range of weight losing percentage of the composites were 6–36%, in which MXCBCFeO gave a maximum of 35.71% while 8.85% of weight loss occurred in MXCNTC25 (Figure 9a,c and Table S7). The MC and FC degradations of the whole temperature range were 6.84% and 6.4%, respectively, and the corresponding degradation beginning temperatures were 190 °C and 255.5 °C, respectively (Figure 9a,c and Table S7). The functionalization of MC by using chitiooligosaccharide increased the thermal stability. This was attributed to the fact that the chitiooligosaccharide contains lots of the hydroxyl groups, thus creating the hydrogen bonding which led to higher thermal stability [68]. According to Raagulan et al., rapid mass changes occurred due to the loss of surface functionalities, water, and decomposition of fabric [5,6,63,64]. In addition, the MXene–carbon fabric composite prohibited degradation up to 235 °C and a Gamage et al. study showed CNTO–carbon fabric composites prohibited degradation until 284 °C. The combination of MXene, CNTO, and nanoparticles considerably diminished the degradation temperature (Table S7). The DTA analysis revealed various peaks raised due the degradation of the composites (Figure 9b,d). Strong peaks were placed in the region of the rapid changing of TGA curve [77]. Further, we observed that the slope of the TGA curve depends on the intensity of the corresponding peaks in DTA analysis, and that the high intensity of the DTA peaks lead to rapid degradation. The more intense peaks were located between 209–934.2 °C temperature range (Figure 9b,d, Figure S9a–m and Table S8). It was obvious that the introduction of the nanomaterials in the fabric network shifted the peaks to lower temperatures and diminished the decomposition temperature.

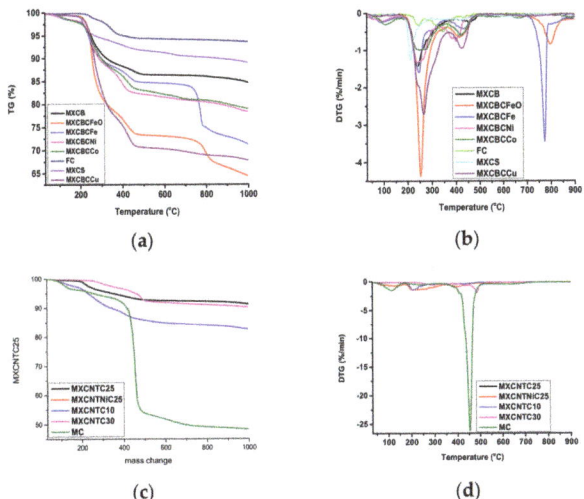

Figure 9. TGA and differential thermal analysis (DTA) analysis of the composites (**a**) TGA of filtration-based composite, (**b**) TGA of filtration-based composite, (**c**) TGA of spray coated composite, and (**d**) DTA of spray coated composite.

4. Conclusions

The filtration-based nanocomposite and spray coated nanocomposite were successfully prepared. The effect of the functionalization of non-woven carbon fabric was analyzed as well. The composites were lightweight and thinner. The density of the composites ranged from 0.108 to 0.288 g·cm^{-3} while the thickness of the composites ranged between 0.266–0.408 mm. The ranges of Ms and Mr were 0.0099–0.86 emu·g^{-1} and 0.00022–0.44 emu·g^{-1}, respectively. The MXCBCNi displayed a higher Ms of 0.86 emu·g^{-1} and a 0.44 emu·g^{-1} of Mr was shown by MXCBCFeO. The MXCCFeO exhibited a maximum Kp of 1.128 while FC was 0.022. The functionalized carbon fabric displayed a hydrophobic nature with a 134.18° contact angle, whereas MXCNTC25 displayed a contact angle of 131.3°. The electric conductivities of the composites varied between 2.65–12 S/cm, whereas the surface resistances were in the range of 2.08–13.98 Ω/sq. The composites showed good thermal stability and resisted complete thermal degradation above 120°. In addition, functionalization increased the thermal stability and prevented degradation up to 255.5°. The maximum EMI-SE shown by MXCS was 50.5 dB, and the EMI shielding range of the composite was 50.5–28.5 dB. The maximum SE$_R$ and SE$_A$ were shown by MXCB and MXCS, respectively. The ranges of SSE and SSE/t were 149.37–401.93 dB cm^3·g^{-1} and 4330.82–12,639.35 dB cm^2·g^{-1}, respectively, where FC showed maximums of both. Hence, the manufactured fabric composite presented high EMI shielding, magnetic behavior, low density, smaller thickness, and flexibility.

Supplementary Materials: The following are available online at http://www.mdpi.com/2079-4991/9/4/519/s1, Figure S1: SEM image of CNTO decorated by (a) Fe3O4 (b) Fe (c) Ni (d) Cu; Figure S2: SEM image of carbon fabric composite of (a) MC (500) (b) MXCB (500) (c) surface of MXCS (10000) (d) MXCBCFeO (500) (e)MXCBCFe (500) (f)MXCBCNi (450) (g) MXCBCCo (150) (h) MXCBCCo (100000) (i) MXXBCCu (45000) (j) MXCNTC30 (100) (k) MXCNTNi25 (1000) (l) MXCNT10 (400); Figure S3: SEM image of (a) MC (×500), (b) MC (×2000), (c) cracks on fiber (×300000), (d) MXene-CNTO coated carbon fabric (×500), (e) MXene-CNTO coated carbon fabric (×300), (f) MXene-CNTO coated fabric (Ni coated fabric) (×300), (g) MXene and CNTO on the surface of the fabric (×120000), (h) Ti3AlC2 (×22000), (i) Ti3C2Tx (×50000) and (j) Ti3C2Tx (×80000); Figure S4: Normalized curve of Raman spectrum of (a) decorated CNT, (b) composites (c) CNT, (d) CNTO, (e) CCu, (f) CCo, (g) CNi, (h) CFe, (i) CFeO, (j) FC (k) comparison of MXene, MAX phase and MC, (l) MC, (m) MXCS, (n) MXCBCCu, (o) MXCBCCo, (p) MXCBCNi, (q) MXCBCFe, (r) MXCBCFeO and (s) MXCB; Figure S5: The c1s fitting curve of the MXene, functionalized fabric and composites; Figure S6: Resistivity profile of the composites; Figure S7: Contact angle of (a) MXCNTNiC25 (b) FC and (c) represent other composites; Figure S8: The magnetization of composites against the applied field at 300 K (a) MXCBCCu, (b) MXCS, (c) FC, (d) MXCBCCo, (e) MXCBCNi, (f) MXCBCFe, (g) MXCBCFe and (h) MXCBCFeO; Figure S9: Comparison of the TG and DTA of all the composites (a) MXCB, (b) MXCBCFeO, (c) MXCBCFe, (d) MXCBCNi, (e) MXCBCCo, (f) MXCS, (g) MXCBCCu, (h) FC, (i) MXCNTC25, (j) MXCNTNiC25, (k) MXCNTC10, (l) MXCNTC30 and (m) MC; Table S1: Elemental percentage of composites from EDX analysis; Table S2: Raman spectra band positions; Table S3: Atomic percentage MXene, MC, CNT, CNTO and composites from XPS analysis; Table S4: Comparison of EMI SE with thickness; Table S5: Comparison of maximum (MAX), minimum (MINI), average (AVE) shielding, SSE and SSE/t of the composite in each case; Table S6: Density of the composites; Table S7: Comparison of mass changes with different temperature range from TGA analysis; Table S8: Comparison of different peak positions range from DTA analysis.

Author Contributions: K.Y.C. and R.B. designed the project; K.R., B.M.K. and J.J.M. performed the experiments; L.R.L. and J.L. analyzed the data; S.B.L. supervised the analysis; K.R. wrote the manuscript.

Funding: This research was supported by the Leading Human Resource Training Program of Regional Neo industry through the National Research Foundation of Korea (NRF) funded by the Ministry of Science, ICT and future Planning (grant number) (NRF-2017H1D5A1043865).

Conflicts of Interest: The author declare no conflict of interest.

References

1. Pawar, S.P.; Rzeczkowski, P.; Potschke, P.; Krause, B.; Bose, S. Does the Processing Method Resulting in Different States of an Interconnected Network of Multiwalled Carbon Nanotubes in Polymeric Blend Nanocomposites Affect EMI Shielding Properties? *ACS Omega* **2018**, *3*, 5771–5782. [CrossRef]
2. Shahzad, F.; Alhabeb, M.; Hatter, C.B.; Anasori, B.; Hong, S.M.; Koo, C.M.; Gogotsi, Y. Electromagnetic interference shielding with 2D transition metal carbides (MXenes). *Science* **2016**, *353*, 1137–1140. [CrossRef] [PubMed]

3. Cao, L.; Liu, G.; Li, J. Damage assessment of long-range rocket system by electromagnetic pulse weapon. *AIP Conf. Proc.* **2017**, *1864*, 020115.
4. Choi, G.; Shahzad, F.; Bahk, Y.M.; Jhon, Y.M.; Park, H.; Alhabeb, M.; Anasori, B.; Kim, D.S.; Koo, C.M.; Gogotsi, Y.; et al. Enhanced Terahertz Shielding of MXenes with Nano-Metamaterials. *Adv. Opt. Mater.* **2018**, *6*, 1701076. [CrossRef]
5. Pothupitiya Gamage, S.J.; Yang, K.; Braveenth, R.; Raagulan, K.; Kim, H.S.; Lee, Y.S.; Yang, C.M.; Moon, J.J.; Chai, K.Y. MWCNT coated free-standing carbon fiber fabric for enhanced performance in EMI shielding with a higher absolute EMI SE. *Materials* **2017**, *10*, 1350. [CrossRef]
6. Raagulan, K.; Braveenth, R.; Jang, H.; Seon Lee, Y.; Yang, C.M.; Mi Kim, B.; Moon, J.; Chai, K. Electromagnetic Shielding by MXene-Graphene-PVDF Composite with Hydrophobic, Lightweight and Flexible Graphene Coated Fabric. *Materials* **2017**, *11*, 1803. [CrossRef] [PubMed]
7. Lu, L.; Xing, D.; Teh, K.S.; Liu, H.; Xie, Y.; Liu, X.; Tang, Y. Structural effects in a composite nonwoven fabric on EMI shielding. *Mater. Des.* **2018**, *120*, 354–362. [CrossRef]
8. Al-Saleh, M.H.; Sundararaj, U. Electromagnetic interference shielding mechanisms of CNT/polymer composites. *Carbon* **2009**, *47*, 1738–1746. [CrossRef]
9. Yadav, R.S.; Kuřitka, I.; Vilcakova, J.; Skoda, D.; Urbánek, P.; Machovsky, M.; Masař, M.; Kalina, L.; Havlica, J. Lightweight NiFe$_2$O$_4$-Reduced Graphene Oxide-Elastomer Nanocomposite flexible sheet for electromagnetic interference shielding application. *Compos. B Eng.* **2019**, *166*, 95–111. [CrossRef]
10. Ravindren, R.; Mondal, S.; Nath, K.; Das, N.C. Synergistic effect of double percolated co-supportive MWCNT-CB conductive network for high-performance EMI shielding application. *Polym. Adv. Technol.* **2019**, 1–12. [CrossRef]
11. Xu, Z.; Liang, M.; He, X.; Long, Q.; Yu, J.; Xie, K.; Liao, L. The preparation of carbonized silk cocoon-Co-graphene composite and its enhanced electromagnetic interference shielding performance. *Compos. Part A Appl. Sci. Manuf.* **2019**, *119*, 111–118. [CrossRef]
12. Wu, H.Y.; Jia, L.C.; Yan, D.X.; Gao, J.F.; Zhang, X.P.; Ren, P.G.; Li, Z.M. Simultaneously improved electromagnetic interference shielding and mechanical performance of segregated carbon nanotube/polypropylene composite via solid phase molding. *Compos. Sci. Technol.* **2018**, *156*, 87–94. [CrossRef]
13. Ma, X.; Shen, B.; Zhang, L.; Liu, Y.; Zhai, W.; Zheng, W. Porous superhydrophobic polymer/carbon composites for lightweight and self-cleaning EMI shielding application. *Compos. Sci. Technol.* **2018**, *158*, 86–93. [CrossRef]
14. Lu, S.; Shao, J.; Ma, K.; Chen, D.; Wang, X.; Zhang, L.; Meng, Q.; Ma, J. Flexible, mechanically resilient carbon nanotube composite films for high-efficiency electromagnetic interference shielding. *Carbon* **2018**, *136*, 387–394. [CrossRef]
15. Mondal, S.; Das, P.; Ganguly, S.; Ravindren, R.; Remanan, S.; Bhawal, P.; Das, T.K.; Das, N.C. Thermal-air ageing treatment on mechanical, electrical, and electromagnetic interference shielding properties of lightweight carbon nanotube based polymer nanocomposites. *Compos. Pt A. Appl. Sci. Manuf.* **2018**, *107*, 447–460. [CrossRef]
16. Lu, D.; Mo, Z.; Liang, B.; Yang, L.; He, Z.; Zhu, H.; Tang, Z.; Gui, X. Flexible, lightweight carbon nanotube sponges and composites for high-performance electromagnetic interference shielding. *Carbon* **2018**, *133*, 457–463. [CrossRef]
17. Liu, Y.F.; Feng, L.M.; Chen, Y.F.; Shi, Y.D.; Chen, X.D.; Wang, M. Segregated polypropylene/cross-linked poly (ethylene-co-1-octene)/multi-walled carbon nanotube nanocomposites with low percolation threshold and dominated negative temperature coefficient effect: Towards electromagnetic interference shielding and thermistors. *Compos. Sci. Technol.* **2018**, *159*, 152–161.
18. Alhabeb, M.; Maleski, K.; Anasori, B.; Lelyukh, P.; Clark, L.; Sin, S.; Gogotsi, Y. Guidelines for Synthesis and Processing of Two-Dimensional Titanium Carbide (Ti3C2Tx MXene). *Chem. Mater.* **2017**, *29*, 7633–7644. [CrossRef]
19. He, P.; Cao, M.; Shu, J.C.; Cai, Y.Z.; Wang, X.; Zhao, Q.; Yuan, J. Atomic Layer Tailoring Titanium Carbide MXene to Tune Transport and Polarization for Utilization of Electromagnetic Energy beyond Solar and Chemical Energy. *ACS Appl. Mater. Interfaces* **2019**. [CrossRef]
20. Xiang, C.; Guo, R.; Lin, S.; Jiang, S.; Lan, J.; Wang, C.; Cui, C.; Xiao, H.; Zhang, Y. Lightweight and ultrathin TiO$_2$-Ti$_3$C$_2$T$_X$/graphene film with electromagnetic interference shielding. *Chem. Eng. J.* **2019**, *360*, 1158–1166. [CrossRef]

21. Zhou, Z.; Liu, J.; Zhang, X.; Tian, D.; Zhan, Z.; Lu, C. Ultrathin MXene/Calcium Alginate Aerogel Film for High-Performance Electromagnetic Interference Shielding. *Adv. Mater. Interfaces* **2019**, 1802040–1802049. [CrossRef]
22. Cui, C.; Xiang, C.; Geng, L.; Lai, X.; Guo, R.; Zhang, Y.; Xiao, H.; Lan, J.; Lin, S.; Jiang, S. Flexible and ultrathin electrospun regenerate cellulose nanofibers and d-$Ti_3C_2T_x$ (MXene) composite film for electromagnetic interference shielding. *J. Alloys Compd.* **2019**, *785*, 1246–1255. [CrossRef]
23. Li, X.; Yin, X.; Liang, S.; Li, M.; Cheng, L.; Zhang, L. 2D carbide MXene Ti_2CT_X as a novel high-performance electromagnetic interference shielding material. *Carbon* **2019**, *146*, 210–217. [CrossRef]
24. Raagulan, K.; Braveenth, R.; Jang, H.J.; Lee, Y.S.; Yang, C.M.; Kim, B.M.; Moon, J.J.; Chai, K.Y. Fabrication of Nonwetting Flexible Free-Standing MXene-Carbon Fabric for Electromagnetic Shielding in S-Band Region. *Bull. Korean Chem. Soc.* **2018**, *39*, 1412–1419. [CrossRef]
25. Georgakilas, V.; Gournis, D.; Tzitzios, V.; Pasquato, L.; Guldi, D.M.; Prato, M. Decorating carbon nanotubes with metal or semiconductor nanoparticles. *J. Mater. Chem.* **2017**, *17*, 2679–2694. [CrossRef]
26. Megiel, E. Surface modification using TEMPO and its derivatives. *Adv. Colloid Interface Sci.* **2017**, *250*, 158–184. [CrossRef] [PubMed]
27. Young, K.D. Bacterial morphology: Why have different shapes? *Curr. Opin. Microbiol.* **2017**, *10*, 596–600. [CrossRef] [PubMed]
28. Allen, J.W.; Sihanonth, P.; Gartz, J.; Toro, G. An ethnopharmacological and ethnomycological update on the occurrence, use, cultivation, chemical analysis, and SEM photography of neurotropic fungi from Thailand, Cambodia and other regions of South and Southeast Asia, Indonesia and Bali. *Ethnomycological Journals: Sacred Mushroom Studies*, **2012**, *9*, 1–129.
29. Clark, T.R.; Leonard, N.D.; Zhao, J.X.; Brodie, J.; McCook, L.J.; Wachenfeld, D.R.; Nguyen, A.D.; Markham, H.L.; Pandolfi, J.M. Historical photographs revisited: A case study for dating and characterizing recent loss of coral cover on the inshore Great Barrier Reef. *Sci. Rep.* **2016**, *6*, 19285. [CrossRef]
30. Yan, P.; Zhang, R.; Jia, J.; Wu, C.; Zhou, A.; Xu, J.; Zhang, X. Enhanced supercapacitive performance of delaminated two-dimensional titanium carbide/carbon nanotube composites in alkaline electrolyte. *J. Power Sources* **2015**, *284*, 38–43. [CrossRef]
31. Zhang, H.B.; Zhou, Y.C.; Bao, Y.W.; Li, M.S. Abnormal thermal shock behavior of Ti_3SiC_2 and Ti_3AlC_2. *J. Mater. Res.* **2006**, *21*, 2401–2407. [CrossRef]
32. Ivasyshyn, A.; Ostash, O.; Prikhna, T.; Podhurska, V.; Basyuk, T. Oxidation Resistance of Materials Based on Ti3AlC2 Nanolaminate at 600 °C in Air. *Nanoscale Res. Lett.* **2016**, *11*, 358. [CrossRef] [PubMed]
33. Wang, L.; Zhang, H.; Wang, B.; Shen, C.; Zhang, C.; Hu, Q.; Zhou, A.; Liu, B. Synthesis and electrochemical performance of $Ti_3C_2T_x$ with hydrothermal process. *Electron. Mater. Lett.* **2016**, *12*, 702–710. [CrossRef]
34. Han, M.; Yin, X.; Wu, H.; Hou, Z.; Song, C.; Li, X.; Zhang, L.; Cheng, L. Ti_3C_2 MXenes with modified surface for high-performance electromagnetic absorption and shielding in the X-band. *ACS Appl. Mater. Interfaces* **2016**, *8*, 21011–21019. [CrossRef]
35. Qian, A.; Hyeon, S.E.; Seo, J.Y.; Chung, C.H. Capacitance changes associated with cation-transport in free-standing flexible $Ti_3C_2T_x$(TO,F,OH) MXene film electrodes. *Electrochim. Acta* **2018**, *266*, 86–93. [CrossRef]
36. Saleh, T.A.; Agarwal, S.; Gupta, V.K. Synthesis of $MWCNT/MnO_2$ and their application for simultaneous oxidation of arsenite and sorption of arsenate. *Appl. Catal. B Environ.* **2011**, *106*, 46–53. [CrossRef]
37. Jia-Jia, C.; Xin, J.; Qiu-Jie, S.; Chong, W.; Qian, Z.; Ming-Sen, Z.; Quan-Feng, D. The preparation of nano-sulfur/MWCNTs and its electrochemical performance. *Electrochim. Acta* **2010**, *55*, 8062–8066. [CrossRef]
38. Gupta, V.K.; Agarwal, S.; Saleh, T.A. Synthesis and characterization of alumina-coated carbon nanotubes and their application for lead removal. *J. Hazard. Mater.* **2011**, *185*, 17–23. [CrossRef]
39. Ma, X.; Yu, J.; Wang, N. Glycerol plasticized-starch/multiwall carbon nanotube composites for electroactive polymers. *Compos. Sci. Technol.* **2008**, *68*, 268–273. [CrossRef]
40. Gupta, V.; Saleh, T.A. *Syntheses of carbon nanotube-metal oxides composites; adsorption and photo-degradation*. In Carbon Nanotubes-From Research to Applications; InTech: Shanghai, China, 2011.
41. użny, W.; Bańka, E. Relations between the structure and electric conductivity of polyaniline protonated with camphorsulfonic acid. *Macromolecules* **2000**, *33*, 425–429. [CrossRef]
42. Lorencová, L.; Bertok, T.; Dosekova, E.; Holazová, A.; Paprckova, D.; Vikartovská, A.; Sasinková, V.; Filip, J.; Kasák, P.; Jerigová, M.; et al. Electrochemical performance of Ti3C2Tx MXene in aqueous media: Towards ultrasensitive H_2O_2 sensing. *Electrochim. Acta* **2017**, *235*, 471–479. [CrossRef] [PubMed]

43. Wang, X.; Garnero, C.; Rochard, G.; Magne, D.; Morisset, S.; Hurand, S.; Chartier, P.; Rousseau, J.; Cabioc'h, T.; Coutanceau, C.; et al. A new etching environment (FeF$_3$/HCl) for the synthesis of two-dimensional titanium carbide MXenes: A route towards selective reactivity vs. water. *J. Mater. Chem. A* **2017**, *5*, 22012–22023. [CrossRef]
44. Yan, J.; Ren, C.E.; Maleski, K.; Hatter, C.B.; Anasori, B.; Urbankowski, P.; Sarycheva, A.; Gogotsi, Y. Flexible MXene/graphene films for ultrafast supercapacitors with outstanding volumetric capacitance. *Adv. Funct. Mater.* **2017**, *27*, 1701264. [CrossRef]
45. Sui, X.M.; Giordani, S.; Prato, M.; Wagner, H.D. Effect of carbon nanotube surface modification on dispersion and structural properties of electrospun fibers. *Appl. Phys. Lett.* **2009**, *95*, 233113. [CrossRef]
46. Mishra, A.K.; Ramaprabhu, S. Magnetite decorated multiwalled carbon nanotube based supercapacitor for arsenic removal and desalination of seawater. *J. Phys. Chem. C* **2017**, *114*, 2583–2590. [CrossRef]
47. He, Y.; Huang, L.; Cai, J.S.; Zheng, X.M.; Sun, S.G. 2010. Structure and electrochemical performance of nanostructured Fe$_3$O$_4$/carbon nanotube composites as anodes for lithium ion batteries. *Electrochimica Acta* **2010**, *55*, 1140–1144. [CrossRef]
48. Grassi, G.; Scala, A.; Piperno, A.; Iannazzo, D.; Lanza, M.; Milone, C.; Pistone, A.; Galvagno, S. A facile and ecofriendly functionalization of multiwalled carbon nanotubes by an old mesoionic compound. *Chem. Comm.* **2012**, *48*, 6836–6838. [CrossRef]
49. Mali, S.S.; Betty, C.A.; Bhosale, P.N.; Patil, P.S. Synthesis, characterization of hydrothermally grown MWCNT–TiO$_2$ photoelectrodes and their visible light absorption properties. *ECS J. Solid State Sci. Technol.* **2012**, *1*, M15–M23. [CrossRef]
50. Dong, C.K.; Li, X.; Zhang, Y.; Qi, J.Y.; Yuan, Y.F. Fe$_3$O$_4$ nanoparticles decorated multi-walled carbon nanotubes and their sorption properties. *Chem. Res. Chin. Univ.* **2009**, *25*, 936–940.
51. Jiao, Z.; Qiu, J. Microwave absorption performance of iron oxide/multiwalled carbon nanotubes nanohybrids prepared by electrostatic attraction. *J. Mater. Sci.* **2018**, *53*, 3640–3646. [CrossRef]
52. Crane, R.A.; Scott, T.B. The Effect of Vacuum Annealing of Magnetite and Zero-Valent Iron Nanoparticles on the Removal of Aqueous Uranium. *J. Nanotechnol.* **2013**, *2013*, 11. [CrossRef]
53. Jiang, Z.; Xie, J.; Jiang, D.; Wei, X.; Chen, M. Modifiers-assisted formation of nickel nanoparticles and their catalytic application to p-nitrophenol reduction. *CrystEngComm* **2013**, *15*, 560–569. [CrossRef]
54. Balela, M.D.L.; Yagi, S.; Lockman, Z.; Aziz, A.; Amorsolo, A.V.; Matsubara, E. 2009. Electroless deposition of ferromagnetic cobalt nanoparticles in propylene glycol. *J. Electrochem. Soc.* **2009**, *156*, E139–E142. [CrossRef]
55. Betancourt-Galindo, R.; Reyes-Rodriguez, P.Y.; Puente-Urbina, B.A.; Avila-Orta, C.A.; Rodríguez-Fernández, O.S.; Cadenas-Pliego, G.; Lira-Saldivar, R.H.; García-Cerda, L.A. Synthesis of copper nanoparticles by thermal decomposition and their antimicrobial properties. *J. Nanomater.* **2014**, *2014*, 10. [CrossRef]
56. Graat, P.; Somers, M.A. Quantitative analysis of overlapping XPS peaks by spectrum reconstruction: Determination of the thickness and composition of thin iron oxide films. Surface and Interface Analysis: An International Journal devoted to the development and application of techniques for the analysis of surfaces. *Interfaces Thin Films* **1998**, *26*, 773–782.
57. Lee, T.W.; Lee, S.E.; Jeong, Y.G. Carbon nanotube/cellulose papers with high performance in electric heating and electromagnetic interference shielding. *Compos. Sci. Technol.* **2016**, *131*, 77–87. [CrossRef]
58. Datsyuk, V.; Kalyva, M.; Papagelis, K.; Parthenios, J.; Tasis, D.; Siokou, A.; Kallitsis, I.; Galiotis, C. Chemical oxidation of multiwalled carbon nanotubes. *Carbon* **2008**, *46*, 833–840. [CrossRef]
59. Halim, J. Synthesis and Characterization of 2D Nanocrystals and Thin Films of Transition Metal Carbides (MXenes). Doctoral Dissertation, Linköping University Electronic Press, Linköping, Sweden, 2014.
60. Shen, C.; Wang, L.; Zhou, A.; Wang, B.; Wang, X.; Lian, W.; Hu, Q.; Qin, G.; Liu, X. Synthesis and Electrochemical Properties of Two-Dimensional RGO/Ti$_3$C$_2$T$_x$ Nanocomposites. *Nanomaterials* **2018**, *8*, 80. [CrossRef]
61. Shah, S.A.; Habib, T.; Gao, H.; Gao, P.; Sun, W.; Green, M.J.; Radovic, M. Template-free 3D titanium carbide (Ti$_3$C$_2$T$_x$) MXene particles crumpled by capillary forces. *Chem. Commun.* **2017**, *53*, 400–403. [CrossRef] [PubMed]
62. Liu, J.; Zhang, H.B.; Sun, R.; Liu, Y.; Liu, Z.; Zhou, A.; Yu, Z.Z. Hydrophobic, Flexible, and Lightweight MXene Foams for High-Performance Electromagnetic-Interference Shielding. *Adv. Mater.* **2017**, *29*, 1702367. [CrossRef] [PubMed]

63. Ibrahim, S.F.; El-Amoudy, E.S.; Shady, K.E. Thermal analysis and characterization of some cellulosic fabrics dyed by a new natural dye and mordanted with different mordants. *Int. J. Chem.* **2011**, *3*, 40. [CrossRef]
64. Taira, M.; Araki, Y. DTG thermal analyses and viscosity measurements of three commercial agar impression materials. *J. Oral Rehabil.* **2002**, *29*, 697–701. [CrossRef] [PubMed]
65. Kushwaha, A.; Aslam, M. Roughness enhanced surface defects and photoconductivity of acid etched ZnO nanowires. In Proceedings of the 2012 International Conference on Emerging Electronics, Mumbai, India, 15–17 December 2012; pp. 1–4.
66. Zhao, M.Q.; Ren, C.E.; Ling, Z.; Lukatskaya, M.R.; Zhang, C.; Van Aken, K.L.; Barsoum, M.W.; Gogotsi, Y. Flexible MXene/carbon nanotube composite paper with high volumetric capacitance. *Adv. Mater.* **2015**, *27*, 339–345. [CrossRef] [PubMed]
67. Kim, B.C.; Innis, P.C.; Wallace, G.G.; Low, C.T.J.; Walsh, F.C.; Cho, W.J.; Yu, K.H. Electrically conductive coatings of nickel and polypyrrole/poly (2-methoxyaniline-5-sulfonic acid) on nylon Lycra®textiles. *Prog. Org. Coat.* **2013**, *76*, 1296–1301. [CrossRef]
68. Zhang, L.; Zhang, Y. Fabrication and magnetic properties of Fe_3O_4 nanowire arrays in different diameters. *Journal of Magnetism and Magnetic. Materials* **2009**, *321*, L15–L20.
69. Iida, H.; Takayanagi, K.; Nakanishi, T.; Osaka, T. Synthesis of Fe_3O_4 nanoparticles with various sizes and magnetic properties by controlled hydrolysis. *J. Colloid Interface Sci.* **2007**, *314*, 274–280. [CrossRef] [PubMed]
70. Chang, H.; Su, H.T. Synthesis and magnetic properties of Ni nanoparticles. *Rev. Adv. Mater. Sci.* **2008**, *18*, 667–675.
71. Alborzi, Z.; Hassanzadeh, A.; Golzan, M.M. Superparamagnetic behavior of the magnetic hysteresis loop in the Fe2O3@ Pt core-shell nanoparticles. *Int. J. Nanosci. Nanotechnol.* **2012**, *8*, 93–98.
72. Hu, C.; Gao, Z.; Yang, X. Fabrication and magnetic properties of Fe3O4 octahedra. *Chem. Phys. Lett.* **2006**, *429*, 513–517. [CrossRef]
73. Xing, D.; Lu, L.; Teh, K.S.; Wan, Z.; Xie, Y.; Tang, Y. Highly flexible and ultra-thin Ni-plated carbon-fabric/polycarbonate film for enhanced electromagnetic interference shielding. *Carbon* **2018**, *132*, 32–41. [CrossRef]
74. Xu, Y.; Yang, Y.; Duan, H.; Gao, J.; Yan, D.X.; Zhao, G.; Liu, Y. Flexible and highly conductive sandwich nylon/nickel film for ultra-efficient electromagnetic interference shielding. *Appl. Surf. Sci.* **2018**, *455*, 856–863. [CrossRef]
75. Bagotia, N.; Mohite, H.; Tanaliya, N.; Sharma, D.K. A comparative study of electrical, EMI shielding and thermal properties of graphene and multiwalled carbon nanotube filled polystyrene nanocomposites. *Polym. Compos.* **2018**, *39*, E1041–E1051. [CrossRef]
76. Cao, W.T.; Chen, F.F.; Zhu, Y.J.; Zhang, Y.G.; Jiang, Y.Y.; Ma, M.G.; Chen, F. Binary Strengthening and Toughening of MXene/Cellulose Nanofiber Composite Paper with Nacre-Inspired Structure and Superior Electromagnetic Interference Shielding Properties. *ACS Nano* **2018**, *12*, 4583–4593. [CrossRef] [PubMed]
77. Lodhi, G.; Kim, Y.S.; Hwang, J.W.; Kim, S.K.; Jeon, Y.J.; Je, J.Y.; Ahn, C.B.; Moon, S.H.; Jeon, B.T.; Park, P.J. Chitooligosaccharide and its derivatives: Preparation and biological applications. *BioMed Res. Int.* **2014**, *2014*, 654913. [CrossRef] [PubMed]

© 2019 by the authors. Licensee MDPI, Basel, Switzerland. This article is an open access article distributed under the terms and conditions of the Creative Commons Attribution (CC BY) license (http://creativecommons.org/licenses/by/4.0/).

Article

Improved Catalytic Durability of Pt-Particle/ABS for H_2O_2 Decomposition in Contact Lens Cleaning

Yuji Ohkubo [1,*], Tomonori Aoki [1], Satoshi Seino [1], Osamu Mori [2], Issaku Ito [2], Katsuyoshi Endo [1] and Kazuya Yamamura [1]

1. Graduate School of Engineering, Osaka University, Suita, Osaka 565-0871, Japan; t-aoki@div1.upst.eng.osaka-u.ac.jp (T.A.); seino@mit.eng.osaka-u.ac.jp (S.S.); endo@upst.eng.osaka-u.ac.jp (K.E.); yamamura@prec.eng.osaka-u.ac.jp (K.Y.)
2. Menicon Co., Ltd., Kasugai, Aichi 487-0032, Japan; o-mori@menicon.co.jp (O.M.); issaku-ito@menicon.co.jp (I.I.)
* Correspondence: okubo@upst.eng.osaka-u.ac.jp; Tel.: +81-6-6879-7294

Received: 29 January 2019; Accepted: 18 February 2019; Published: 3 March 2019

Abstract: In a previous study, Pt nanoparticles were supported on a substrate of acrylonitrile–butadiene–styrene copolymer (ABS) to give the ABS surface catalytic activity for H_2O_2 decomposition during contact lens cleaning. Although the Pt-particle/ABS catalysts exhibited considerably high specific catalytic activity for H_2O_2 decomposition, the catalytic activity decreased with increasing numbers of repeated usage, which meant the durability of the catalytic activity was low. Therefore, to improve the catalytic durability in this study, we proposed two types of pretreatments, as well as a combination of these treatments before supporting Pt nanoparticles on the ABS substrate. In the first method, the ABS substrate was etched, and in the second method, the surface charge of the ABS substrate was controlled. A combination of etching and surface charge control was also applied as a third method. The effects of these pretreatments on the surface morphology, surface chemical composition, deposition behavior of Pt particles, and Pt loading weight were investigated by scanning electron microscopy (SEM), X-ray photoelectron spectroscopy (XPS), cross-sectional SEM, and inductively coupled plasma atomic emission spectroscopy (ICP-AES), respectively. Both etching and controlling the surface charge effectively improved the catalytic durability for H_2O_2 decomposition. In addition, the combination treatment was the most effective.

Keywords: catalytic durability; nanoparticle; supported catalyst; radical reactions; platinum (Pt); H_2O_2 decomposition; contact lens cleaning

---

1. Introduction

The number of contact lens wearers is estimated to be approximately 140 million all over the world [1]. Contact lens materials have been improved based on demands from the wearers [2]. Contact lenses are divided into two types: disposable and extended wear. Extended-wear contact lenses can be used repeatedly, which offers long-term cost advantages. However, to prevent eye troubles, a repeatable-use-type contact lens requires daily cleaning and sterilization. Three methods are used to clean and sterilize contact lenses: boiling, and cleaning in either H_2O_2 or multipurpose solution (MPS). When cleaning with a MPS, a single solution plays the roles of cleaning, sterilizing, and preserving lenses. Thus, MPS cleaning is a simple method, and about 70% of contact lens wearers currently use an MPS to clean their lenses [3]. However, if contact lens wearers are not careful while using the MPS, eye troubles are likely to occur due to inadequate sterilization. Therefore, the number of wearers using H_2O_2 cleaning, which has higher sterilization performance, has gradually increased in recent years [3]. In H_2O_2 cleaning, a 35,000 ppm H_2O_2 solution is used to clean and sterilize contact lenses. Although the H_2O_2 solution exhibits high sterilizing performance, it involves the risk of eyes becoming bloodshot or painful, and can even lead to blindness if the H_2O_2 solution enters the eyes without decomposing. Thus, a Pt catalyst is used to promote H_2O_2 decomposition to lower the residual H_2O_2 concentration below 100 ppm [4]. A Pt film plated on an acrylonitrile–butadiene–styrene copolymer (ABS) container using electroless plating is usually used to catalyze H_2O_2 decomposition. Thus, each ABS container needs a Pt loading weight of 1.5 mg. However, Pt is an expensive material. In addition, the Pt-film/ABS container must be thrown away after repeated use for a month. Therefore, a technique that decreases the amount of Pt used to clean contact lenses is needed.

In a previous report [5], we proposed replacing the Pt film with Pt nanoparticles. Several methods can synthesize and immobilize metal nanoparticles, for example: impregnation [6–8], polyol [9–11], and sonolytic methods [12–15]. These methods have the disadvantages of high processing temperature, long processing time, nonuniform deposition of the metal ions, and low producibility. Therefore, we selected a radiolytic synthesis method that uses a high-energy electron beam (EB) to synthesize and immobilize Pt nanoparticles on the ABS container. This method is called the electron-beam irradiation reduction method (EBIRM), and it offers the advantages of a low processing temperature, short processing time, highly uniform deposition, and high producibility [16–19]. We successfully decreased the Pt loading weight from 1.5 mg/substrate for Pt-film/ABS to 5.9 µg/substrate for Pt-particle/ABS and synthesized a Pt-particle/ABS catalyst having considerably higher specific catalytic activity for H_2O_2 decomposition than the Pt-film/ABS catalyst. However, the catalytic activity of the Pt-particle/ABS catalyst decreased with increasing number of repeated uses, although the catalytic activity of the Pt-film/ABS catalyst did not change. The decrease in catalytic activity of the Pt-particle/ABS catalyst was caused by decreasing the Pt loading weight with increasing numbers of repeated uses, not by poisoning Pt particles [5]. This problem of catalytic durability for H_2O_2 decomposition remains unsolved. Thus, in this study, we attempted to improve the catalytic durability using two pretreatments before EB irradiation—etching and controlling the surface charge—and a combination of both. The effects of pretreatments on ABS surface, catalytic activity, and catalytic durability were investigated.

2. Results and Discussion

2.1. Effect of Pretreatment on ABS Substrate

To examine the effect of etching on the surface morphology of the ABS substrate, etched ABS surfaces not containing Pt particles were observed using a scanning electron microscope (SEM). Figure 1 shows the SEM images of the surface of the ABS substrate before and after etching (the samples are hereafter labeled with their pretreatment and whether they contain Pt). Although no holes were present before etching, as shown in Figure 1a, many holes with a diameter of 100–500 nm appeared after etching, as shown in Figure 1b. The etching process was confirmed to dissolve butadiene rubber, thereby increasing the surface area of the ABS substrate.

To examine the effects of etching and surface charge control on the chemical composition of an ABS substrate, the chemical compositions of the pretreated ABS surfaces not containing Pt particles were investigated using X-ray photoelectron spectroscopy (XPS). Figure 2a,b shows the C1s-XPS spectra of the surface of the ABS substrate before and after etching. When the ABS substrate was etched, the intensity of the peak indexed to C–H and C–C (285 eV) decreased whereas the intensity of the peaks indexed to C=O–O (289 eV), C=O (287.5 eV), C–N and C–O (286.5 eV) increased. These results indicate that etching not only dissolves butadiene rubber, but also introduces oxygen-containing functional groups. Figure 2a,c shows the C1s-XPS spectra of the surface of the ABS substrate before and after the surface charge control treatment. When the surface charge of the ABS substrate was controlled, the intensity of the peak indexed to C–H and C–C (285 eV) decreased, whereas that of the peaks indexed to C–N and C–O (286.5 eV) increased, indicating that the ABS surface was covered with surface charge controllers. When both etching and surface charge control were performed, the C1s-XPS spectrum for ABS-Etch&Charge was shaped similarly to that for ABS-Charge, as shown in Figure 2c,d, respectively. This similarity indicates that the ABS-Etch&Charge substrate was also covered with surface charge controllers.

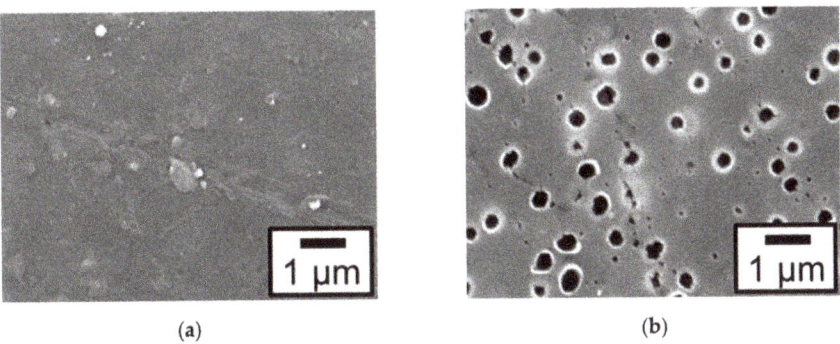

Figure 1. Scanning electron microscope (SEM) images of the surface of the acrylonitrile–butadiene–styrene copolymer (ABS) substrate not containing Pt particles before and after etching: (a) ABS-untreated and (b) ABS-Etch.

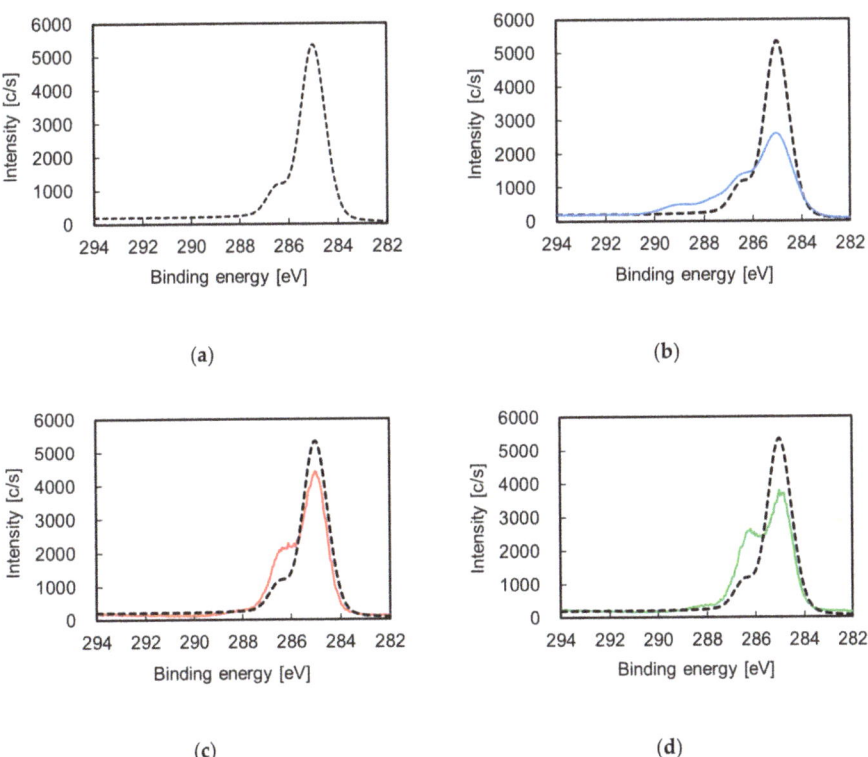

Figure 2. C1s-X-ray photoelectron spectroscopy (XPS) spectra of the ABS surface before (dotted lines) and after surface (solid colored lines): (a) ABS-untreated, (b) ABS-Etch, (c) ABS-Charge, and (d) ABS-Etch&Charge.

The effects of etching and surface charge control on the deposition behavior of Pt particles on four types of the ABS samples were examined. Figure 3 shows the field-emission (FE) SEM images of the surface morphology for the four types of Pt/ABS samples: Pt/ABS-untreated, Pt/ABS-Etch, Pt/ABS-Charge, and Pt/ABS-Etch&Charge. Main Pt particles with diameter of <20 nm and partial Pt particles with a diameter of 20–60 nm were observed on a surface of all four types of Pt/ABS samples. The size of the Pt particles was almost the same whether the surface was etched or its surface charge was controlled.

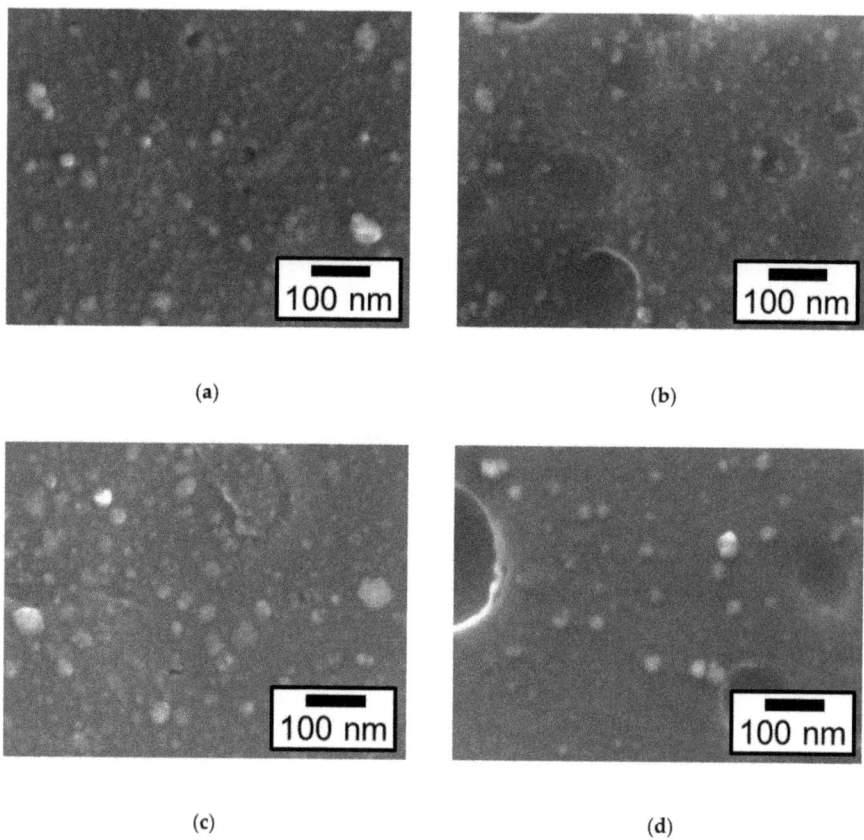

Figure 3. Field-emission (FE) SEM images of the surface morphology of four types of Pt/ABS samples: (a) Pt/ABS-untreated, (b) Pt/ABS-Etch, (c) Pt/ABS-Charge, and (d) Pt/ABS-Etch&Charge.

Cross-sectional FE-SEM images confirmed where Pt particles were deposited on etched ABS samples. Figure 4 shows the cross-sectional backscattered electron images of the Pt/ABS-Etch and Pt/ABS-Etch&Charge samples. The Pt particles were observed both on the ABS surface and in the holes opened by etching. These results indicated that etching increased not only the specific surface area of ABS, but also the number of sites for Pt deposition. In addition, the deposition behavior of Pt particles in the holes was confirmed to be almost the same as that of Pt particles on the ABS surface.

The effects of etching and surface charge control on the Pt loading weight of Pt/ABS samples were also examined. Figure 5 shows the Pt loading weights of the four types of Pt/ABS samples. The Pt loading weights for the Pt/ABS-Etch, Pt/ABS-Charge, and Pt/ABS-Etch&Charge samples were higher than that for the Pt/ABS-untreated sample. This result indicated that both types of pretreatments and their combination increased the Pt loading weight. In addition, the Pt loading weights for the

Pt/ABS-Etch and Pt/ABS-Etch&Charge samples were higher than that for the Pt/ABS-Charge sample. When an ABS substrate is etched, the butadiene rubber component dissolves, which results in a larger surface area. This larger surface area would increase the Pt loading weight because of the increase in sites for the immobilization of Pt nanoparticles. The Pt loading weight per unit area of the ABS substrate covered with an electroless-plated Pt film (Pt-film/ABS), which was calculated in the previous report [5], and the maximum Pt loading weight per unit area of Pt/ABS samples were 2240 and 18.2 ng/mm^2, respectively. Thus, the amount of Pt consumed for the Pt/ABS samples prepared in this study was at least 120 times less than that for the Pt-film/ABS.

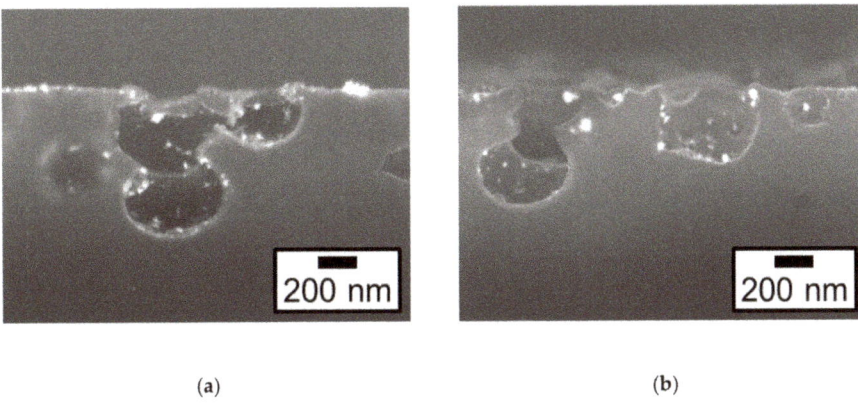

Figure 4. Cross-sectional backscattered electron images for etched samples: (a) Pt/ABS-Etch and (b) Pt/ABS-Etch&Charge.

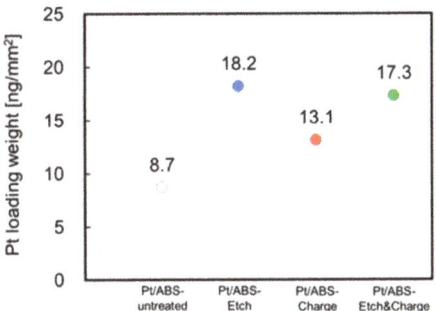

Figure 5. Pt loading weights of four types of Pt/ABS samples prepared using an electron-beam irradiation reduction method (EBIRM): Pt/ABS-untreated, Pt/ABS-Etch, Pt/ABS-Charge, and Pt/ABS-Etch&Charge.

2.2. Catalytic Activity for H_2O_2 Decomposition

To evaluate the catalytic activity for H_2O_2 decomposition, the residual H_2O_2 concentration was measured after employing the four types of Pt/ABS samples. Thus, the untreated ABS, Pt/ABS-untreated, Pt/ABS-Etch, Pt/ABS-Charge, and Pt/ABS-Etch&Charge samples were immersed in a 35,000 ppm H_2O_2 solution for 360 min. The catalytic activities of the four types of Pt/ABS samples are compared in Figure 6. The untreated ABS sample did not decompose H_2O_2 at all within 360 min, whereas all the Pt-supported ABS samples significantly decreased the residual H_2O_2 concentration from 35,000 to less than 400 ppm. Moreover, the residual H_2O_2 concentrations for the Pt/ABS-Etch, Pt/ABS-Charge, and Pt/ABS-Etch&Charge samples became lower than that of the Pt/ABS-untreated sample. Therefore, the two types of pretreatments and their combination improved the catalytic activity for H_2O_2 decomposition. The difference in catalytic activity could be explained by the increase in Pt loading weight. The Pt/ABS-Etch&Charge sample exhibited the highest catalytic activity for H_2O_2 decomposition, successfully reaching the target value of 100 ppm.

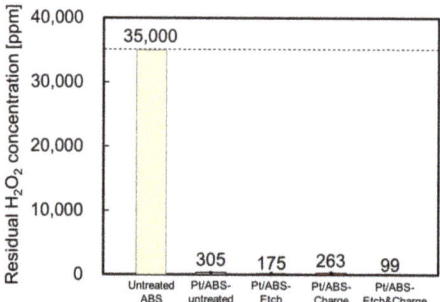

Figure 6. Catalytic activity of the untreated ABS and Pt/ABS samples with various pretreatment: residual H_2O_2 concentration after immersion for 360 min.

2.3. Catalytic Durability in H_2O_2 Decomposition

To examine the effects of etching and surface charge control on the catalytic durability, the relation between the number of repeated uses and residual H_2O_2 concentration was examined. Figure 7 shows the catalytic durability of the four types of Pt/ABS samples. After the Pt/ABS-untreated catalyst was used 10 times, the residual H_2O_2 concentration was 3056 ppm. This result suggests that much of the Pt remained on Pt/ABS-untreated, and more than 90% of H_2O_2 was decomposed after using it 10 times. However, this residual H_2O_2 concentration increased from 305 to 3056 ppm after repeated usage, thus demonstrating insufficient durability. For the pretreated Pt/ABS samples, the residual H_2O_2 concentrations after using Pt/ABS-Etch, Pt/ABS-Charge, and Pt/ABS-Etch&Charge samples 10 times were 851, 713, and 479 ppm, respectively. Thus, the residual H_2O_2 concentration decreased in the order: Pt/ABS-untreated > Pt/ABS-Etch > Pt/ABS-Charge > Pt/ABS-Etch&Charge. This result indicates that both etching and surface charge control effectively improved the catalytic activity for H_2O_2 decomposition. Moreover, the combination of etching and surface charge control was the most effective. However, the residual H_2O_2 concentration for Pt/ABS-Etch&Charge gradually increased with the number of repeated uses, suggesting that the catalytic durability was insufficient for use in practical applications, although the catalytic durability steadily improved upon etching and surface charge control. Therefore, the desorption of Pt nanoparticles must be further prevented.

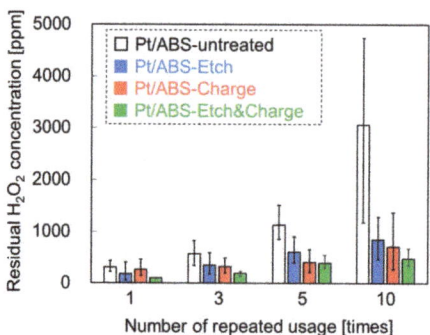

Figure 7. Catalytic durability of the Pt/ABS samples: relation between the number of repeated uses and the residual H_2O_2 concentration.

3. Materials and Methods

3.1. Pretreatment and Synthesis of Pt-Particle/ABS

Pt nanoparticles were immobilized on an ABS substrate using EBIRM according to previous studies [16,20]. The methods for washing the ABS substrate and the radiolytic synthesis of Pt-particle/ABS samples were the same as those reported in the previous article [5]. The main difference between this and previous reports was the use of pretreatments to improve the catalytic durability for H_2O_2 decomposition.

A commercially available 1-mm-thick ABS sheet (2-9229-01, AS-ONE, Nishi-ku, Osaka, Japan) with dimensions of 20 mm × 15 mm × 1 mm was used as the ABS substrate. First, ABS substrates were sequentially washed with ethanol (99.5%, Kishida Chemical, Chuo-ku, Osaka, Japan) and pure water for 10 min each using an ultrasonic cleaner (USK-1R, AS-ONE). Then, they were dried using an N_2 gun (99.99%, Iwatani Fine Gas, Amagasaki, Hyogo, Japan). Prior to immobilizing the Pt nanoparticles, the washed substrates were pretreated via either etching, surface charge control, or both. Table 1 shows the sample conditions and IDs.

Table 1. Sample condition and IDs.

Sample ID	Etching	Surface Charge Modification	EBIRM (EB Irradiation)
ABS-untreated	—	—	—
ABS-Etch	○	—	—
ABS-Charge	—	○	—
ABS-Etch&Charge	○	○	—
Pt/ABS-untreated	—	—	○
Pt/ABS-Etch	○	—	○
Pt/ABS-Charge	—	○	○
Pt/ABS-Etch&Charge	○	○	○

A potassium permanganate solution ($KMnO_4$, 0.2 M, Fujifilm Wako Pure Chemical, Chuo-ku, Osaka, Japan) and concentrated sulfuric acid (H_2SO_4; 97%, Kishida Chemical) were used for etching. An etching solution of molar ratio $KMnO_4/H_2SO_4 = 0.16/3.6$ prepared according to patent [21] was used to dissolve butadiene rubber on the ABS surface. The ABS substrates were immersed in this etching solution for 20 min at room temperature. Then, the etched ABS substrates were washed with pure water for 10 min using an ultrasonic cleaner, followed by drying with an N_2 gun.

Hexadecyltrimethylammonium chloride (Condiriser FR Conc, Okuno Chemical Industries, Chuo-ku, Osaka, Japan) was used to control surface charge, as shown in Figure 8. The hydrocarbon part adsorbs onto the ABS surface, whereas the ammonium end group modifies the surface charge of the ABS surface to be positive. Hexachloroplatinic acid hexahydrate ($H_2PtCl_6 \cdot 6H_2O$; 98.5%, Wako Pure Chemical Industries) was used as the Pt precursor. $H_2PtCl_6 \cdot 6H_2O$ becomes $PtCl_6^{2-}$ in aqueous solution—that is, it becomes negatively charged—which is why Condiriser FR was selected. A 5% v/v solution of Condiriser FR was prepared, and ABS substrates were immersed in this surface charge controller solution for 5 min at 40 °C while stirring at 300 rpm using a magnetic hot stirrer (RCT Basic, IKA, Staufen, Baden-Württemberg, Germany) and a PTFE stirring bar. Excess surface charge controllers were washed away from the ABS substrates with pure water for 20 s. Then, the substrates were dried naturally at room temperature.

To immobilize the Pt nanoparticles, 4 mM solutions of H_2PtCl_6 were separately prepared in cylindrical polystyrene (PS) containers (diameter = 33 mm and height = 16 mm), and 2-propanol (IPA; 99.7%, Kishida Chemical) was added to the solutions to be controlled at 1% v/v. Then, the pretreated ABS substrates were immersed in the Pt precursor solutions. These PS containers with the Pt precursor solutions and the pretreated ABS substrates were then irradiated for 7 s with a high-energy EB of 4.8 MeV using the Dynamitron® accelerator at SHI-ATEX Co. Ltd., in Osaka, Japan. After EB irradiation, the substrates were removed from the solution and were washed with pure water using an ultrasonic cleaner for 10 min to remove the unsupported Pt nanoparticles. Finally, they were dried using the N$_2$ gun. Figure 9 schematically shows the entire process.

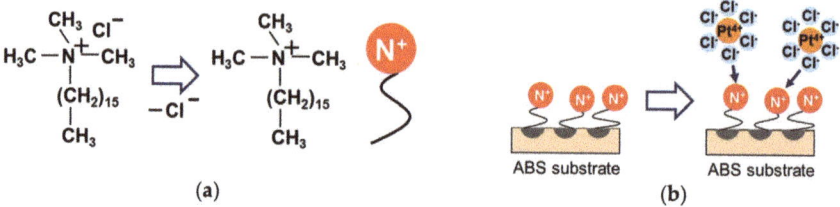

Figure 8. (a) Chemical formula of the surface charge controller (hexadecyltrimethylammonium chloride) and (b) schematic of electrostatic interactions between the surface charge controllers and Pt precursors (PtCl$_6^{2-}$).

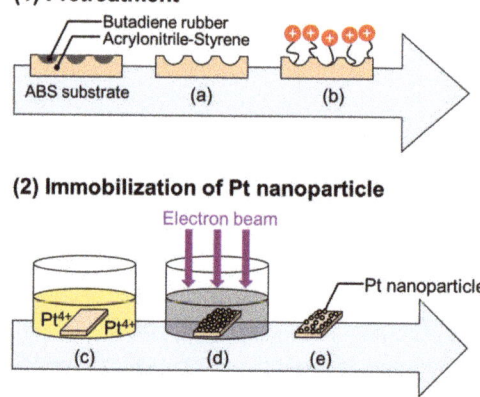

Figure 9. Schematic of the processes for pretreating and preparing a Pt-particle/ABS-Etch&Charge sample using an EBIRM through pretreatment: (a) KMnO$_4$/H$_2$SO$_4$ etching to dissolve butadiene rubber on the ABS surface; (b) surface charge control; (c) immersion of the pretreated ABS substrate in the Pt precursor solution; (d) irradiation with an electron beam; and (e) removing it from the solution, washing using an ultrasonic cleaner, and drying by blowing with N$_2$ gas.

3.2. Characterization

To confirm that butadiene rubber dissolved from the ABS surface, the ABS surface was observed before and after etching using a SEM (JCM-6000, JEOL, Akishima, Tokyo, Japan) at an accelerating voltage of 10 kV. Prior to observation, a thin layer of Au was sputtered on the ABS surfaces using a Smart Coat DII-29010SCTR (JEOL) to prevent from electrostatic charge buildup during SEM observation.

To investigate the effects of etching and surface charge control on the chemical composition of the ABS surface, XPS measurement was performed using a Quantum 2000 (Ulvac-Phi, Chigasaki, Kanagawa, Japan) attached to an Al-Kα source at 15 kV. The area of X-ray irradiation was ⌀100 µm, the pass energy was 23.50 eV, and the step size was 0.05 eV. The XPS spectra were recorded at take-off angles of 45°. A low-speed EB and an Ar ion beam were irradiated on the measured samples during the XPS measurement to neutralize their charges.

To investigate the deposition behavior of the Pt particles on the ABS surface, the ABS surfaces of the four types of Pt/ABS samples were observed using a FE-SEM (JSM-7800F, JEOL) at an accelerating voltage of 8 kV. Prior to observation, Os was coated on the Pt/ABS surfaces via plasma chemical vapor deposition using an Osmium Coater HPC-20 (VACUUM DEVICE, Mito, Ibaragi, Japan) to prevent electrostatic charge buildup during FE-SEM observation. Cross-sectional samples of Pt/ABS-Etch and Pt/ABS-Etch&Charge were prepared using a cross section polisher (IB-09020CP, JEOL) with a broad Ar ion beam source. Cross-sectional backscattered electron images were also obtained using the same FE-SEM (JSM-7800F, JEOL).

To measure the Pt loading weight on the Pt/ABS samples, inductively coupled plasma atomic emission spectrometry (ICP-AES; ICPE-9000, Shimadzu, Kyoto, Kyoto, Japan) was utilized. Pt nanoparticles on the Pt/ABS substrates were dissolved using aqua regia (volume ratio of $HCl/HNO_3 = 3/1$). Then, the diluted aqua regia solutions were sprayed into a plasma torch in the ICPE-9000. The amount of Pt in the Pt/ABS samples was calculated from a calibration curve of a Pt standard solution (1000 ppm, Wako Pure Chemical Industries), as shown in the Supporting Information of the previous report [5].

H_2O_2 decomposition is accelerated by the platinum catalysts, then water and oxygen gas are generated, as shown in Equation (1):

$$2H_2O_2 \rightarrow 2H_2O + O_2 \tag{1}$$

The generation of O_2 bubbles were observed during the H_2O_2 decomposition test in the present study as well as the previous study [5]. It was clear that H_2O_2 decomposition was accelerated by the Pt/ABS samples. To evaluate the catalytic activity for H_2O_2 decomposition, the residual H_2O_2 concentration was measured after immersing the Pt/ABS samples in a 35,000 ppm diluted solution of H_2O_2 (30% w/w, Kishida Chemical) at 25 °C for 360 min in an incubator (i-CUBE FCI-280, AS-ONE). In the previous study, a H_2O_2 decomposition curve was obtained by collecting the data of the residual H_2O_2 concentration with different H_2O_2 decomposition times of 2, 5, 10, 20, 30, 60, 120, 240, and 360 min, and it was confirmed that the residual H_2O_2 concentration steadily decreased with increasing H_2O_2 decomposition times [5]. Therefore, in the present study, the residual H_2O_2 concentration was measured after immersion of the Pt/ABS samples for only 360 min. The method for measuring the H_2O_2 concentration was the same that reported in the previous article [5]. A 5% w/w diluted solution of titanium sulfate (Ti$(SO_4)_2$; 30% w/w, Wako Pure Chemical Industries) was added to the H_2O_2 solution to color the H_2O_2 solution. Then, the optical absorbance of the colored H_2O_2 solution was measured using a deuterium-halogen and tungsten lamp (DH-2000, Ocean Optics, Largo, FL, USA), fiber multichannel spectrometer (HR-4000, Ocean Optics), and optical fiber (P600-1-UV/VIS, Ocean Optics). The absorbance at 407 nm was used to calculate the residual H_2O_2 concentration from the calibration curve, as shown in the Supporting Information of the previous report [5].

To evaluate the catalytic durability for H_2O_2 decomposition, the residual H_2O_2 concentration was measured after Pt/ABS samples were repetitively used 1, 3, 5, and 10 times.

4. Conclusions

We prepared four types of Pt/ABS catalysts with EBIRM and investigated the effects of two types of pretreatments—etching, surface charge control, and the combination of both—on the ABS surface, catalytic activity, Pt loading weight, and durability of these catalysts. Etching increased the Pt loading weight because of the increase in surface area of the ABS substrate, which in turn increased the catalytic activity. Etching also increased the catalytic durability, which could be attributed to the holes created by etching, which partially prevented Pt particles from detaching from the ABS surface. Surface charge control increased the Pt loading weight, which increased both the catalytic activity and durability. These improvements could be explained by electrostatic interactions between the Pt nanoparticles and surface charge controllers on the ABS substrate. The effects of etching on the Pt loading weight and catalytic activity were larger than those of the surface charge control. In contrast, the effect of surface charge control on the catalytic durability was higher than that of etching. Finally, the combination of etching and surface charge control most effectively improved both the catalytic activity and durability. Thus, we successfully improved the catalytic durability through either etching, surface charge control, or both before EB irradiation. Although the catalytic durability was insufficient for cleaning contact lenses in practical applications, these pretreatments would be useful for improving the adhesion between metal nanoparticles and resin substrates or microparticles except in severe conditions such as in H_2O_2 solution.

Author Contributions: Y.O., K.E. and K.Y. supervised the work. T.A. and S.S. prepared the Pt-particle/ABS samples. Y.O. and T.A. performed SEM observation, AFM observation, and XPS analysis. Y.O., T.A. and S.S. measured and calculated the Pt loading weights of the samples using ICP-AES. T.A. evaluated the catalytic activity and catalytic durability. O.M. and I.I. helped the evaluations. All authors contributed to the scientific discussion and manuscript preparation. Y.O. wrote the manuscript.

Funding: The research was funded by the Japan Science and Technology Agency, grant number JST No. MP27215667957 and JST No. VP29117941540.

Acknowledgments: We thank the staff of the SHI-ATEX Co. Ltd. for their assistance with the electron-beam irradiation experiments. We thank the staff of the Okuno Chemical Industries because hexadecyltrimethylammonium chloride (Condiriser FR Conc) was provided by them. We also thank the staff of the JEOL for their assistance with the SEM observations.

Conflicts of Interest: The authors declare no conflict of interest.

References

1. Muntz, A.; Subbaraman, L.N.; Sorbara, L.; Jones, L. Tear exchange and contact lenses: A review. *J. Optom.* **2015**, *8*, 2–11. [CrossRef] [PubMed]
2. Musgrave, C.S.A.; Fang, F. Contact lens materials: A materials science perspective. *Materials* **2019**, *12*, 261. [CrossRef] [PubMed]
3. Nichols, J.J. Continuing upward trends in daily disposable prescribing and other key segments maintained a healthy industry. *Contact Lens Spectr.* **2018**, *33*, 20–25.
4. Paugh, J.R.; Brennan, N.A.; Efron, N. Ocular response to hydrogen peroxide. *Am. J. Optom. Physiol. Opt.* **1988**, *65*, 91–98. [CrossRef] [PubMed]
5. Ohkubo, Y.; Aoki, T.; Seino, S.; Mori, O.; Ito, I.; Endo, K.; Yamamura, K. Radiolytic synthesis of Pt-particle/ABS catalysts for H_2O_2 decomposition in contact lens cleaning. *Nanomaterials* **2017**, *7*, 235. [CrossRef] [PubMed]
6. Liao, P.C.; Carberry, J.J.; Fleisch, T.H.; Wolf, E.E. CO oxidation activity and XPS studies of PtCu γ-Al_2O_3 bimetallic catalysts. *J. Catal.* **1982**, *74*, 307–316. [CrossRef]
7. Mohamed, R.M. Characterization and catalytic properties of nano-sized Pt metal catalyst on TiO_2-SiO_2 synthesized by photo-assisted deposition and impregnation methods. *J. Mater. Process. Technol.* **2009**, *209*, 577–583. [CrossRef]
8. Rahsepar, M.; Pakshir, M.; Piao, Y.; Kim, H. Preparation of highly active 40 wt % Pt on multiwalled carbonnanotube by improved impregnation method for fuel cell applications. *Fuel Cells* **2012**, *12*, 827–834. [CrossRef]
9. Toshima, N.; Wang, Y. Preparation and catalysis of novel colloidal dispersions of copper/noble metal bimetallic clusters. *Langmuir* **1994**, *10*, 4574–4580. [CrossRef]
10. Toshima, N.; Wang, Y. Polymer-protected Cu/Pd bimetallic clusters. *Adv. Mater.* **1994**, *6*, 245–247. [CrossRef]
11. Daimon, H.; Kurobe, Y. Size reduction of PtRu catalyst particle deposited on carbon support by addition of non-metallic elements. *Catal. Today* **2006**, *111*, 182–187. [CrossRef]
12. Mizukoshi, Y.; Oshima, R.; Maeda, Y.; Nagata, Y. Preparation of platinum nanoparticles by sonochemical reduction of the Pt(II) ion. *Langmuir* **1999**, *15*, 2733–2737. [CrossRef]
13. Nitani, H.; Yuya, M.; Ono, T.; Nakagawa, T.; Seino, S.; Okitsu, K.; Mizukoshi, Y.; Emura, S.; Yamamoto, T.A. Sonochemically synthesized core-shell structured Au-Pd nanoparticles supported on γ-Fe_2O_3 particles. *J. Nanopart. Res.* **2006**, *8*, 951–958. [CrossRef]
14. Mizukoshi, Y.; Tsuru, Y.; Tominaga, A.; Seino, S.; Masahashi, N.; Tanabe, S.; Yamamoto, T.A. Sonochemical immobilization of noble metal nanoparticles on the surface of maghemite: Mechanism and morphological control of the products. *Ultrason. Sonochem.* **2008**, *15*, 875–880. [CrossRef] [PubMed]

15. Ziylan-Yavas, A.; Mizukoshi, Y.; Maeda, Y.; Ince, N.H. Supporting of pristine TiO_2 with noble metals to enhance the oxidation and mineralization of paracetamol by sonolysis and sonophotolysis. *Appl. Catal. B Environ.* **2015**, *172–173*, 7–17. [CrossRef]
16. Seino, S.; Kinoshita, T.; Nakagawa, T.; Kojima, T.; Taniguchi, R.; Okuda, S.; Yamamoto, T.A. Radiation induced synthesis of gold/iron-oxide composite nanoparticles using high energy electron beam. *J. Nanopart. Res.* **2008**, *10*, 1071–1076. [CrossRef]
17. Kageyama, S.; Sugano, Y.; Hamaguchi, Y.; Kugai, J.; Ohkubo, Y.; Seino, S.; Nakagawa, T.; Ichikawa, S.; Yamamoto, T.A. Pt/TiO_2 composite nanoparticles synthesized by electron beam irradiation for preferential CO oxidation. *Mater. Res. Bull.* **2013**, *48*, 1347–1351. [CrossRef]
18. Ohkubo, Y.; Seino, S.; Kageyama, S.; Kugai, J.; Nakagawa, T.; Ueno, K.; Yamamoto, T.A. Effect of decrease in the size of Pt nanoparticles using sodium phosphinate on electrochemically active surface area. *J. Nanopart. Res.* **2014**, *16*, 2237. [CrossRef]
19. Ohkubo, Y.; Kageyama, S.; Seino, S.; Nakagawa, T.; Kugai, J.; Ueno, K.; Yamamoto, T.A. Mass production of highly loaded and highly dispersed PtRu/C catalysts for methanol oxidation using an electron-beam irradiation reduction method. *J. Exp. Nanosci.* **2016**, *11*, 123–137. [CrossRef]
20. Belloni, J. Nucleation, growth and properties of nanoclusters studied by radiation chemistry: Application to catalysis. *Catal. Today* **2006**, *113*, 141–156. [CrossRef]
21. Kanao, Y. Japan Patent P2010-159457A. Available online: https://www.j-platpat.inpit.go.jp/web/PU/JPA_H22159457/57BCE0F5B3F81A184B6B6C63C93F1815 (accessed on 22 July 2010).

© 2019 by the authors. Licensee MDPI, Basel, Switzerland. This article is an open access article distributed under the terms and conditions of the Creative Commons Attribution (CC BY) license (http://creativecommons.org/licenses/by/4.0/).

Article

Effect of SiO₂ Nanoparticles on the Performance of PVdF-HFP/Ionic Liquid Separator for Lithium-Ion Batteries

Stefano Caimi, Antoine Klaue, Hua Wu * and Massimo Morbidelli *

Department of Chemistry and Applied Biosciences, Institute for Chemical and Bioengineering, ETH Zurich, 8093 Zurich, Switzerland; stefano.caimi@chem.ethz.ch (S.C.); antoine.klaue@chem.ethz.ch (A.K.)
* Correspondence: hua.wu@chem.ethz.ch (H.W.); massimo.morbidelli@chem.ethz.ch (M.M.); Tel.: +41-446-320-635 (H.W.); +41-446-323-034 (M.M.)

Received: 10 October 2018; Accepted: 5 November 2018; Published: 8 November 2018

Abstract: Safety concerns related to the use of potentially explosive, liquid organic electrolytes in commercial high-power lithium-ion batteries are constantly rising. One promising alternative is to use thermally stable ionic liquids (ILs) as conductive media, which are however, limited by low ionic conductivity at room temperature. This can be improved by adding fillers, such as silica or alumina nanoparticles (NPs), in the polymer matrix that hosts the IL. To maximize the effect of such NPs, they have to be uniformly dispersed in the matrix while keeping their size as small as possible. In this work, starting from a water dispersion of silica NPs, we present a novel method to incorporate silica NPs at the nanoscale level (<200 nm) into PVdF-HFP polymer clusters, which are then blended with the IL solution and hot-pressed to form separators suitable for battery applications. The effect of different amounts of silica in the polymer matrix on the ionic conductivity and cyclability of the separator is investigated. A membrane containing 10 wt.% of silica (with respect to the polymer) was shown to maximize the performance of the separator, with a room temperature ionic conductivity of of 1.22 mS cm^{-1}. The assembled half-coin cell with LiFePO₄ and Li as the cathode and the anode exhibited a capacity retention of more than 80% at a current density of 2C and 60 °C.

Keywords: lithium-ion battery; ionic-liquid-based separator; hot-pressing; inorganic nanoparticle; nanocomposite; fractal cluster

---

1. Introduction

Recent large-scale power applications of lithium-ion batteries (LIB) including hybrid vehicles and smart grids require long cycle life, low impact on the environment and high reliability and safety [1–6]. Current technologies are based on the use of organic liquid electrolytes, which guarantee high ionic conductivity at low temperatures and long cycle stability. However, they are also highly volatile, toxic and thermally unstable and may leak out of the battery under abnormal operations [7–14]. One of the most promising alternatives to replace liquid electrolytes is the employment of ionic liquids (ILs), which possess high ion density and are characterized by high thermal stability [6,15–24]. Among the several existing ILs, those based on pyrrolidinium, and in particular Pyr13TFSI, are often considered since they exhibit low viscosities and are chemically inert towards the cell components [10,19,25,26]. The main drawback of the use of ILs is the limited ionic conductivity at low temperature, especially when mixed with the host polymer to form the separator. One possibility to increase the ionic conductivity at low temperatures is the addition of inorganic nanoparticles (NPs) such as SiO₂, Al₂O₃, TiO₂ and CeO₂, to the polymer matrix [27–31]. These fillers can improve the conductivity by reducing the polymer crystallinity and by interacting with the ionic species in the electrolyte through Lewis acid–base interactions [29,32–35]. Moreover, it is well established that the addition of inorganic NPs into the polymer matrix improves its mechanical stability, thus preventing thermal shrinkage and mechanical breakdown of the separator [35–37]. In order to maximize the effect of the added NPs, it is essential to disperse them in the polymer matrix uniformly and at the nanoscale level [38,39]. To achieve this, in this work, we start from a dispersion of silica NPs in water and mix it with an aqueous dispersion of PVdF-HFP NPs. The binary dispersion is then subjected to shear-driven gelation by passing through a microchannel where, if present alone, the polymer NPs undergo gelation whereas the silica NPs are shear-inactive (i.e., they are stable and do not aggregate). As the process occurs in few milliseconds the silica NPs cannot escape from the polymer gel network and remain entrapped and dispersed uniformly in the polymer matrix [40–42]. Poly(vinylidenefluoride-co-hexafluoropropylene) (PVdF-HFP) is chosen as it possesses high dielectric constant, it is chemically compatible with the electrode materials, and it is characterized by low crystallinity [27,28,39,43]. The method to form a freestanding, uniform and transparent membrane through hot-pressing, starting from the polymer/filler clusters and the IL has been developed earlier in our group and described elsewhere [44]. The effect of the presence of silica NPs in the IL-based membrane is investigated in terms of ionic conductivity and electrochemical cyclability.

2. Materials and Methods

2.1. Materials

The following chemicals have been employed without further treatments: Sodium dodecyl sulphate (SDS, purity 99%) and N-Propyl-N-Methylpyrrolidinium bis(trifluoromethane-sulfonyl)-imide (Pyr1308b, purity 99.5%) were purchased from Apollo Scientific (Bredbury, UK) and Solvionic (Toulouse, France), respectively. The water dispersion of PVdF-HFP NPs, the amorphous silica powder Tixosil 365 and bis(trifluoro methane)sulfonamide lithium salt (LiTFSI) were provided by Solvay (Bollate, Italy). The ion-exchange resin (Dowex MR-3) was purchased from Sigma-Aldrich (Steinheim, Germany). The cathode material lithium iron phosphate (LFP) is commercial grade Life Power P2 from Clariant (Muttenz, Switzerland). Before assembling, LFP and the IL-based separator were dried overnight under vacuum at 130 °C and 60 °C, respectively.

2.2. Methods

To form a suitable dispersion, Millipore water is added to the silica powder to reach a solid fraction of 30% and the mixture is mechanically stirred and repeatedly sonicated using a digital sonifier from Branson. Eventually, the dispersion is centrifuged at 2000 rpm for 10 min to remove the remaining large clusters. The binary dispersion of PVdF-HFP and SiO_2 NPs is prepared by adding the silica NP dispersion dropwise to the PVdF-HFP NP dispersion under agitation. As the addition of SiO_2 NPs may destabilize the PVdF-HFP dispersion due to a repartition of the adsorbed surfactant between the two species, some additional SDS is added (0.2% with respect to the solid content of the dispersion). The content of the added silica is evaluated as mass percentage with respect to the mass of polymer.

A high-shear device, HC-5000 (Microfluidics, Westwood, MA, USA), connected to a L30Z microchannel with a width of 300 μm, a rectangular cross section of $5.26 \cdot 10^{-8}$ m^2 and a length of 5.8 mm was used to perform shear-driven gelation of the PVdF-HFP/SiO_2 NPs. The operating conditions for the gelation and subsequent drying to obtain the polymer-silica clusters (PSiCs) are reported elsewhere [44].

The IL solution consisted of a 0.5 M solution of LiTFSI salt dissolved into Pyr13TFSI.

The obtained clusters were mixed with the IL solution at a mass fraction of PSiC/IL equal to 30/70 wt.%, following the procedure previously reported [44]. The slurry was then transferred between two aluminum sheets in a preheated hydraulic hand-press (Rondol, Strasbourg, France) and hot-pressed at 120 °C and 10 kN to form the separator. The cooling phase was performed while holding the pressure. The formed separator is here referred to as the PSiCIL membrane.

Small-angle light scattering (SALS) measurements were taken using a Mastersizer 2000 (Malvern, UK) equipped with a laser having a wavelength λ = 633 nm to characterize the obtained clusters in terms of their average radius of gyration, $\langle R_g \rangle$, and their fractal dimension, d_f, following the procedure reported elsewhere [45,46]. Measures of dynamic light scattering and zeta potential were conducted using Zetasizer Nano ZS 3600 from Malvern Instruments (Malvern, UK).

Differential scanning calorimetry (DSC) measurements were conducted using Q1000 instrument (TA Instruments, New Castle, DE, USA) using 40 µL crucibles in aluminum and a heating and cooling rate of 5 °C min^{-1} in a nitrogen atmosphere in the temperature range from 80 to 200 °C. The solid content is measured using a HG53 Halogen Moisture Analyzer (Mettler-Toledo, Columbus, OH, USA). Powder XRD measurements were carried out with a X'Pert PRO-MPD diffractometer (Malvern PANalytical, Malvern, UK). Data were recorded in the 5–70° 2θ range with an angular step size of 0.05° and a counting time of 0.26 s per step. The peaks at 2θ = 18.2, 20.0, 26.6 and 38.8 correspond to the (100), (020), (110) and (021) crystalline peaks of PVdF-HFP, respectively [47].

AC (alternating current) impedance spectroscopy was used to measure the ionic conductivity of the PSiCIL membranes using a conductivity cell consisting of two stainless steel blocking electrodes. The measurement was carried out under PEIS conditions (impedance under potentiostatic mode), ΔV = 5 mV and frequency range from 300 kHz to 1 Hz. The resistance of the polymer electrolyte was measured and the ionic conductivity (σ) was obtained as follows:

$$\sigma = \frac{d}{R_b S} \tag{1}$$

where d is the thickness of the separator, R_b the bulk resistance and S the area of the stainless steel electrode.

SEM images were taken with a Zeiss Leo 1530 (Zeiss, Oberkochen, Germany) microscopy with a field emission gun of 5 kV and platinum coating. TEM images were performed using a Morgagni 268 from FEI equipped with a tungsten emitter operated at 100 kV.

The battery consisted of CR2032 coin cells assembled using lithium iron phosphate (LFP) and lithium metal as the cathode and the counter electrode, respectively, and the PSiCIL membrane as the separator. The electrochemical tests were performed with a Land CT2001A battery tester and with a mass loading of the active material per cell equal to 4 mg. The battery assembly was performed under argon atmosphere. The cells were cycled in the voltage range 2.5–4 V (vs. Li/Li$^+$) at 60 °C, at current densities from 0.1 to 2C. For charge/discharge performance characterization, 1C is defined as 170 mAh g^{-1}.

3. Results and Discussion

3.1. Preparation of the PSiCIL Separators

The procedure to obtain the silica dispersion is reported in the experimental section. Typical properties of the obtained dispersion of SiO$_2$ NPs in water are summarized in Table 1.

Table 1. Properties of the dispersion of SiO$_2$ NPs in water.

	Average Clusters Diameter [nm]	PDI [—]	Zeta Potential [mV]
Tixosil 365	160	0.19	−47.9

For the application at hand, the silica NPs have to be well-dispersed in water and should be negatively charged to maintain their stability while mixed with the negatively charged PVdF-HFP NPs. As reported in Table 1, these requirements are met by the silica NPs, showing an average diameter smaller than 200 nm and a negative potential larger than 40 mV (absolute value). In order to investigate the effect of pH, Figure 1a reports the average diameter and the zeta potential of the silica NPs in the pH range 6–12. To better appreciate the morphology of the silica NPs, Figure 1b reports a TEM image of the dried silica dispersion. From Figure 1a, it is seen that the dispersed SiO_2 NPs have a constant average diameter smaller than 200 nm and are negatively charged with the zeta potential ranging from −48 to −42 mV. Moreover, from Figure 1b, it is possible to recognize that the each silica NP is a nanocluster made of 10 nm silica primary particles.

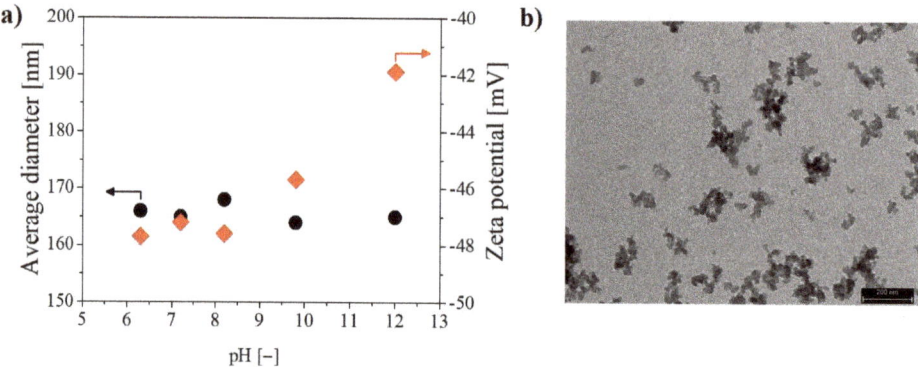

Figure 1. (a) Average diameter and zeta potential of the silica NPs as a function of pH in the silica dispersion. (b) TEM image of the dried silica NPs (scale 200 nm).

The chemical and physical properties of the used PVdF-HFP NP dispersion are reported in the Supplementary Materials (Table S1). The anionic surfactant was extracted from the polymer dispersion by repeated washing with ion-exchange resin Dowex MR-3 to reduce the colloidal stability of the polymer NPs and facilitate the gelation under shear.

The polymer and filler dispersions are mixed as described in the experimental section and the binary system is subjected to intense shear by forcing it to pass through a microchannel so as to have rapid gelation of the polymer NPs with the typical fractal characteristics. Since the formation of the polymer gel network occurs in few milliseconds and the silica NPs are shear-inactive (i.e., they do not aggregate under the given shear rate), the fillers, silica NPs, have no time to escape and remains entrapped in the formed matrix at a nanoscale level. In order to have complete capture of the silica nanoparticles, it is of utmost importance to obtain a compact gel after a single passage through the microchannel, as discussed elsewhere [40,48]. This depends on the solid content of the polymer/silica dispersion: the higher the solid content, the greater the compactness of the formed gel. In order to analyze the distribution of the silica NPs inside the polymer matrix, SEM pictures of the gel obtained from the microchannel are shown in Figure 2.

From Figure 2 it is evident that the shear-induced gelation is capable of dispersing uniformly and randomly the silica NPs as fillers within the polymer matrix, where the silica NPs with respect to the PVdF-HFP NPs are clearly and easily distinguishable (whiter and smaller ones, encircled in violet in Figure 2). It is worth noticing that the fillers are uniformly dispersed at the nanoscale level (i.e., the silica NPs are smaller than 200 nm). Moreover, it is possible to observe that also the PVdF-HFP NPs maintain their original identity. In order to investigate the physical properties of the pure polymer and of the composite material, DSC is performed. Figure 3 reports the results of the heating and cooling ramps of the dried gel at different contents of SiO_2. The melting and crystallization temperatures as well as the crystallinity of the composite derived from the DSC heating and cooling curves are reported in Table 2.

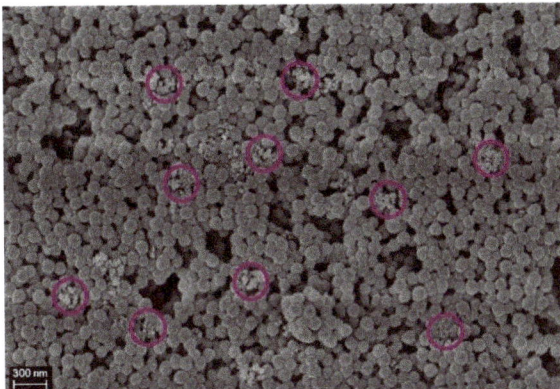

Figure 2. SEM picture of the silica NPs entrapped in the polymer matrix after a single passage through the microchannel (scale 300 nm). Some silica NPs are encircled in violet to be better visualized.

Table 2. Properties of the composite material at increasing amount of SiO_2, derived from the DSC curves in Figure 3. T_m: melting temperature. T_c: crystallization temperature. ΔH_m: enthalpy of melting. X_c: crystallinity.

	T_m [°C]	T_c [°C]	ΔH_m [J g^{-1}]	X_c [%]
Pure	133.6	97.6	22.19	21.2
10% SiO_2	132.8	100	18.21	17.4
20% SiO_2	130.5	99.7	16.92	16.2
30% SiO_2	130.6	99.6	15.66	14.9

The results reported in Table 2 show that the amount of SiO_2 NPs affects only slightly the melting and crystallization temperatures. The melting temperature of the pure polymer PVdF-HFP results in approximately 134 °C as derived from the maximum of the heating curve and progressively decreases with increasing amount of SiO_2 to reach approximately 131 °C at 30 wt.% content of silica. The crystallization temperature, on the other hand, is measured from the minimum of the cooling curve, which is approximately 98 °C for the pure polymer and 100 °C for the composite material, independently of the amount of silica. Furthermore, the enthalpy of melting and the crystallinity significantly decrease with increasing content of SiO_2 NPs. These results are expected and are related to the hindered reorganization of the polymer chains due to the cross-linking centers formed by the interaction of Lewis acid groups with the polar groups (i.e., the -F atoms of the polymer chains). This interaction can stabilize the amorphous structure and facilitates the transport of Li$^+$ ions (i.e., the ionic conductivity), as observed by several authors [29,33,34,49,50].

In order to measure the crystallinity of the polymer and to investigate the effect of the introduction of the SiO_2 NPs on it, XRD measurements are performed and the results are shown in Figure 4.

It is seen that the spectrum of the pure polymer confirms the partial crystallization of PVdF units in the copolymer and gives a semi-crystalline structure of PVdF-HFP [27]. Moreover, the crystallinity of the polymer has been considerably reduced upon the addition of silica NPs. The intensity of the crystalline peaks, indeed, decreases and broadens when increasing the amount of SiO_2. This reduction in crystallinity is attributed to the changes of the chain conformation due to the presence of the silica NPs, which again facilitate higher ionic conduction [51,52].

As described in the experimental section, the dried PSiCs containing different percentages of SiO_2 are mixed with the IL solution at a mass fraction PSiC/IL equal to 30/70 wt.% and left at rest overnight to allow full impregnation of the pores of the PSiCs by the IL solution, before being hot-pressed. After the hot-pressing, it is possible to obtain a freestanding, homogeneous and transparent 50-μm thick IL-based membrane (referred as PSiCIL membrane) containing the silica NPs at a nanoscale level. A SEM picture showing the preservation of the internal morphology after hot-pressing has been reported in our previous work [44]. To prevent water absorption, the PSiCIL membranes are stored in a nitrogen atmosphere.

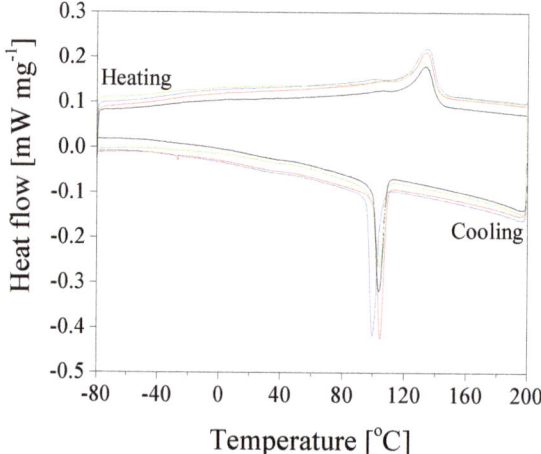

Figure 3. DSC of the composite gel obtained after one passage through the microchannel at different percentages of the fillers, silica NPs. Blue curve: PVdF-HFP; Red curve: PVdF-HFP + 10% SiO_2; Black curve: PVdF-HFP + 20% SiO_2; Green curve: PVdF-HFP + 30% SiO_2.

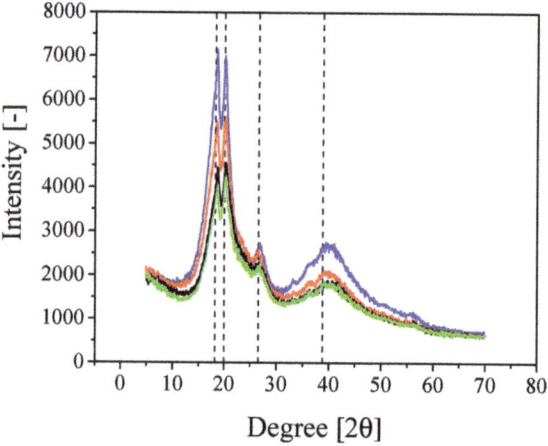

Figure 4. XRD of the composite gel obtained after one passage through the microchannel at different percentages of the fillers, silica NPs. Blue curve: PVdF-HFP; Red curve: PVdF-HFP + 10% SiO_2; Black curve: PVdF-HFP + 20% SiO_2; Green curve: PVdF-HFP + 30% SiO_2.

3.2. Electrochemical Properties of the PSiCIL Separators

The ionic conductivity of the PSiCIL membranes was measured by AC impedance spectroscopy in the temperature range from 25 to 80 °C, at increasing percentages of silica (with respect to the polymer) in the PSiC, and the results are reported in Figure 5 and Table S2 in the Supplementary Materials. As can be seen in Figure 5, the ionic conductivity improves as the temperature increases. This can be understood by taking into account the effect of the temperature on the viscosity of the IL, which decreases as the temperature rises. The introduction of SiO$_2$ NPs within the polymer matrix substantially improves the ionic conductivity of the PSiCIL membrane. In particular, the room temperature conductivity increases from 0.51 mS cm^{-1} for the pure polymer to 1.04, 1.22 and 0.71 mS cm^{-1} with 5%, 10%, and 15% of SiO$_2$, respectively. The same trend is observed in the entire temperature range, where the membrane containing 10% of SiO$_2$ NPs shows the highest values of ionic conductivity reaching 1.77, 2.51, 2.95 and 3.23 mS cm^{-1} at 40, 55, 70 and 80 °C, respectively (the conductivity data for the other samples are summarized in Table S2 in the Supplementary Materials). Interestingly, it is observed that the increase in conductivity is not a linear function of the SiO$_2$ content. The conductivity values corresponding to the SiO$_2$ content of 15% are lower than those at an SiO$_2$ content of 5% and 10%. This behavior has been previously observed in the literature and attributed to the fact that at low filler concentrations the interaction between polymer matrix and SiO$_2$ NPs facilitates the transport of Li$^+$ ions. However, when the SiO$_2$ concentration reaches a certain level, the dilution effect predominates and the ionic conductivity decreases [27,53]. As reported by Stephan et al. [27], the highest conductivity is reached when the filler content ranges from 8 to 10 wt.%. The values of the ionic conductivity obtained in this work are in line with those previously reported in the literature for polymer/IL separators [44,54].

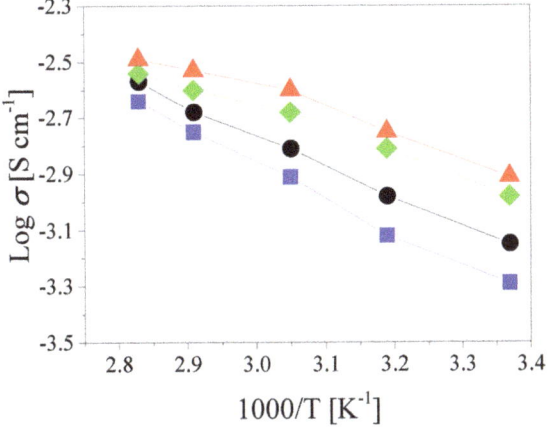

Figure 5. Ionic conductivity at 25, 40, 55, 70 and 80 °C of the PSiCIL membranes containing 70 wt.% of IL and 0% (blue squares) [44], 5% (green diamonds), 10% (red triangles), and 15% (black circles) of SiO$_2$, respectively.

The membrane with an SiO$_2$ content of 10%, showing the highest ionic conductivity, was used as separator to assemble CR2032 coin cells, with LFP and lithium metal as electrodes. The corresponding discharge capacity was measured as a function of the applied current density and compared in Figure 6 with the values obtained using the membrane made of the PVdF-HFP/IL composite without silica [44]. It is seen that the discharge capacities of the two membranes are almost identical up to a current density of 1C, with a capacity retention higher than 90% with respect to the initial cycles at 0.1C. At a current density of 2C, a clear difference between the two membranes is observed, where the normalized discharge capacity is 79.5% and 83.3% for the separators with 0 and 10% of SiO$_2$, respectively. Such a superior high capacity retention can be attributed to the fractal structure of the polymer clusters and to the bicontinuous morphology of the separator. Indeed, the high and well-controlled porosity formed via shear-gelation is preserved during the hot-pressing phase, where the IL solution fully impregnates the pores forming a multitude of channels through which the ions can flow, thus showing limited loss of capacity at high current density. On the other hand, the positive contribution of the silica NPs can be attributed to two factors. Firstly, the silica NPs might hinder the reorganization, even if already limited, of the polymer particles during the membrane formation process, thus leaving more channels open to the ions transfer. Secondly, as discussed in the previous paragraph, the performance at high C-rates are improved because of the same silica/polymer interactions which favor the ionic conductivity. At low current densities this is not observed as the existing channels, given the fractal geometry of the polymer clusters, are well-developed and the ion transfer is not limited by them. It is worth mentioning that after the cycles at 2C, the battery is tested again at 0.2C showing a recovery for the membranes containing 0 and 10% of SiO$_2$ of 97.5% and 99%, respectively, thus showing limited performance loss during the cycles at high current densities. It is also worth pointing out that the cycle efficiency remained close to unity during all cycles. Moreover, the very low dependence of the discharge capacity on the applied current density is not commonly observed when considering earlier literature results [36,52,55].

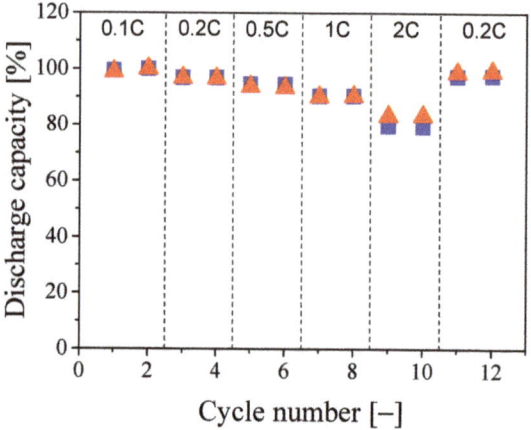

Figure 6. Discharge capacity, normalized with respect to the initial cycle at 0.1C, at 60 °C of the PSiCIL membranes containing 70 wt.% of IL with 0% (blue squares) and 10% (red triangles) of SiO$_2$, respectively.

4. Conclusions

In this work, we have analyzed the effect of dispersing silica NPs into PVdF-HFP/IL membranes on the ionic conductivity and discharge capacity of lithium-ion batteries. In particular, starting from the corresponding powder, we have formed a stable water dispersion of silica NPs, which could be mixed with a PVdF-HFP NP dispersion, to form a binary dispersion which was then subjected to intense shear-driven gelation. As the gelation occurs extremely fast, the silica NPs cannot escape during the gel network formation and remain entrapped and dispersed into the polymer matrix at the nanoscale level. The introduction of silica NPs into the polymer matrix was shown via DSC and XRD to reduce the crystallinity of the polymer, thus stabilizing the amorphous structure and facilitating the transport of Li$^+$ ions.

The so-produced PVdF-HFP-SiO$_2$ composite clusters (PSiCs) were mixed with an IL solution and hot-pressed to form a membrane, so as to analyze the effect of the silica NPs on its electrochemical performance. It was observed that the ionic conductivity increases as the SiO$_2$ content increases. The ionic conductivity reaches a maximum at an SiO$_2$ content of 10%, being 1.22 mS cm^{-1} at room temperature, and then decreases as the SiO$_2$ content further increases. The membrane formed with 10% SiO$_2$ was used to assemble coin cells and tested for cyclability at different C-rates at 60 °C. At low current densities, no significant differences between the membranes with 0% and 10% silica were observed and the measured discharge capacities at 1C were higher than 90% of the ones measured at 0.1C, showing excellent capacity retention even at high current densities. At 2C, the membrane containing 10% silica performed better, showing a discharge capacity of 83.3%, compared to 79.5% of the membrane containing no silica. This can be attributed to the positive effect of the dispersed SiO_2 NPs, which, on one side hinder the reorganization of the polymer NPs, thus reducing the crystallinity and increasing the amorphous phase, and on the other side, favor the transfer of ions because of their interaction with the polymer matrix.

Supplementary Materials: The following are available online at http://www.mdpi.com/2079-4991/8/11/926/s1, Table S1: Properties of the aqueous dispersion of PVdF-HFP NPs, Table S2: Ionic conductivity results.

Author Contributions: Conceptualization, S.C., H.W. and M.M.; Investigation, S.C., A.K.; Writing—Original Drat Preparation, S.C.; Writing—Review & Editing, S.C., A.K., H.W. and M.M.; Supervision, S.C., H.W. and M.M.

Funding: This research was funded by the Swiss National Science Foundation grant number 200020_165917.

Acknowledgments: The authors would like to express their special thanks to Lu Jin for the TEM image, Christine Hamon and Riccardo Pieri for their hints, and to Maurizio Biso for the electrochemical tests. The PVdF-HFP NP dispersion was supplied by Solvay (Italy).

Conflicts of Interest: The authors declare no conflict of interest. The founding sponsors had no role in the design of the study; in the collection, analyses, or interpretation of data; in the writing of the manuscript, or in the decision to publish the results.

References

1. Armand, M.; Tarascon, J.M. Building better batteries. *Nature* **2008**, *451*, 652–657. [CrossRef] [PubMed]
2. Scrosati, B.; Garche, J. Lithium batteries: Status, prospects and future. *J. Power Sources* **2010**, *195*, 2419–2430. [CrossRef]
3. Scrosati, B.; Hassoun, J.; Sun, Y.K. Lithium-ion batteries. A look into the future. *Energy Environ. Sci.* **2011**, *4*, 3287–3295. [CrossRef]
4. Dunn, B.; Kamath, H.; Tarascon, J.M. Electrical energy storage for the grid: A battery of choices. *Science* **2011**, *334*, 928–935. [CrossRef] [PubMed]
5. Thackeray, M.M.; Wolverton, C.; Isaacs, E.D. Electrical energy storage for transportation-approaching the limits of, and going beyond, lithium-ion batteries. *Energy Environ. Sci.* **2012**, *5*, 7854–7863. [CrossRef]
6. Kalhoff, J.; Eshetu, G.G.; Bresser, D.; Passerini, S. Safer Electrolytes for Lithium-Ion Batteries: state of the Art and Perspectives. *ChemSusChem* **2015**, *8*, 2154–2175. [CrossRef] [PubMed]
7. Hammami, A.; Raymond, N.; Armand, M. Lithium-ion batteries: Runaway risk of forming toxic compounds. *Nature* **2003**, *424*, 635–636. [CrossRef] [PubMed]
8. Xu, K. Nonaqueous liquid electrolytes for lithium-based rechargeable batteries. *Chem. Rev.* **2004**, *104*, 4303–4418. [CrossRef] [PubMed]
9. Lu, Y.; Moganty, S.S.; Schaefer, J.L.; Archer, L.A. Ionic liquid-nanoparticle hybrid electrolytes. *J. Mater. Chem.* **2012**, *22*, 4066–4072. [CrossRef]
10. Eshetu, G.G.; Grugeon, S.; Laruelle, S.; Boyanov, S.; Lecocq, A.; Bertrand, J.P.; Marlair, G. In-depth safety-focused analysis of solvents used in electrolytes for large scale lithium-ion batteries. *Phys. Chem. Chem. Phys.* **2013**, *15*, 9145–9155. [CrossRef] [PubMed]
11. Kuo, P.L.; Tsao, C.H.; Hsu, C.H.; Chen, S.T.; Hsu, H.M. A new strategy for preparing oligomeric ionic liquid gel polymer electrolytes for high-performance and nonflammable lithium ion batteries. *J. Membr. Sci.* **2016**, *499*, 462–469. [CrossRef]
12. Rollins, H.W.; Harrup, M.K.; Dufek, E.J.; Jamison, D.K.; Sazhin, S.V.; Gering, K.L.; Daubaras, D.L. Fluorinated phosphazene co-solvents for improved thermal and safety performance in lithium-ion battery electrolytes. *J. Power Sources* **2014**, *263*, 66–74. [CrossRef]
13. Ye, Y.S.; Rick, J.; Hwang, B.J. Ionic liquid polymer electrolytes. *J. Mater. Chem. A* **2013**, *1*, 2719–2743. [CrossRef]
14. MacFarlane, D.R.; Tachikawa, N.; Forsyth, M.; Pringle, J.M.; Howlett, P.C.; Elliott, G.D.; Davis, J.H.; Watanabe, M.; Simon, P.; Angell, C.A. Energy applications of ionic liquids. *Energy Environ. Sci.* **2014**, *7*, 232–250. [CrossRef]
15. Wasserscheid, P.; Keim, W. Ionic liquids-new "solutions" for transition metal catalysis. *Angew. Chem. Int. Ed.* **2000**, *39*, 3772–3789. [CrossRef]

16. Anderson, J.L.; Ding, J.; Welton, T.; Armstrong, D.W. Characterizing ionic liquids on the basis of multiple solvation interactions. *J. Am. Chem. Soc.* **2002**, *124*, 14247–14254. [CrossRef] [PubMed]
17. Armand, M.; Endres, F.; MacFarlane, D.R.; Ohno, H.; Scrosati, B. Ionic-liquid materials for the electrochemical challenges of the future. *Nat. Mater.* **2009**, *8*, 621–629. [CrossRef] [PubMed]
18. Borgel, V.; Markevich, E.; Aurbach, D.; Semrau, G.; Schmidt, M. On the application of ionic liquids for rechargeable Li batteries: High voltage systems. *J. Power Sources* **2009**, *189*, 331–336. [CrossRef]
19. Lux, S.F.; Schmuck, M.; Appetecchi, G.B.; Passerini, S.; Winter, M.; Balducci, A. Lithium insertion in graphite from ternary ionic liquid–lithium salt electrolytes: II. Evaluation of specific capacity and cycling efficiency and stability at room temperature. *J. Power Sources* **2009**, *192*, 606–611. [CrossRef]
20. Schmitz, R.W.; Murmann, P.; Schmitz, R.; Müller, R.; Krämer, L.; Kasnatscheew, J.; Isken, P.; Niehoff, P.; Nowak, S.; Roeschenthaler, G.V. Investigations on novel electrolytes, solvents and SEI additives for use in lithium-ion batteries: systematic electrochemical characterization and detailed analysis by spectroscopic methods. *Prog. Solid State Chem.* **2014**, *42*, 65–84. [CrossRef]
21. Di Leo, R.A.; Marschilok, A.C.; Takeuchi, K.J.; Takeuchi, E.S. Battery electrolytes based on saturated ring ionic liquids: Physical and electrochemical properties. *Electrochim. Acta* **2013**, *109*, 27–32. [CrossRef]
22. Watanabe, M.; Thomas, M.L.; Zhang, S.; Ueno, K.; Yasuda, T.; Dokko, K. Application of ionic liquids to energy storage and conversion materials and devices. *Chem. Rev.* **2017**, *117*, 7190–7239. [CrossRef] [PubMed]
23. Cao, X.; He, X.; Wang, J.; Liu, H.; Röser, S.; Rad, B.R.; Evertz, M.; Streipert, B.; Li, J.; Wagner, R. High voltage LiNi$_{0.5}$Mn$_{1.5}$O$_4$/Li$_4$Ti$_5$O$_{12}$ lithium ion cells at elevated temperatures: carbonate-versus ionic liquid-based electrolytes. *ACS Appl. Mater. Interfaces* **2016**, *8*, 25971–25978. [CrossRef] [PubMed]
24. Nádherná, M.; Reiter, J.; Moškon, J.; Dominko, R. Lithium bis (fluorosulfonyl) imide-PYR$_{14}$TFSI ionic liquid electrolyte compatible with graphite. *J. Power Sources* **2011**, *196*, 7700–7706. [CrossRef]
25. Kalhoff, J.; Bresser, D.; Bolloli, M.; Alloin, F.; Sanchez, J.; Passerini, S. Enabling LiTFSI-based electrolytes for safer lithium-ion batteries by using linear fluorinated carbonates as (co)solvent. *ChemSusChem* **2014**, *7*, 2939–2946. [CrossRef] [PubMed]
26. Aravindan, V.; Gnanaraj, J.; Madhavi, S.; Liu, H. Lithium-ion conducting electrolyte salts for lithium batteries. *Chem. Eur. J.* **2011**, *17*, 14326–14346. [CrossRef] [PubMed]
27. Stephan, A.M.; Nahm, K. Review on composite polymer electrolytes for lithium batteries. *Polymer* **2006**, *47*, 5952–5964. [CrossRef]
28. Stolarska, M.; Niedzicki, L.; Borkowska, R.; Zalewska, A.; Wieczorek, W. Structure, transport properties and interfacial stability of PVdF/HFP electrolytes containing modified inorganic filler. *Electrochim. Acta* **2007**, *53*, 1512–1517. [CrossRef]
29. Chung, S.; Wang, Y.; Persi, L.; Croce, F.; Greenbaum, S.; Scrosati, B.; Plichta, E. Enhancement of ion transport in polymer electrolytes by addition of nanoscale inorganic oxides. *J. Power Sources* **2001**, *97*, 644–648. [CrossRef]
30. Wachtler, M.; Ostrovskii, D.; Jacobsson, P.; Scrosati, B. A study on PVdF-based SiO$_2$-containing composite gel-type polymer electrolytes for lithium batteries. *Electrochim. Acta* **2004**, *50*, 357–361. [CrossRef]
31. Vijayakumar, G.; Karthick, S.; Priya, A.S.; Ramalingam, S.; Subramania, A. Effect of nanoscale CeO$_2$ on PVDF-HFP-based nanocomposite porous polymer electrolytes for Li-ion batteries. *J. Solid State Electrochem.* **2008**, *12*, 1135–1141. [CrossRef]
32. Wu, C.G.; Lu, M.I.; Tsai, C.C.; Chuang, H.J. PVdF-HFP/metal oxide nanocomposites: the matrices for high-conducting, low-leakage porous polymer electrolytes. *J. Power Sources* **2006**, *159*, 295–300. [CrossRef]
33. Raghavan, P.; Zhao, X.; Kim, J.K.; Manuel, J.; Chauhan, G.S.; Ahn, J.H.; Nah, C. Ionic conductivity and electrochemical properties of nanocomposite polymer electrolytes based on electrospun poly (vinylidene fluoride-co-hexafluoropropylene) with nano-sized ceramic fillers. *Electrochim. Acta* **2008**, *54*, 228–234. [CrossRef]
34. Raghavan, P.; Choi, J.W.; Ahn, J.H.; Cheruvally, G.; Chauhan, G.S.; Ahn, H.J.; Nah, C. Novel electrospun poly (vinylidene fluoride-co-hexafluoropropylene)-in situ SiO$_2$ composite membrane-based polymer electrolyte for lithium batteries. *J. Power Sources* **2008**, *184*, 437–443. [CrossRef]
35. Croce, F.; Appetecchi, G.; Persi, L.; Scrosati, B. Nanocomposite polymer electrolytes for lithium batteries. *Nature* **1998**, *394*, 456. [CrossRef]

36. Jeong, H.S.; Lee, S.Y. Closely packed SiO$_2$ nanoparticles/poly (vinylidene fluoride-hexafluoropropylene) layers-coated polyethylene separators for lithium-ion batteries. *J. Power Sources* **2011**, *196*, 6716–6722. [CrossRef]
37. Deimede, V.; Elmasides, C. Separators for Lithium-Ion Batteries: a Review on the Production Processes and Recent Developments. *Energy Technol.* **2015**, *3*, 453–468. [CrossRef]
38. Bishop, K.J.; Wilmer, C.E.; Soh, S.; Grzybowski, B.A. Nanoscale forces and their uses in self-assembly. *Small* **2009**, *5*, 1600–1630. [CrossRef] [PubMed]
39. Kumar, A.; Deka, M. Nanofiber Reinforced Composite Polymer Electrolyte Membranes. In *Nanofibers*; INTECH: Rijeka, Croatia, 2010; pp. 13–38.
40. Sheng, X.; Xie, D.; Zhang, X.; Zhong, L.; Wu, H.; Morbidelli, M. Uniform distribution of graphene oxide sheets into a poly-vinylidene fluoride nanoparticle matrix through shear-driven aggregation. *Soft Matter* **2016**, *12*, 5876–5882. [CrossRef] [PubMed]
41. Wu, H.; Zaccone, A.; Tsoutsoura, A.; Lattuada, M.; Morbidelli, M. High shear-induced gelation of charge-stabilized colloids in a microchannel without adding electrolytes. *Langmuir* **2009**, *25*, 4715–4723. [CrossRef] [PubMed]
42. Wu, H.; Lattuada, M.; Sandkühler, P.; Sefcik, J.; Morbidelli, M. Role of sedimentation and buoyancy on the kinetics of diffusion limited colloidal aggregation. *Langmuir* **2003**, *19*, 10710–10718. [CrossRef]
43. Stephan, A.M.; Nahm, K.S.; Kumar, T.P.; Kulandainathan, M.A.; Ravi, G.; Wilson, J. Nanofiller incorporated poly (vinylidene fluoride–hexafluoropropylene)(PVdF-HFP) composite electrolytes for lithium batteries. *J. Power Sources* **2006**, *159*, 1316–1321. [CrossRef]
44. Caimi, S.; Wu, H.; Morbidelli, M. PVdF-HFP and Ionic Liquid-Based, Freestanding Thin Separator for Lithium-Ion Batteries. *ACS Appl. Energy Mater.* **2018**, *1*, 5224–5232. [CrossRef]
45. Caimi, S.; Cingolani, A.; Jaquet, B.; Siggel, M.; Lattuada, M.; Morbidelli, M. Tracking of fluorescently labeled polymer particles reveals surface effects during shear-controlled aggregation. *Langmuir* **2017**, *33*, 14038–14044. [CrossRef] [PubMed]
46. Cingolani, A.; Baur, D.; Caimi, S.; Storti, G.; Morbidelli, M. Preparation of Perfusive Chromatographic Materials via Shear-Induced Reactive Gelation. *J. Chromatogr. A* **2018**, *1538*, 25–33. [CrossRef] [PubMed]
47. Saikia, D.; Kumar, A. Ionic conduction in P(VDF-HFP)/PVDF-(PC+DEC)-LiClO$_4$ polymer gel electrolytes. *Electrochim. Acta* **2004**, *49*, 2581–2589. [CrossRef]
48. Meng, X.; Wu, H.; Storti, G.; Morbidelli, M. Effect of dispersed polymeric nanoparticles on the bulk polymerization of methyl methacrylate. *Macromolecules* **2016**, *49*, 7758–7766. [CrossRef]
49. Li, Z.; Su, G.; Wang, X.; Gao, D. Micro-porous P(VDF-HFP)-based polymer electrolyte filled with Al$_2$O$_3$ nanoparticles. *Solid State Ion.* **2005**, *176*, 1903–1908. [CrossRef]
50. Wieczorek, W.; Stevens, J.; Florjańczyk, Z. Composite polyether based solid electrolytes. The Lewis acid–base approach. *Solid State Ion.* **1996**, *85*, 67–72. [CrossRef]
51. Liu, Y.; Lee, J.; Hong, L. In situ preparation of poly (ethylene oxide)-SiO$_2$ composite polymer electrolytes. *J. Power Sources* **2004**, *129*, 303–311. [CrossRef]
52. Scrosati, B.; Croce, F.; Panero, S. Progress in lithium polymer battery R&D. *J. Power Sources* **2001**, *100*, 93–100.
53. Croce, F.; Persi, L.; Scrosati, B.; Serraino-Fiory, F.; Plichta, E.; Hendrickson, M. Role of the ceramic fillers in enhancing the transport properties of composite polymer electrolytes. *Electrochim. Acta* **2001**, *46*, 2457–2461. [CrossRef]
54. Ferrari, S.; Quartarone, E.; Mustarelli, P.; Magistris, A.; Fagnoni, M.; Protti, S.; Gerbaldi, C.; Spinella, A. Lithium ion conducting PVdF-HFP composite gel electrolytes based on N-methoxyethyl-N-methylpyrrolidinium bis (trifluoromethanesulfonyl)-imide ionic liquid. *J. Power Sources* **2010**, *195*, 559–566. [CrossRef]
55. Wu, F.; Chen, N.; Chen, R.; Zhu, Q.; Qian, J.; Li, L. "Liquid-in-Solid" and "Solid-in-Liquid" Electrolytes with High Rate Capacity and Long Cycling Life for Lithium-Ion Batteries. *Chem. Mater.* **2016**, *28*, 848–856. [CrossRef]

© 2018 by the authors. Licensee MDPI, Basel, Switzerland. This article is an open access article distributed under the terms and conditions of the Creative Commons Attribution (CC BY) license (http://creativecommons.org/licenses/by/4.0/).

Article

Multifunctional Platform Based on Electroactive Polymers and Silica Nanoparticles for Tissue Engineering Applications

Sylvie Ribeiro [1,2], Tânia Ribeiro [3,4], Clarisse Ribeiro [1,5,*], Daniela M. Correia [6,7], José P. Sequeira Farinha [3,4], Andreia Castro Gomes [2], Carlos Baleizão [3,4] and Senentxu Lanceros-Méndez [7,8]

1. Centro/Departamento de Física, Universidade do Minho, 4710-057 Braga, Portugal; sribeiro@fisica.uminho.pt
2. Centre of Molecular and Environmental Biology (CBMA), Universidade do Minho, Campus de Gualtar, 4710-057 Braga, Portugal; agomes@bio.uminho.pt
3. Centro de Química-Física Molecular and Institute of Nanosciences and Nanotechnology, Instituto Superior Técnico, Universidade de Lisboa, 1049-001 Lisboa, Portugal; tania.ribeiro@tecnico.ulisboa.pt (T.R.); farinha@tecnico.ulisboa.pt (J.P.S.F.); carlos.baleizao@tecnico.ulisboa.pt (C.B.)
4. Centro de Química Estrutural, Instituto Superior Técnico, Universidade de Lisboa, 1049-001 Lisboa, Portugal
5. CEB—Centre of Biological Engineering, Universidade do Minho, Campus de Gualtar, 4710 057 Braga, Portugal
6. Chemical Department and CQ-VR, Universidade de Trás-os-Montes e Alto Douro, 5001-801 Vila Real, Portugal; d.correia@fisica.uminho.pt
7. BCMaterials, Basque Centre for Materials, Applications and Nanostructures, UPV/EHU Science Park, 48940 Leioa, Spain; senentxu.lanceros@bcmaterials.net
8. IKERBASQUE, Basque Foundation for Science, 48013 Bilbao, Spain
* Correspondence: cribeiro@fisica.uminho.pt; Tel.: +351-253-406-074

Received: 25 October 2018; Accepted: 6 November 2018; Published: 9 November 2018

Abstract: Poly(vinylidene fluoride) nanocomposites processed with different morphologies, such as porous and non-porous films and fibres, have been prepared with silica nanoparticles (SiNPs) of varying diameter (17, 100, 160 and 300 nm), which in turn have encapsulated perylenediimide (PDI), a fluorescent molecule. The structural, morphological, optical, thermal, and mechanical properties of the nanocomposites, with SiNP filler concentration up to 16 wt %, were evaluated. Furthermore, cytotoxicity and cell proliferation studies were performed. All SiNPs are negatively charged independently of the pH and more stable from pH 5 upwards. The introduction of SiNPs within the polymer matrix increases the contact angle independently of the nanoparticle diameter. Moreover, the smallest ones (17 nm) also improve the PVDF Young's modulus. The filler diameter, physico-chemical, thermal and mechanical properties of the polymer matrix were not significantly affected. Finally, the SiNPs' inclusion does not induce cytotoxicity in murine myoblasts (C2C12) after 72 h of contact and proliferation studies reveal that the prepared composites represent a suitable platform for tissue engineering applications, as they allow us to combine the biocompatibility and piezoelectricity of the polymer with the possible functionalization and drug encapsulation and release of the SiNP.

Keywords: nanostructures; polymer matrix composites (PMCs); mechanical properties; thermal properties

1. Introduction

The development of advanced multifunctional materials is essential for the development of society [1]. Nanocomposites are among the most important materials for an increasing number of applications due to the possibility of designing materials with tailored properties meeting specific application demands in areas ranging from the automotive industry [2,3] to food packaging [4,5] and tissue engineering [6,7], among others. The introduction of inorganic nanomaterials into polymers allows for the combination of the rigidity and high thermal stability of the inorganic material with the ductibility, flexibility and processability of the organic polymers [8], as well as the introduction/tuning of further functionalities such as magnetic [9], mechanical [10] or electrical properties [11,12]. Typical nanomaterials include nanoparticles, nanotubes, nanofibres, fullerenes and nanowires [13]. Among nanomaterials, silica is widely present in the environment and has several key features [14].

Properties of silica nanoparticles (SiNPs), such as high mechanical strength, permeability, thermal and chemical stability, relatively low refractive index, high surface area, and the fact of being used for coatings of other particles, such as magnetic and quantum dots [15–17], make these nanoparticles highly interesting for various applications [18]. Furthermore, their biocompatibility and the different possibilities of functionalizing them are the basis of their potential for biomedicine and tissue engineering applications [19]. Silica nanostructures have been extensively used as supports or carriers in drug delivery [20,21], nanomedicine [22,23] and bioanalysis [24]. Their characteristics can be tuned during synthesis to obtain a wide range of particle diameters ranging from 20 to 500 nm, different pore sizes and the incorporation of molecules such as drugs or fluorophores [24], as well as magnetic nanostructures [25]. Mesoporous silica nanoparticles (MSNPs) [20,26] have attracted particular attention for their functionalization versatility. Silica-based mesoporous nanoparticles, due to the strong Si-O bond compared to niosomes, liposomes and dendrimers, are more resistant to degradation and mechanical stress, obviating the need for any external stabilization of the MSNPs [27,28].

With respect to tissue engineering, different tissues require different microenvironments for suitable regeneration [29]. Thus, muscle tissue has electromechanical responses and needs electrical stimulation to support ionic exchange, mainly sodium by calcium ion [30]. In this context, electroactive polymers such as magnetoelectric [31,32], piezoelectric and conductive polymers [33], among others [34], show strong potential for tissue engineering applications. Among the different electroactive polymers, piezoelectric polymers have already shown their suitability for tissue engineering [6,29] due to their capacity to vary surface charge when a mechanical load is applied or vice versa. These materials can play a significant role because electric stimulation can be found in many living tissues of the human body, namely nerves [35] and bones [36,37], and it can provide the electromechanical solicitations for muscle tissue [38]. Poly(vinylidene fluoride) (PVDF) is the biocompatible piezoelectric polymer with the highest piezoelectric response. It can crystallize in four differentiated crystalline phases, α, β, γ, δ, with the β-phase being the one with the highest piezoelectric coefficient. Furthermore, it can be processed in different morphologies, including fibres, spheres, membranes and 3D scaffolds [29,39], providing a suitable platform for tissue engineering.

In order to further exploit the applicability of PVDF in regenerative medicine, polymer nanocomposites based on PVDF using silica nanoparticles with different diameters were prepared, improving the electroactive characteristics of PVDF with the aforementioned characteristics of MSNPs for biomedical applications. Together with the physico-chemical characteristics of the developed composites, their biocompatibility was evaluated in murine myoblast cells.

2. Materials and Methods

2.1. Materials

PVDF (Solef 1010) was purchased from Solvay, N,N-dimethylformamide (DMF) from Merck. Absolute ethanol (EtOH, Panreac, Barcelona, Spain, 99.5%), ammonium hydroxide solution (NH_4OH, 28% in water, Fluka, Carnaxide, Portugal) and tetraethyl orthosilicate (TEOS, Sigma-Aldrich, Sintra, Portugal, 99%) were used as received. Deionized water from a Millipore (Oeiras, Portugal) system

Milli-Q $\geq$ 18 MΩ cm was used in the synthesis of the silica nanoparticles. Perylenediimide derivative (PDI) was synthesized according to the literature [40].

2.2. Silica Nanoparticles

2.2.1. Preparation of the Silica Nanoparticles

Fluorescent silica nanoparticles, doped with PDI were prepared by a modified Stöber method [41,42]. Water, absolute ethanol, and PDI (previously dispersed in ethanol, 1×10^{-6} M) were mixed and after 30 min the ammonia solution was added to the mixture, followed by TEOS. The reaction was kept under stirring at constant temperature for 24 h. After that time, the nanoparticles were recovered and washed with ethanol (three cycles of centrifugation). The nanoparticles were redispersed in ethanol and dried at 50 °C in a ventilated oven. The experimental details are provided in Table 1.

Table 1. Experimental details used for the preparation of the SiNPs.

Particle Diameter (nm)	EtOH (g)	H$_2$O (g)	PDI Solution (mL)	NH$_3$ (mL)	TEOS (mL)	Reaction Temperature (°C)
17	84.13	7.99	3	1.51	4.46	50
100	105.73	4.65	4	6.68	9.00	30
160	53.18	11.03	4	2.67	4.46	50
300	53.18	11.03	4	2.67	4.46	30

2.2.2. Characterization of the SiNPs

Transmission electron microscopy: Transmission electron microscopy (TEM) images were obtained on a Hitachi (Krefeld, Germany) transmission electron microscope (model H-8100 with a LaB6 filament) with an acceleration voltage of 200 kV. One drop of the dispersion of particles in ethanol was placed on a carbon grid and dried in air before observation. The images were processed with the Fiji software (Madison, WI, USA).

Zeta potential: The surface charge of the nanoparticles was estimated with the use of zeta potential (Zetasizer NANO ZS-ZEN3600, Malvern). The zeta potential of the fluorescent SiNPs with different diameters were evaluated at different pH values (3, 5, 7, 11, 13). To adjust the pH, it was used a solutions of HCl (1M) and NaOH (1M). The average value and standard deviation for each sample were obtained from six measurements.

2.3. Nanocomposite Samples

2.3.1. Preparation of the SiNPs/PVDF Nanocomposites

SiNPs/PVDF nanocomposites with 16 wt % of SiNPs were prepared by dispersing the respective mass of SiNPs in the DMF solvent within an ultrasound bath for 4 h at room temperature. The filler concentration was selected based in [43], as it shows a suitable filler content without compromising the mechanical characteristics of the polymer matrix and allowing a suitable dispersion of the filler. After we obtained a good dispersion of the nanoparticles, PVDF was added ay a concentration of 15% (w/w) and the solution was magnetically stirred at room temperature until the complete dissolution of the polymer. The materials were then prepared by different production methods [39].

First, SiNPs/PVDF samples (porous and non-porous films) were prepared by solution casting on a clean glass substrate and, in some cases, melted at different temperatures for different times (Table 2). The different preparation conditions allowed us to tailor the porosity and to study the possibility of the nucleation of the electroactive β-phase of the polymer by the fillers [44]. The thickness of the films ranges from 30 to 50 μm.

Table 2. Denomination, relevant preparation conditions and morphology of the PVDF and nanocomposite samples prepared in this work.

Morphology	Temperature (°C)	Time to Melt/Dry	Diameter of the Nanoparticles	Samples Morphology (P: Porous; NP: Non-Porous)	Denomination
Films (F)	90	30	—	NP	F90-NP
			17	NP	F90-17NP
			100	NP	F90-100NP
			160	NP	F90-160NP
			300	NP	F90-300NP
	210	10	17	NP	F210-17NP
	Room temperature (Trt)	—	17	P	FTrt-17P
Oriented fibres (O)	—	—		P	O-17P
Random fibres (R)	—	—		P	R-17P

For SiNPs/PVDF electrospun fibre mats, the solution was placed in a plastic syringe (10 mL) fitted with a steel needle with inner diameter of 0.5 mm. After an optimization procedure, electrospinning was conducted with a high-voltage power supply from Glasman (model PS/FC30P04, Radeberg, Germany) at 14 kV with a feed rate of 0.5 mL·h^{-1} (with a syringe pump from Syringepump, Porto, Portugal). The electrospun fibres were collected in an aluminium plate (placed 20 cm from the needle) and in a rotating drum (1500 rpm) to obtain random and oriented nanofibres, respectively.

Table 2 summarizes the main characteristics of the samples and the corresponding denomination that refers the type of sample and processing temperature, the nanoparticle diameter and the composite morphology. For example, F90-17NP is a film (F) obtained at 90 °C (90) with nanoparticles with a diameter of 17 nm (17), which is non-porous (NP).

2.3.2. Characterization of the Nanocomposite Samples

Scanning electron microscopy: A desktop scanning electron microscope (SEM) (Phenom ProX, Eindhoven, The Netherlands) was used to observe the morphology and microstructure of the PVDF and SiNPs/PVDF nanocomposites. This technique was also used to observe the cell morphology seeded on the different fibrous samples. All the samples were added to the aluminium pin stubs with electrically conductive carbon adhesive tape (PELCO TabsTM, Redding, CA, USA). The aluminium pin stub was then placed on a phenom Charge Reduction sample Holder. All results were acquired using the ProSuite software (Waarschoot, Belgium). The images were obtained with an acceleration voltage of 10 kV.

Laser scanning confocal fluorescence microscopy: Laser scanning confocal fluorescence microscopy (LSCFM) images were obtained with a Leica TCS SP5 laser scanning microscope (Leica Microsystems CMS GmbH, Mannheim, Germany) using an inverted microscope (DMI6000), a HCX PL APO CS 10× dry immersion objective (10× magnification and 0.4 numerical aperture) and a HC PL FLUOTAR 50× dry immersion objective (50× magnification and 0.8 numerical aperture). Imaging used the 488 nm line of an argon ion laser.

Contact angle measurements: Water contact angle (CA) measurements (sessile drop in dynamic mode) were performed at room temperature in a Data Physics OCA20 (Data Physics, Filderstadt, Germany) setup using ultrapure water as the test liquid. The samples wettability was determined by using water drops (3 µL) placed onto the surface of the samples. Each sample was measured at six different locations and the mean contact angle and standard deviation were calculated.

Fourier transform infrared spectroscopy: Fourier transform infrared spectroscopy (FTIR) measurements in attenuated total reflectance (ATR) were performed at room temperature, using a Nicolet Nexus 670 FTIR-spectrophotometer (ThermoFisher Scientific, Porto Salvo, Portugal) with

Smart Orbit Accessory equipment (ThermoFisher Scientific, Porto Salvo, Portugal). The analysis was performed from 4000 to 600 cm^{-1}, after 64 scans with a resolution of 4 cm^{-1}. The spectra of each sample was used to determine the relative content of the electroactive β-phase in the composite samples, by using the method presented in [44]. In short, the ®-phase content (F®) was calculated by Equation (1):

$$F_\beta = \frac{A_\beta}{\left(\frac{K_\beta}{K_\alpha}\right) \times A_\alpha + A_\beta},\qquad(1)$$

where A_β are the absorbance at 840 cm^{-1} and $K_\beta = 7.7 \times 10^4$ cm^2·mol^{-1} is the absorption coefficients and correspond to the β phase. A_α is the absorbance at 760 cm^{-1} and $K_\alpha = 6.1 \times 10^4$ cm^2·mol^{-1} is the absorption coefficient, and correspond to the α phase.

Thermal properties: Differential scanning calorimetry (DSC) was carried out with a DSC 6000 Perkin Elmer (Mettler Toledo, Columbus, OH, USA) instrument. The samples were heated from 30 to 200 °C at a rate of 10 °C·min^{-1} under a flowing nitrogen atmosphere. Samples were cut from the middle region of the samples and placed in aluminium pans.

From the melting in the DSC thermograms, the degree crystallinity (X_c) of the samples was calculated by the following equation [44]:

$$X_c = \frac{\Delta H_f}{x\Delta H_\alpha + y\Delta H_\beta},\qquad(2)$$

where ΔH_f is the melting enthalpy of the sample, x and y represent the α and β phase contents present in the sample, respectively, and ΔH_α and ΔH_β are the melting enthalpies for a 100% α-PVDF (93.04 J·g^{-1}) and β-PVDF (104.4 J·g^{-1}) crystalline samples respectively.

Mechanical characterization: Mechanical measurements were performed with a universal testing machine (Shimadzu model AG-IS, Kyoto, Japan) at room temperature, in tensile mode at a test velocity of 1 mm·min^{-1}, with a load cell of 50 N. The tests were performed on rectangular samples (30 × 10 mm) with a thickness between 30 and 50 µm (Fischer Dualscope 603-478, digital micrometer, Windsor, CT, USA). The mechanical parameters were calculated from the average of triplicate measurements. Hook's law was used to obtain the effective Young's modulus (E) of PVDF and SiNPs/PVDF nanocomposite samples in the linear zone of elasticity between 0 and 1% strain.

2.4. Cell Culture Experiments

2.4.1. Sample Sterilization

The samples were sterilized by multiple immersions into 70% ethanol for 30 min each and to remove any residual solvent, they were washed five times in a phosphate buffered saline (PBS) 1× solution for 5 min each. Each side of the samples was then exposed to ultraviolet (UV) light for 1 h.

2.4.2. Cell Culture

Murine myoblasts (C2C12 cell line) were cultivated in Dulbecco's Modified Eagle's Medium (DMEM, Gibco, Porto Salvo, Portugal) with 4.5 g·L^{-1} containing 10% of Foetal Bovine Serum (FBS, Biochrom, Cambridge, UK) and 1% of Penicillin/Streptomycin (P/S, Biochrom). The cells were grown in a 75 cm^2 cell-culture flask at 37 °C in a humidified air containing 5% CO_2 atmosphere. Every two days, the culture medium was changed. The cells were trypsinized with 0.05% trypsin-EDTA when they reached 60–70% confluence. For the cytotoxicity assays, SiNPs/PVDF nanocomposites with different morphologies were cut according to the ISO_10993-12. The extraction ratio (surface area or mass/volume) was 6 cm^2·mL^{-1}. To analyse cell morphology and viability, the materials were cut into 6 mm diameter. PVDF films without nanoparticles were used as the control.

2.4.3. Cytotoxicity Assay by the Indirect Contact

C2C12 cells were seeded at the density of 2×10^4 cells·mL^{-1} in 96-well tissue culture polystyrene plates. Cells were allowed to attach for 24 h, after which the culture medium was removed and the conditioned medium (the medium that was in contact with the samples) was added to the wells (100 µL). Afterwards, the cells were incubated for 24 or 72 h, and the number of viable cells was quantified by (3-(4,5-Dimethylthiazol-2-yl)-2,5-diphenyltetrazolium bromide) (MTT) assay. The cells received MTT solution (5 mg·mL^{-1} in PBS dissolved in DMEM in proportion of 10%) and were incubated in the dark at 37 °C for 2 h. The medium was then removed and 100 µL of DMSO/well were added to dissolve the precipitated formazan. The quantification was determined by measuring the absorbance at 570 nm using a microplate reader. All quantitative results were obtained from four replicate samples and controls and were analysed as the average of viability ± standard deviation (SD).

2.4.4. Direct Contact and Proliferation

Since MTT interferes with the materials, we chose the MTS as having the same theoretical basis but a soluble reaction product. C2C12 cells (4000) were seeded on each sample. After 24 h and 72 h, the viable cell number was determined using the (3-(4,5-dimethylthiazol-2-yl)-5-(3-carboxymethoxyphenyl) -2-(4-sulfophenyl)-2H-tetrazolium) (MTS) assay. At the desired time points, the MTS reagent was added into each well in a 1:5 proportion of DMEM medium, and incubated at 37 °C for 2 h. The absorbance was detected at 490 nm with a microplate reader. Experimental data were obtained from four replicates.

2.4.5. Immunofluorescence Staining

Using the same time points as in the proliferation assays, the nanocomposite samples were subjected to immunofluorescence staining to analyse the cytoskeleton morphology of the cells, also verifying the cell viability and adhesion. At each time point, the medium of each well was removed, the samples were washed with PBS and the cells fixed with 4% formaldehyde for 10 min at 37 °C in a 5% CO_2 incubator. After fixation, the samples were washed with PBS 1× (three times) and incubated for 45 min at room temperature in 0.1 µg mL^{-1} of green phalloidin (Sigma-Aldrich, Sintra, Portugal). Then, the samples were incubated for 5 min with 1 µg mL^{-1} of 4,6-diamidino-2-phenylindole (DAPI, Sigma). Afterwards, the samples were washed again with PBS 1× (three times) and one time with distillate water. Finally, the samples were visualized with fluorescence microscopy (Olympus BX51 Microscope, Lisboa, Portugal).

3. Results and Discussion

3.1. Silica Nanoparticles

3.1.1. Morphology and Size of the Nanoparticles

The morphology and the size of the SiNPs were analysed from TEM images (Figure 1). The spherical nanoparticles prepared by the Stober method [45] were prepared in four different diameters: 17 ± 2, 100 ± 18, 160 ± 17 and 300 ± 37 nm. The corresponding histograms are presented as insets in Figure 1.

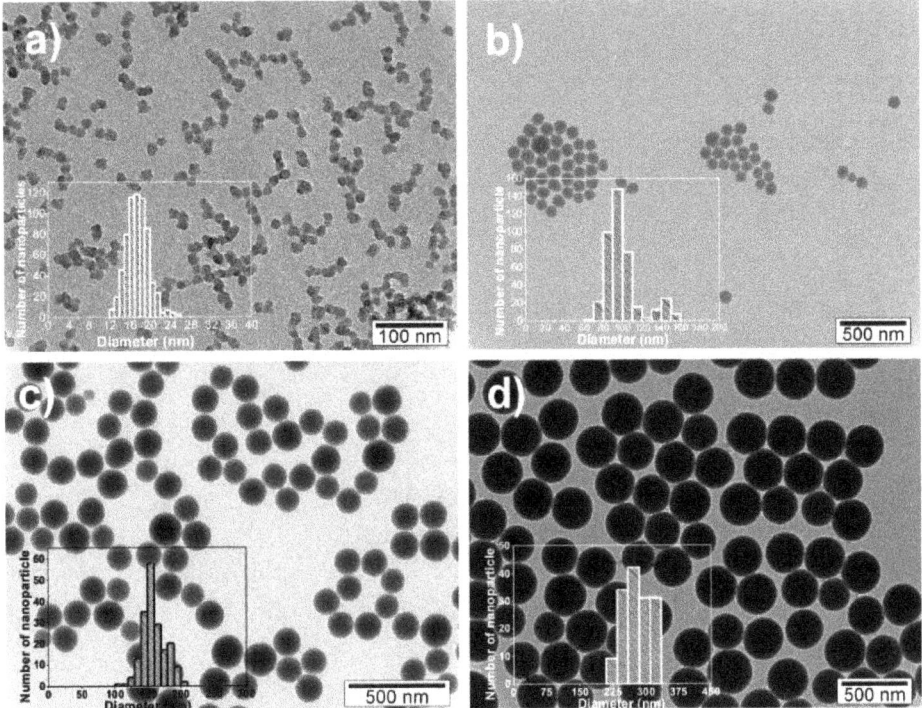

Figure 1. TEM images of SiNPs-PDI with different particle size: (a) 17 ± 2 nm, (b) 100 ± 1 m, (c) 160 ± 1 m and (d) 300 ± 3 m.

3.1.2. Surface Charge of the Nanoparticles

Figure 2 shows the zeta potential of aqueous dispersions of the different SiNPs at different pH to analyse the periphery charge of the particles.

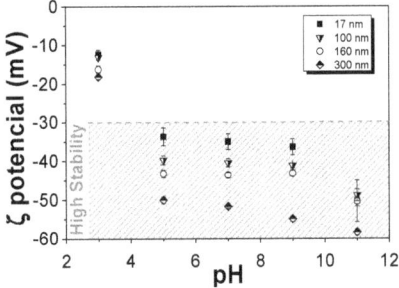

Figure 2. Zeta potential of the different SiNPs nanoparticles at different pH.

The particles are considered more stable with a zeta potential above +30 mV or below −30 mV. This fact is due to the electrostatic repulsions between the nanoparticles that prevent their aggregation. Figure 2 shows that all nanoparticles are more stable at pH ≥ 5, independently of their average diameter. On the other hand, nanoparticles with higher average diameters are more stable. The isoelectric point of SiNPs is close to pH 2 so, from this pH upwards, the silica nanoparticles are negatively charged in acidic, neutral and basic environments, which can be taken advantage of as it has been demonstrated that the interactions between negatively charged nanoparticle surfaces and the positive charge density of the CH$_2$ groups of the PVDF polymer can promote the nucleation of the electroactive β-phase of the polymer [46].

3.2. SiNPs/PVDF Nanocomposite Samples

3.2.1. Morphology of the Nanocomposites

The morphology of the nanocomposites was assessed by SEM. Figure 3 shows the different morphologies obtained after the different processing methods as well as the variations due to the introduction of fillers with different diameters. Figure 3 shows the cross section (Figure 3a–c) of the nanocomposites and electrospun fibres samples (Figure 3d) with 16 wt % of SiNPs. Figure 3a,b present the differences between the samples obtained at 90 °C with SiNPs of different diameters, showing that the higher diameter particles are well-dispersed in the PVDF polymer matrix, in contrast to the SiNPs with lower diameter that present particle agglomerates. Furthermore, a small porosity is observed (Figure 3a), which is in agreement with the literature [47].

It is important to note that the nanoparticles act as nucleation agents for crystallization in PVDF composites [48], which can be verified with the results obtained, indicating a good interfacial interaction between the PVDF chains and silica nanoparticles.

Figure 3a,c shows the differences in composite morphology due to the crystallization process. The samples obtained at 90 °C (Figure 3a,b) present a slightly more porous morphology than the ones obtained at 210 °C (Figure 3c).

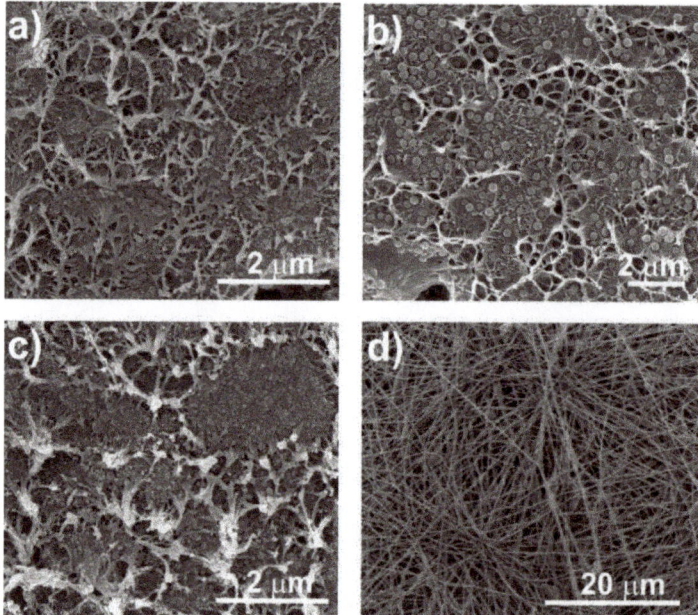

Figure 3. Cross section SEM micrographs of SiNPs/PVDF nanocomposite samples with nanoparticles of different diameters and different processing conditions: (**a**) F90-17NP, (**b**) F90-300NP, (**c**) F210-17NP, (**d**) R-17P.

Once the SiNPs of 17 nm do not show a suitable dispersion in the films, electrospinning was used in order to produce fibres with well-dispersed particles. Relative to the fibres (Figure 3d), smooth randomly oriented fibres with encapsulated particles are observed, with no particles at the surface.

This result is confirmed by the confocal images represented in Figure 4. It was observed that the introduction of the particles increases the fibre diameter (243 ± 89 nm to 339 ± 92 nm). Oriented fibres with SiNPs were also produced (data not shown), verifying the particles' encapsulation within the fibres and a fibre diameter of 683 ± 140 nm. The increase of fibre diameter with the incorporation of the SiNPs is attributed to the higher viscosity of the solution, with also hinders fibre stretching by the applied field. The higher diameter of the oriented fibres relative to the randomly oriented fibres is attributed to the merging of aligned fibrils that crystallize simultaneously [49].

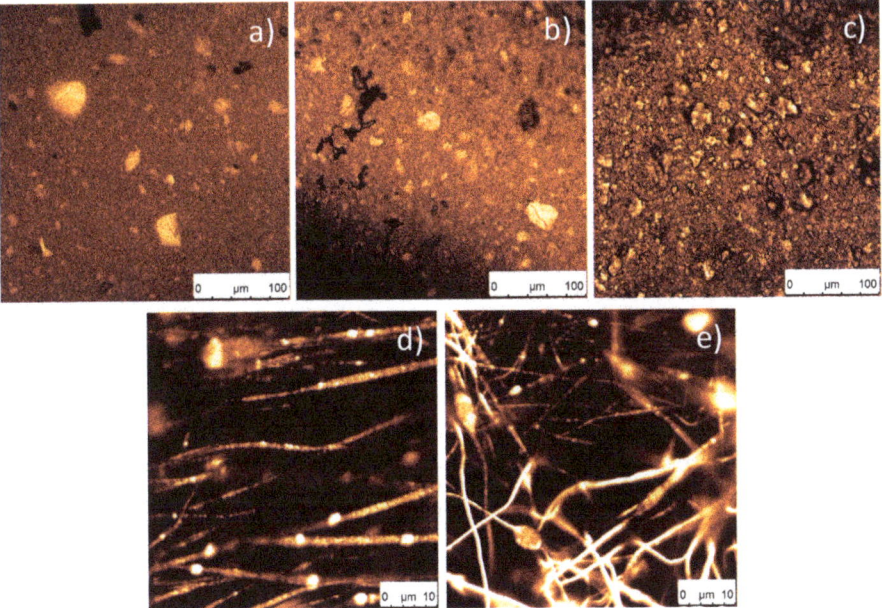

Figure 4. Representative confocal images of SiNPs/PVDF nanocomposites with different morphologies: (a) F210-17NP, (b) F90-17NP, (c) Ftrt-17P, (d) O-17P and (e) R-17P.

3.2.2. Confocal Fluorescence Microscopy of the Nanocomposites

The incorporation of PDI in the silica nanoparticles can increase their application range, in particular, for biomedical applications, as it allows their tracking and localization [42,50]. In Figure 4, the green identifies the fluorescence of the nanoparticles; a higher colour intensity indicates a higher number of nanoparticles present. Figure 4a–c shows that, as the processing temperature decreases, a larger aggregation of nanoparticles is observed. In Figure 4a, where the temperature is higher, more homogeneous samples were obtained.

Relative to the oriented and random fibres (Figure 4d,e, respectively), it is observed that the nanoparticles are present and included within the fibres.

3.2.3. Wettability of the Nanocomposites

Material surface characteristics are essential in determining cell response in tissue engineering applications. For this reason, the static CA was measured on the different SiNPs/PVDF nanocomposites and the values are presented in Figure 5.

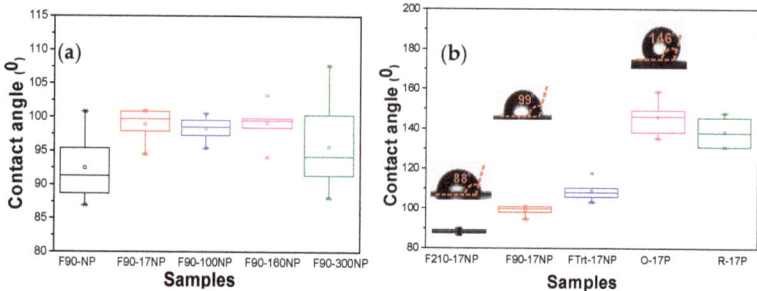

Figure 5. Contact angle of the SiNPs/PVDF nanocomposites: (**a**) PVDF with the SiNPs with different diameters processed at 90 °C and (**b**) SiNPs/PVDF samples with silica nanoparticles (17 nm) with different morphologies.

The introduction of the Si nanoparticles increases the CA values, independently of the diameter of the silica nanoparticles [18], to around 100° excepting for the samples with silica nanoparticles with the highest diameter (F90-300NP). This increase is attributed to the hydrophobic properties of the silica nanoparticles [18]. Samples with nanoparticles with the highest diameter show a higher range of CA values, which is explained by the variation in the diameter of the nanoparticles, as observed in Figure 1. Regarding Figure 5b, the CA for the composite samples with the smallest silica nanoparticles show that the CA of PVDF fibres increases significantly compared to the one of PVDF films, and the CA of the oriented PVDF fibres is slightly higher than the one for randomly oriented PVDF fibres, showing a contact angle of 146.0 ± 7.2°. These results support the idea that the increase in the hydrophobicity of electrospun samples is mainly related to the membrane morphology [8], with the fibres being significantly more hydrophobic than films. In the case of PVDF films, the CA is also higher for films with higher porosity, as already reported for pristine films [43].

3.2.4. Structural Properties and Electroactive Phase Content of the Nanocomposites

FTIR-ATR spectra allow us to identify and quantify (Equation (2)) the polymer phase present in the samples and, therefore, to evaluate possible modifications induced by the introduction of silica nanoparticles (Figure 6).

Figure 6a shows the FTIR spectra of the different samples prepared at 90 °C as well as the corresponding quantification of the β-phase content (Figure 6c, calculated after Equation (1)). The characteristic bands of β PVDF (840 cm^{-1}) is present in all samples, with low traces of α-PVDF (bands at 766, 855 cm^{-1}), with the exception of F210-17NP. This is mainly attributed to the processing temperature [47], which mainly governs the solvent evaporation kinetics and the polymer crystallization in the β phase for processing at temperatures below 90 °C [44]. The introduction of SiNPs in PVDF does not significantly change the β-phase content, independently of the SiNPs content and average diameter. The β-phase value of pristine PVDF is 83 ± 3.3% and for the nanocomposites F90-17NP, F90-100NP, F90-160NP and F90-300NP, is 62 ± 2.5, 91 ± 3.6, 79 ± 3.1 and 74 ± 3, respectively (Figure 6c). On the other hand, Figure 6d shows that, depending on the nanocomposites' morphology, the polymer crystallizes in different phases, mainly due to the different processing conditions. Thus, electrospinning involves room-temperature solvent evaporation and polymer stretching during jet formation, both favourable conditions for the crystallization of the polymer fibres in the β phase [49]. With respect to the films, the F210-17NP nanocomposite, which is processed by a melting and recrystallization process, crystallizes in the α-phase and shows that the addition of SiNPs does not induce the nucleation of the electroactive β-phase of the polymer, as observed in previous study with Fe$_3$O$_4$ spherical nanoparticles [51]. On the other hand, the porous samples, as well as the fibres, are prepared after solvent evaporation at room temperature, conditions leading to the crystallization in the β-phase. This fact is not affected by the introduction of the nanoparticles. Thus, it is concluded that the presence of the nanoparticles does not induce strong interactions with the polymer chain, leading to the nucleation of a specific phase, as observed with other fillers such as $CoFe_2O_4$ [52] and NaY zeolite [43]. Thus, processing temperature and solvent/polymer ratio remain the main factors determining polymer phase content in those composites [39,44].

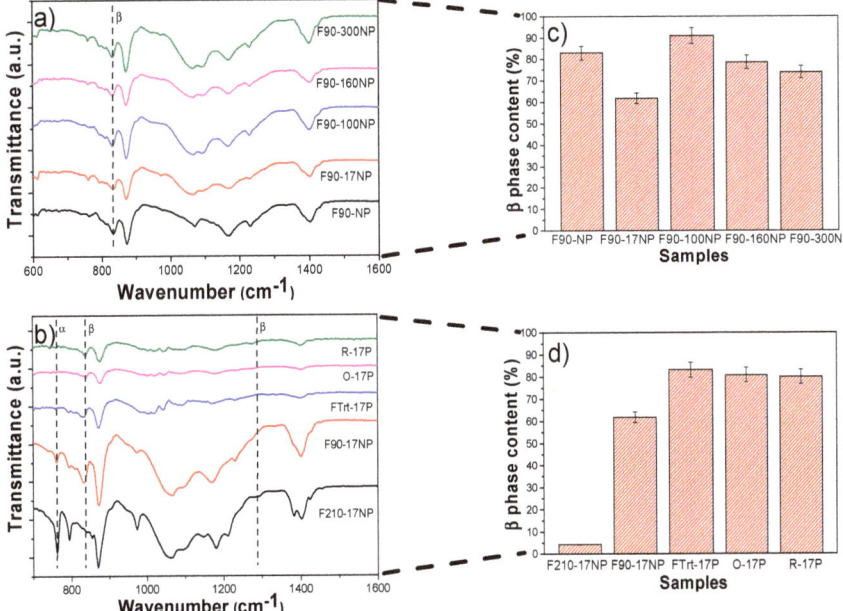

Figure 6. FTIR spectra of (**a**) neat PVDF and SiNPs/PVDF nanocomposites with silica nanoparticles of different diameters processed at 90 °C and (**b**) different morphologies of SiNPs nanocomposites prepared with the smallest nanoparticles. The β-phase content for the different sample is represented in (**c**,**d**).

3.2.5. Thermal Behaviour of the Nanocomposites

The DSC scans allow us to determine the melting temperature and the degree of polymer crystallinity (Figure 7).

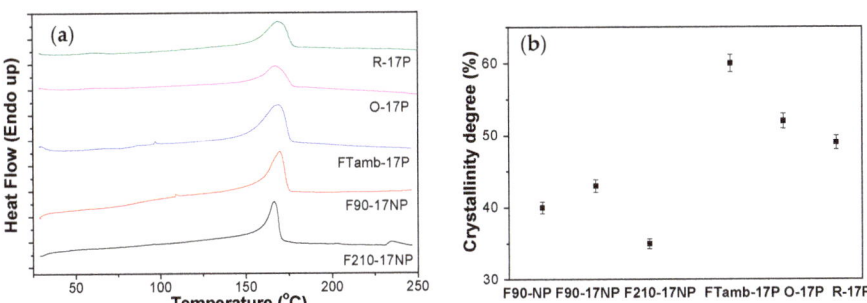

Figure 7. (**a**) DSC thermographs and (**b**) degree of crystallinity of the SiNPs/PVDF nanocomposites with different morphologies and with the fillers of lowest average diameter.

All the samples show an endothermic peak around 168 °C corresponding to the polymer melting of the crystalline phase [44], thus, processing conditions and the incorporation of the filler do not affect the melting temperature. The degree of crystallinity was calculated (Equation (2)) from the enthalpy of the melting peak of the DSC thermograms. It was noticed that the samples prepared by solvent evaporation at 90 °C and after melting and recrystallization showed a lower degree of crystallinity than the samples prepared by solvent evaporation at room temperature, which also includes electrospun samples (Figure 7b). The pristine PVDF film processed at 90 °C shows a degree of crystallinity of ≈40%, which slightly increases with the introduction of the SiNPs and with the size of the SiNPs, being 43% for F90-17NP and 55% for F90-160NP (data not shown). Relative to the different morphologies (Figure 7a), the endothermic peak value is lower for the sample processed at 210 °C, indicating a lower degree of crystallinity if the sample, attributed to the fillers acting as defects during the crystallization from the melt [53]. Inclusion of the nanoparticles in the fibres does not significantly alters the crystallinity degree of the O-17P (52%) and R 17P (49%) with respect to the pristine polymer-oriented fibres (50% [8]).

The latter is ascribed to the combined effect of solvent evaporation at room temperature and stretching during the crystallization process that overcomes the effect of the presence of NP.

3.2.6. Mechanical Properties of the Nanocomposites

The mechanical properties of the materials are essential parameters to design a scaffold suitable for tissues with different mechanical characteristics. The characteristic mechanical strain-stress curves of samples with different morphology, filler type and content are presented in Figure 8.

Figure 8a shows the stress-strain curves for the nanocomposites prepared with fillers with different average diameter after a melting process and Figure 8b refers to the nanocomposites with the same SiNPs (17 nm) after different processing conditions. Independently of the filler average diameter or processing conditions all samples show the typical mechanical behaviour of PVDF [54] characterized by the elastic region, yielding and plastic region, i.e., the typical behaviour of a thermoplastic elastomer.

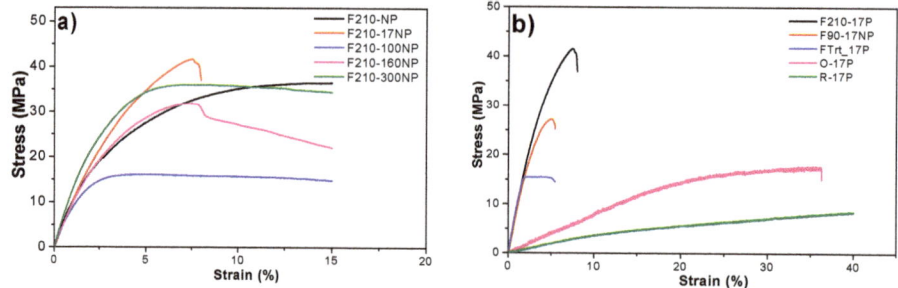

Figure 8. Stress-strain curves for (a) SiNPs/PVDF nanocomposites with different SiNPs average diameters within the PVDF matrix and (b) for nanocomposites obtained after different processing conditions.

The Young's modulus of the samples was calculated from the linear zone of elasticity between 0 and 1% strain, as presented in Figure 9.

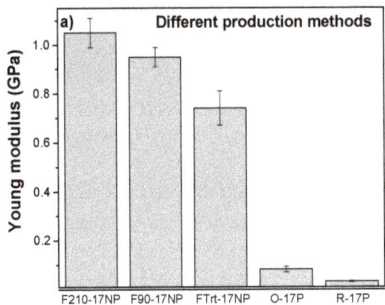 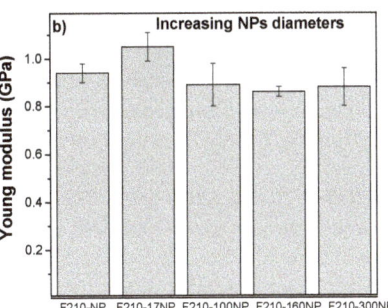

Figure 9. Young's modulus of the SiNPs/PVDF nanocomposites varying (**a**) the processing method and (**b**) the average diameters of the SiNPs. The values are shown as mean ± SD.

The characteristic features of the strain-stress curves are similar for all the materials, demonstrating that the mechanical characteristics are not strongly dependent on nanoparticle diameter. Furthermore, the introduction of particles with different diameters does not significantly affect the Young's modulus of the pristine PVDF (F210-NP) −0.94 ± 0.04 GPa. However, a slight improvement in the Young's modulus is observed for samples prepared with smaller silica nanoparticles (F210-17NP): 1.05 ± 0.06 GPa; this is in line with reports showing that the modulus increases as the particle size decreases [55]. Relative to the different production methods for the polymer films, F210-NP, F90-17NP and Frt-17P, it is observed that the more porous the structure, the lower the Young's modulus, 0.83 ± 0.16 GPa for FTrt-17P. On the other hand, oriented fibres (O-17P) show a higher Young's modulus (0.082 ± 0.012 GPa) than the random fibre samples (R-17P) (0.032 ± 0.002 GPa) due to the larger number of fibres along the stretch direction [8].

Relative to the other samples, the production method has a relevant influence on their mechanical response, as the samples prepared at room temperature by solvent evaporation showed a lower Young's modulus than those obtained at 210 °C due to the porous nature of the former and the compact structure of the latter, as was also visible in the SEM images (Figure 3).

3.3. Cell Culture Studies

In order to explore the potential use of the developed materials in tissue engineering applications, it is necessary to evaluate the putative cytotoxicity of the samples. The study of metabolic activity of C2C12 myoblasts, evaluated with the MTS assay, was applied to all samples and the results for 24 and 72 h are presented in Figure 10. Thus, the effects associated with introducing a fluorescent SiNP with different sizes are analysed, as well as the effect of the different microstructures/morphologies.

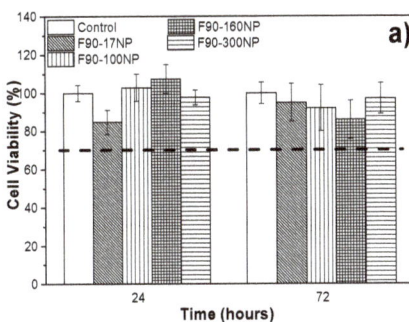 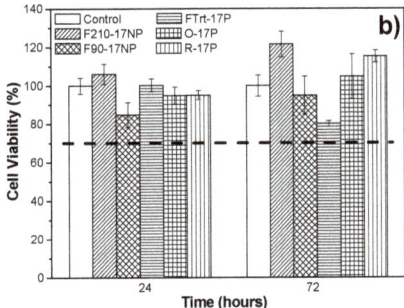

Figure 10. Cytotoxicity indirect test of (**a**) samples prepared with nanoparticles of different diameters and prepared by solvent evaporation at 90 °C and (**b**) samples prepared with SiNPs of 17 nm diameter after different processing methods and therefore with different morphologies.

It has already been reported that PVDF is biocompatible and shows no cytotoxicity to C2C12 cells for 24 or 72 h [29,38]. The SiNPs are also biocompatible for many cells including C2C12 myoblasts [56–58]. It is important to notice that in the polymer composites silica nanoparticles are within the polymer films, avoiding any possible cytotoxic effects from the particles themselves. This is confirmed by the results of the cytotoxic assays of Figure 10. Once PVDF is a non-biodegradable polymer, there is also no risk of the particles leaching out from the films.

Thus, Figure 10 shows that none of the samples are cytotoxic, independently of the nanoparticle diameter and of the material morphology. It is important to note that, despite both materials being biocompatible, the result is not evident, as polymer-filler interface effects or solvent retained in the nanoparticles or in the interface areas, can lead to cytotoxic effects. According to the ISO standard 10993-5, samples are considered cytotoxic when cells suffer a viability reduction larger than 30%. The measured cell viability values are all higher than 70%, confirming the cytocompatibility of the SiNPs/PVDF nanocomposites.

C2C12 myoblasts were used in previous studies to analyse cell proliferation of cultures grown on porous [59] and non-porous [38] PVDF films as well as fibres [38], with the verification that C2C12 cells proliferate better on piezoelectric β-PVDF "poled" samples. The samples obtained in this work were studied to determine the suitability for tissue engineering applications, namely muscle tissue.

MTS (Figure 11), immunofluorescence (Figure 12) and SEM (Figure 13) assays were used to assess cell viability and morphology in the different samples. Relative to the proliferation results (Figure 11), the cell viability has been obtained in relation to the sample of F90-NP at 24 h.

$$\text{Cell Viability}(\%) = \left(\frac{\text{Absorbance of samples at 72 h}}{\text{Absorbance of F90-NP at 24 h}} \times 100 \right) - \text{cell viability of F90-NP at 24 h} \quad (3)$$

Figure 10 shows that the cell viability of the samples increases after 72 h of cell culture, independently of the SiNPs' diameters (Figure 11a) and the morphology of the materials (Figure 11b), when compared with the sample without particles (F90-NP). No significant differences are observed between the samples and the negative control (F90-NP), revealing that C2C12 myoblast proliferation is not affected by the presence of SiNPs in the PVDF matrix. In fact, it has been reported that SiNPs being included in different polymers improves the cell attachment and proliferation, and enhances cellular processes [60,61], which is in agreement with the obtained results.

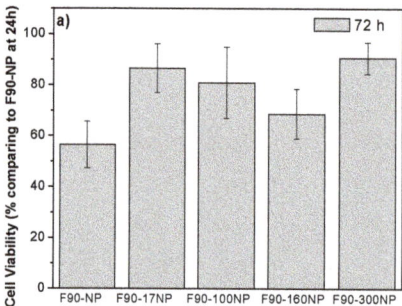 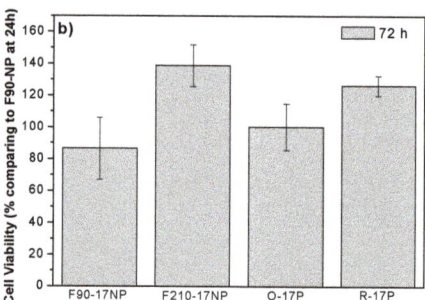

Figure 11. Cell proliferation of C2C12 cells seeded on (**a**) SiNPs/PVDF samples prepared at 90 °C with different sized nanoparticles and (**b**) SiNPs/PVDF samples with different morphologies.

Cell cytoskeleton morphology, viability and adhesion were analysed by fluorescence microscopy for porous and non-porous films and SEM for fibre samples.

Independently of the nanoparticles' diameters and the sample morphology, it is observed that the cell behaviour is similar. Bigger cell agglomerates (and larger nanoparticle agglomerates) are observed with the increasing nanoparticle diameter of the samples (Figure 12a–d). This fact is associated with the interaction between serum proteins and nanoparticles present on the PVDF matrix, as it has been reported that a negative surface charge enhances the adsorption of proteins with isoelectric point more than 5.5 such as immunoglobulin G (IgG) that can be important for C2C12 myoblasts [62,63]. Cell cultures on PVDF fibres prepared with the smaller silica nanoparticles were analysed by SEM and Figure 13 shows the cell morphology of C2C12 cells after 72 h of cell culture on oriented and random PVDF fibre nanocomposites.

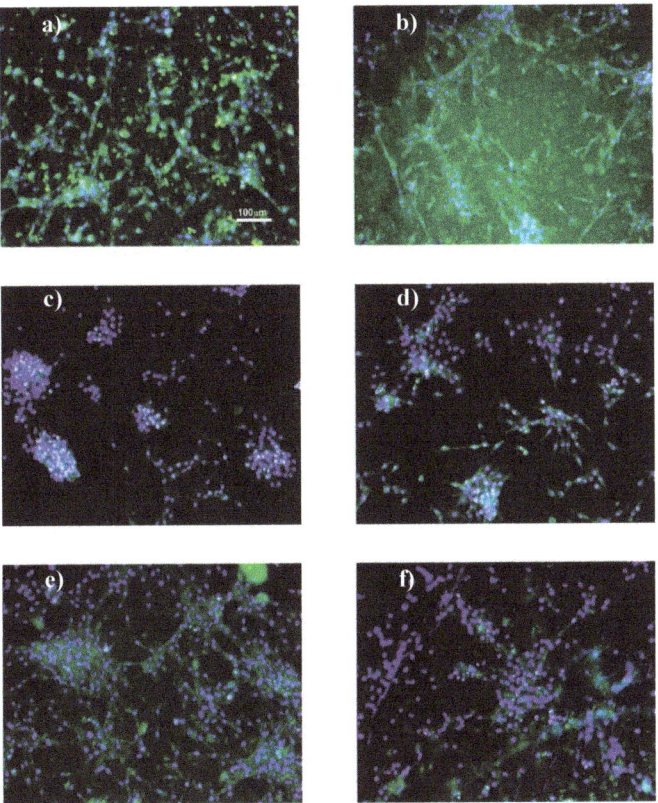

Figure 12. Representative images of C2C12 myoblast culture after 72 h on (**a**) F90-17NP, (**b**) F90-100NP, (**c**) F90-160NP, (**d**) F90-300NP, (**e**) F210-17NP and (**f**) FTrt-17P samples (nucleus stained with DAPI—blue and cytoskeleton stained with FITC—green). Scale bar = 100 μm for all the samples.

These representative images demonstrate that, in the presence of a fibrillar microstructure, the muscle cells orientate their cytoskeleton along the fibres, which is in agreement with the literature [38]. In this way, in the presence of oriented fibres, the cells share a similar architecture to the natural muscle cells in living systems.

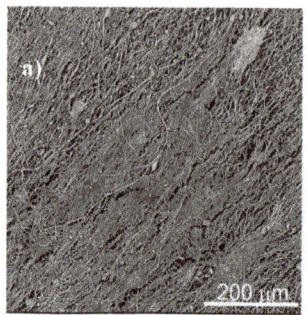 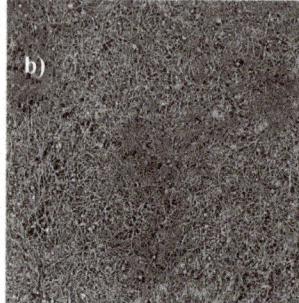

Figure 13. Cell morphology obtained by SEM of C2C12 myoblasts seeded on PVDF fibres: (**a**) O-17P and (**b**) R-17P, after three days of culture. The scale bar is 200 µm for all samples.

Thus, the overall results prove the potential of the use of SiNPs/PVDF piezoelectric nanocomposites for muscle tissue engineering. Physical and chemical stimuli are important factors to obtain tissues with characteristics similar to those of natural living tissues in the human body, developing therefore specific biomimetic microenvironments for different tissues according to their specific biophysico-chemical needs. The developed platform presents nanocomposites with different morphologies (membranes and fibres), piezoelectric β phase and SiNPs diameter (from 17 to 300 nm), which makes it an interesting and complete platform for tissue engineering.

Furthermore, this platform will allow further studies applying mechanical stimuli on the nanocomposites obtained in this work with specific bioreactors [36] applying mechanical and/or mechanoelectrical stimuli. It may also take advantage of the SiNPs' capacity to include specific biomolecules or to develop drug delivery systems, or, more specifically, differentiation factors to promote directed myogenic differentiation. This will not only allow a deeper knowledge of the stimuli necessary for muscle tissue regeneration, but also lead to more effective therapies.

4. Conclusions

Different parameters important for tissue engineering, such as materials morphology, porosity and the PVDF electroactive phase, are modified in the obtained membranes.

Different diameters of silica nanoparticles have been introduced within the PVDF polymer matrix to obtain multifunctional samples for tissue engineering applications.

It is observed that the introduction of the SiNPs fillers in the PVDF matrix decreases its wettability. Furthermore, it is shown that the filler diameter does not significantly affect the properties of the polymer matrix, such as physico-chemical, thermal and mechanical properties.

Cytotoxicity assays with C2C12 cells show no cytotoxicity associated with neat PVDF and composites with different SiNPs diameters and sample morphologies.

Thus, it is demonstrated that the developed platform of PVDF materials with silica nanoparticles demonstrates potential for tissue engineering applications, allowing us to develop electromechanically active microenvironments with different morphologies with SiNPs, allowing protein functionalization and/or controlled release of specific drugs and/or growth or differentiation factors according to the targeted application.

Author Contributions: S.L.-M. conceived and designed the project. S.R. and T.R. contributed to the processing and characterization of the particles. J.P.S.F. and C.B. contributed to the characterization of the nanoparticles. S.R. and D.M.C. contributed to the processing and characterization of the samples in the different morphologies. S.R. was in charge of the cell culture assays and their characterization and interpretation. C.R. contributed to the cell culture assays and the interpretation of the cell culture assays. All authors contributed to the evaluation and interpretation of the data, as well as to the writing of the manuscript. All authors agree with the paper submission.

Funding: This work was supported by the Portuguese Foundation for Science and Technology (FCT) in the framework of the Strategic Funding UID/FIS/04650/2013 and UID/BIA/04050/2013 (POCI-01-0145-FEDER-007569) and project

POCI-01-0145-FEDER-028237 funded by national funds through Fundação para a Ciência e a Tecnologia (FCT) and by the ERDF through the COMPETE2020-Programa Operacional Competitividade e Internacionalização (POCI); and also under the scope of the strategic funding of UID/BIO/04469 unit and COMPETE 2020 (POCI-01-0145-FEDER-006684) and BioTecNorte operation (NORTE-01-0145-FEDER-000004) funded by the European Regional Development Fund under the scope of Norte2020-Programa Operacional Regional do Norte. The authors also thank the FCT for the SFRH/BD/111478/2015 (S.R.), SFRH/BPD/96707/2013 (T.R.), SFRH/BPD/90870/2012 (C.R.) and SFRH/BPD/121526/2016 (D.C) grants. The authors acknowledge funding from the Spanish Ministry of Economy and Competitiveness (MINECO) through the project MAT2016-76039-C4-3-R (AEI/FEDER, UE) and from the Basque Government Industry and Education Departments under the ELKARTEK, HAZITEK and PIBA (PIBA-2018-06) programs, respectively.

Acknowledgments: The SEM measurements have been conducted at the Centre of Biological Engineering (CEB), Braga, Portugal. The authors thank CEB for offering access to their instruments and expertise.

Conflicts of Interest: The authors declare no conflict of interest.

References

1. Camargo, P.H.C.; Satyanarayana, K.G.; Wypych, F. Nanocomposites: Synthesis, structure, properties and new application opportunities. *Mater. Res.-Ibero-Am. J. Mater.* **2009**, *12*, 1–39. [CrossRef]
2. Muller, K.; Bugnicourt, E.; Latorre, M.; Jorda, M.; Sanz, Y.E.; Lagaron, J.M.; Miesbauer, O.; Bianchin, A.; Hankin, S.; Bolz, U.; et al. Review on the processing and properties of polymer nanocomposites and nanocoatings and their applications in the packaging, automotive and solar energy fields. *Nanomaterials* **2017**, *7*, 74. [CrossRef] [PubMed]
3. Raji, M.; Mekhzoum, M.E.M.; Rodrigue, D.; Qaiss, A.E.K.; Bouhfid, R. Effect of silane functionalization on properties of polypropylene/clay nanocomposites. *Compos. B Eng.* **2018**, *146*, 106–115. [CrossRef]
4. Ribeiro, S.; Costa, P.; Ribeiro, C.; Sencadas, V.; Botelho, G.; Lanceros-Méndez, S. Electrospun styrene-butadiene-styrene elastomer copolymers for tissue engineering applications: Effect of butadiene/styrene ratio, block structure, hydrogenation and carbon nanotube loading on physical properties and cytotoxicity. *Compos. B Eng.* **2014**, *67*, 30–38. [CrossRef]
5. Narayanan, K.B.; Han, S.S. Dual-crosslinked poly(vinyl alcohol)/sodium alginate/silver nanocomposite beads—A promising antimicrobial material. *Food Chem.* **2017**, *234*, 103–110. [CrossRef] [PubMed]
6. Ribeiro, C.; Sencadas, V.; Correia, D.M.; Lanceros-Mendez, S. Piezoelectric polymers as biomaterials for tissue engineering applications. *Colloids Surf. B* **2015**, *136*, 46–55. [CrossRef] [PubMed]
7. Cardoso, V.F.; Correia, D.M.; Ribeiro, C.; Fernandes, M.M.; Lanceros-Méndez, S. Fluorinated polymers as smart materials for advanced biomedical applications. *Polymers* **2018**, *10*, 161. [CrossRef]
8. Maciel, M.M.; Ribeiro, S.; Ribeiro, C.; Francesko, A.; Maceiras, A.; Vilas, J.L.; Lanceros-Méndez, S. Relation between fiber orientation and mechanical properties of nano-engineered poly(vinylidene fluoride) electrospun composite fiber mats. *Compos. B Eng.* **2018**, *139*, 146–154. [CrossRef]
9. Cardoso, V.F.; Francesko, A.; Ribeiro, C.; Bañobre-López, M.; Martins, P.; Lanceros-Mendez, S. Advances in magnetic nanoparticles for biomedical applications. *Adv. Healthc. Mater.* **2017**, *7*, 1700845. [CrossRef] [PubMed]
10. Liu, D.; Pallon, L.K.H.; Pourrahimi, A.M.; Zhang, P.; Diaz, A.; Holler, M.; Schneider, K.; Olsson, R.T.; Hedenqvist, M.S.; Yu, S.; et al. Cavitation in strained polyethylene/aluminium oxide nanocomposites. *Eur. Polym. J.* **2017**, *87*, 255–265. [CrossRef]
11. Xu, D.; Cheng, X.; Banerjee, S.; Huang, S. Dielectric and electromechanical properties of modified cement/polymer based 1–3 connectivity piezoelectric composites containing inorganic fillers. *Compos. Sci. Technol.* **2015**, *114*, 72–78. [CrossRef]
12. Liu, D.; Hoang, A.T.; Pourrahimi, A.M.; Pallon, L.H.; Nilsson, F.; Gubanski, S.M.; Olsson, R.T.; Hedenqvist, M.S.; Gedde, U.W. Influence of nanoparticle surface coating on electrical conductivity of LDPE/Al$_2$O$_3$ nanocomposites for HVDC cable insulations. *IEEE Trans. Dielectr. Electr. Insul.* **2017**, *24*, 1396–1404. [CrossRef]
13. Li, Y.; Yang, X.Y.; Feng, Y.; Yuan, Z.Y.; Su, B.L. One-dimensional metal oxide nanotubes, nanowires, nanoribbons, and nanorods: Synthesis, characterizations, properties and applications. *Crit. Rev. Solid State Mater. Sci.* **2012**, *37*, 1–74. [CrossRef]
14. Chen, L.; Jia, Z.; Tang, Y.; Wu, L.; Luo, Y.; Jia, D. Novel functional silica nanoparticles for rubber vulcanization and reinforcement. *Compos. Sci. Technol.* **2017**, *144*, 11–17. [CrossRef]

15. Meinardi, F.; Ehrenberg, S.; Dhamo, L.; Carulli, F.; Mauri, M.; Bruni, F.; Simonutti, R.; Kortshagen, U.; Brovelli, S. Highly efficient luminescent solar concentrators based on earth-abundant indirect-band gap silicon quantum dots. *Nat. Photonics* **2017**, *11*, 177–185. [CrossRef]
16. Bergren, M.R.; Makarov, N.S.; Ramasamy, K.; Jackson, A.; Guglielmetti, R.; McDaniel, H. High-Performance CuInS$_2$ Quantum Dot Laminated Glass Luminescent Solar Concentrators for Windows. *Energy Lett.* **2018**, *3*, 520–525. [CrossRef]
17. Marinins, A.; Shafagh, R.Z.; van der Wijngaart, W.; Haraldsson, T.; Linnros, L.; Veinot, J.G.C.; Popov, S.; Sychugov, I. Light-Converting Polymer/Si Nanocrystal Composites with Stable 60–70% Quantum Efficiency and Their Glass Laminates. *Appl. Mater. Interfaces* **2017**, *9*, 30267–30272. [CrossRef] [PubMed]
18. Ribeiro, T.; Baleizao, C.; Farinha, J.P.S. Functional films from silica/polymer nanoparticles. *Materials* **2014**, *7*, 3881–3900. [CrossRef] [PubMed]
19. Asefa, T.; Tao, Z.M. Biocompatibility of mesoporous silica nanoparticles. *Chem. Res. Toxicol.* **2012**, *25*, 2265–2284. [CrossRef] [PubMed]
20. Rodrigues, A.S.; Ribeiro, T.; Fernandes, F.; Farinha, J.P.S.; Baleizao, C. Intrinsically fluorescent silica nanocontainers: A promising theranostic platform. *Microsc. Microanal.* **2013**, *19*, 1216–1221. [CrossRef] [PubMed]
21. Jiao, J.; Liu, C.; Li, X.; Liu, J.; Di, D.; Zhang, Y.; Zhao, Q.; Wang, S. Fluorescent carbon dot modified mesoporous silica nanocarriers for redox-responsive controlled drug delivery and bioimaging. *J. Colloid Interface Sci.* **2016**, *483*, 343–352. [CrossRef] [PubMed]
22. Burns, A.; Ow, H.; Wiesner, U. Fluorescent core-shell silica nanoparticles: Towards "Lab on a particle" Architectures for nanobiotechnology. *Chem. Soc. Rev.* **2006**, *35*, 1028–1042. [CrossRef] [PubMed]
23. Chen, F.; Hableel, G.; Zhao, E.R.; Jokerst, J.V. Multifunctional nanomedicine with silica: Role of silica in nanoparticles for theranostic, imaging, and drug monitoring. *J. Colloid Interface Sci.* **2018**, *521*, 261–279. [CrossRef] [PubMed]
24. Yan, J.; Estevez, M.C.; Smith, J.E.; Wang, K.; He, X.; Wang, L.; Tan, W. Dye-doped nanoparticles for bioanalysis. *Nano Today* **2007**, *2*, 44–50. [CrossRef]
25. Cardoso, V.F.; Irusta, S.; Navascues, N.; Lanceros-Mendez, S. Comparative study of sol-gel methods for the facile synthesis of tailored magnetic silica spheres. *Mater. Res. Express* **2016**, *3*, 075402. [CrossRef]
26. Slowing, I.I.; Vivero-Escoto, J.L.; Trewyn, B.G.; Lin, V.S.Y. Mesoporous silica nanoparticles: Structural design and applications. *J. Mater. Chem.* **2010**, *20*, 7924–7937. [CrossRef]
27. Liong, M.; Lu, J.; Kovochich, M.; Xia, T.; Ruehm, S.G.; Nel, A.E.; Tamanoi, F.; Zink, J.I. Multifunctional inorganic nanoparticles for imaging, targeting, and drug delivery. *ACS Nano* **2008**, *2*, 889–896. [CrossRef] [PubMed]
28. Bharti, C.; Nagaich, U.; Pal, A.K.; Gulati, N. Mesoporous silica nanoparticles in target drug delivery system: A review. *Int. J. Pharm. Investig.* **2015**, *5*, 124–133. [CrossRef] [PubMed]
29. Parssinen, J.; Hammaren, H.; Rahikainen, R.; Sencadas, V.; Ribeiro, C.; Vanhatupa, S.; Miettinen, S.; Lanceros-Mendez, S.; Hytonen, V.P. Enhancement of adhesion and promotion of osteogenic differentiation of human adipose stem cells by poled electroactive poly(vinylidene fluoride). *J. Biomed. Mater. Res. A* **2015**, *103*, 919–928. [CrossRef] [PubMed]
30. Gilbert, J.R.; Meissner, G. Sodium-calcium ion-exchange in skeletal-muscle sarcolemmal vesicles. *J. Membr. Biol.* **1982**, *69*, 77–84. [CrossRef] [PubMed]
31. Brito-Pereira, R.; Ribeiro, C.; Lanceros-Mendez, S.; Martins, P. Magnetoelectric response on Terfenol-D/P(VDF-TrFE) two-phase composites. *Compos. B Eng.* **2017**, *120*, 97–102. [CrossRef]
32. Ribeiro, C.; Correia, V.; Martins, P.; Gama, F.M.; Lanceros-Mendez, S. Proving the suitability of magnetoelectric stimuli for tissue engineering applications. *Colloids Surf. B* **2016**, *140*, 430–436. [CrossRef] [PubMed]
33. Ribeiro, S.; Correia, D.M.; Ribeiro, C.; Lanceros-Méndez, S. Electrospun polymeric smart materials for tissue engineering applications. In *Electrospun Biomaterials and Related Technologies*; Almodovar, J., Ed.; Springer International Publishing: Cham, Switzerland, 2017; pp. 251–282.
34. Cardoso, V.F.; Ribeiro, C.; Lanceros-Mendez, S. Metamorphic biomaterials. In *Bioinspired Materials for Medical Applications*; Woodhead Publishing: Cambridge, UK, 2016; pp. 69–99.
35. Lee, Y.S.; Collins, G.; Arinzeh, T.L. Neurite extension of primary neurons on electrospun piezoelectric scaffolds. *Acta Biomater.* **2011**, *7*, 3877–3886. [CrossRef] [PubMed]

36. Ribeiro, C.; Moreira, S.; Correia, V.; Sencadas, V.; Rocha, J.G.; Gama, F.M.; Ribelles, J.L.G.; Lanceros-Mendez, S. Enhanced proliferation of pre-osteoblastic cells by dynamic piezoelectric stimulation. *RSC Adv.* **2012**, *2*, 11504–11509. [CrossRef]
37. Ribeiro, C.; Parssinen, J.; Sencadas, V.; Correia, V.; Miettinen, S.; Hytonen, V.P.; Lanceros-Mendez, S. Dynamic piezoelectric stimulation enhances osteogenic differentiation of human adipose stem cells. *J. Biomed. Mater. Res. A* **2015**, *103*, 2172–2175. [CrossRef] [PubMed]
38. Martins, P.M.; Ribeiro, S.; Ribeiro, C.; Sencadas, V.; Gomes, A.C.; Gama, F.M.; Lanceros-Mendez, S. Effect of poling state and morphology of piezoelectric poly(vinylidene fluoride) membranes for skeletal muscle tissue engineering. *RSC Adv.* **2013**, *3*, 17938–17944. [CrossRef]
39. Ribeiro, C.; Costa, C.M.; Correia, D.M.; Nunes-Pereira, J.; Oliveira, J.; Martins, P.; Gonçalves, R.; Cardoso, V.F.; Lanceros-Méndez, S. Electroactive poly(vinylidene fluoride)-based structures for advanced applications. *Nat. Protoc.* **2018**, *13*, 681–704. [CrossRef] [PubMed]
40. Ribeiro, T.; Baleizão, C.; Farinha, J.P.S. Synthesis and characterization of perylenediimide labeled core-shell hybrid silica-polymer nanoparticles. *J. Phys. Chem. C* **2009**, *113*, 18082–18090. [CrossRef]
41. Santiago, A.M.; Ribeiro, T.; Rodrigues, A.S.; Ribeiro, B.; Frade, R.F.M.; Baleizão, C.; Farinha, J.P.S. Multifunctional hybrid silica nanoparticles with a fluorescent core and active targeting shell for fluorescence imaging biodiagnostic applications. *Eur. J. Inorg. Chem.* **2015**, *2015*, 4579–4587. [CrossRef]
42. Ribeiro, T.; Raja, S.; Rodrigues, A.S.; Fernandes, F.; Baleizão, C.; Farinha, J.P.S. Nir and visible perylenediimide-silica nanoparticles for laser scanning bioimaging. *Dyes Pigm.* **2014**, *110*, 227–234. [CrossRef]
43. Lopes, A.C.; Ribeiro, C.; Sencadas, V.; Botelho, G.; Lanceros-Méndez, S. Effect of filler content on morphology and physical-chemical characteristics of poly(vinylidene fluoride)/NaY zeolite-filled membranes. *J. Mater. Sci.* **2014**, *49*, 3361–3370. [CrossRef]
44. Martins, P.; Lopes, A.C.; Lanceros-Mendez, S. Electroactive phases of poly(vinylidene fluoride): Determination, processing and applications. *Prog. Polym. Sci.* **2014**, *39*, 683–706. [CrossRef]
45. Stober, W.; Fink, A.; Bohn, E. Controlled growth of monodisperse silica spheres in micron size range. *J. Colloid Interface Sci.* **1968**, *26*, 62–69. [CrossRef]
46. Martins, P.; Caparros, C.; Gonçalves, R.; Martins, P.M.; Benelmekki, M.; Botelho, G.; Lanceros-Mendez, S. Role of nanoparticle surface charge on the nucleation of the electroactive β-poly(vinylidene fluoride) nanocomposites for sensor and actuator applications. *J. Phys. Chem. C* **2012**, *116*, 15790–15794. [CrossRef]
47. Ferreira, J.C.C.; Monteiro, T.S.; Lopes, A.C.; Costa, C.M.; Silva, M.M.; Machado, A.V.; Lanceros-Mendez, S. Variation of the physicochemical and morphological characteristics of solvent casted poly(vinylidene fluoride) along its binary phase diagram with dimethylformamide. *J. Non-Cryst. Solids* **2015**, *412*, 16–23. [CrossRef]
48. Sencadas, V.; Martins, P.; Pitães, A.; Benelmekki, M.; Gómez Ribelles, J.L.; Lanceros-Mendez, S. Influence of ferrite nanoparticle type and content on the crystallization kinetics and electroactive phase nucleation of poly(vinylidene fluoride). *Langmuir* **2011**, *27*, 7241–7249. [CrossRef] [PubMed]
49. Ribeiro, C.; Sencadas, V.; Ribelles, J.L.G.; Lanceros-Méndez, S. Influence of processing conditions on polymorphism and nanofiber morphology of electroactive poly(vinylidene fluoride) electrospun membranes. *Soft Mater.* **2010**, *8*, 274–287. [CrossRef]
50. Ribeiro, T.; Raja, S.; Rodrigues, A.S.; Fernandes, F.; Farinha, J.P.S.; Baleizao, C. High performance nir fluorescent silica nanoparticles for bioimaging. *RSC Adv.* **2013**, *3*, 9171–9174. [CrossRef]
51. Sebastian, M.S.; Larrea, A.; Goncalves, R.; Alejo, T.; Vilas, J.L.; Sebastian, V.; Martins, P.; Lanceros-Mendez, S. Understanding nucleation of the electroactive beta-phase of poly(vinylidene fluoride) by nanostructures. *RSC Adv.* **2016**, *6*, 113007–113015. [CrossRef]
52. Martins, P.; Costa, C.M.; Lanceros-Mendez, S. Nucleation of electroactive β-phase poly(vinilidene fluoride) with $CoFe_2O_4$ and $NiFe_2O_4$ nanofillers: A new method for the preparation of multiferroic nanocomposites. *Appl. Phys. A* **2011**, *103*, 233–237. [CrossRef]
53. Costa, C.M.; Rodrigues, L.C.; Sencadas, V.; Silva, M.M.; Lanceros-Méndez, S. Effect of the microstructure and lithium-ion content in poly[(vinylidene fluoride)-*co*-trifluoroethylene]/lithium perchlorate trihydrate composite membranes for battery applications. *Solid State Ionics* **2012**, *217*, 19–26. [CrossRef]
54. Zeng, F.; Liu, Y.; Sun, Y. Mechanical properties and fracture behavior of poly(vinylidene fluoride)-polyhedral oligomeric silsesquioxane nanocomposites by nanotensile testing. In Proceedings of the 13th International Conference on Fracture 2013 (ICF 2013), Beijing, China, 6–12 June 2013; pp. 4862–4869.

55. Xu, Y.L.; Yu, L.Y.; Han, L.F. Polymer-nanoinorganic particles composite membranes: A brief overview. *Front. Chem. Eng. China* **2009**, *3*, 318–329. [CrossRef]
56. Beck, G.R.; Ha, S.W.; Camalier, C.E.; Yamaguchi, M.; Li, Y.; Lee, J.K.; Weitzmann, M.N. Bioactive silica-based nanoparticles stimulate bone-forming osteoblasts, suppress bone-resorbing osteoclasts, and enhance bone mineral density in vivo. *Nanomedicine* **2012**, *8*, 793–803. [CrossRef] [PubMed]
57. Liu, D.; He, X.X.; Wang, K.M.; He, C.M.; Shi, H.; Jian, L.X. Biocompatible silica nanoparticles-insulin conjugates for mesenchymal stem cell adipogenic differentiation. *Bioconjug. Chem.* **2010**, *21*, 1673–1684. [CrossRef] [PubMed]
58. Poussard, S.; Decossas, M.; Le Bihan, O.; Mornet, S.; Naudin, G.; Lambert, O. Internalization and fate of silica nanoparticles in c2c12 skeletal muscle cells: Evidence of a beneficial effect on myoblast fusion. *Int. J. Nanomed.* **2015**, *10*, 1479–1492.
59. Nunes-Pereira, J.; Ribeiro, S.; Ribeiro, C.; Gombek, C.J.; Gama, F.M.; Gomes, A.C.; Patterson, D.A.; Lanceros-Mendez, S. Poly(vinylidene fluoride) and copolymers as porous membranes for tissue engineering applications. *Polym. Test.* **2015**, *44*, 234–241. [CrossRef]
60. Mehrasa, M.; Asadollahi, M.A.; Nasri-Nasrabadi, B.; Ghaedi, K.; Salehi, H.; Dolatshahi-Pirouz, A.; Arpanaei, A. Incorporation of mesoporous silica nanoparticles into random electrospun plga and plga/gelatin nanofibrous scaffolds enhances mechanical and cell proliferation properties. *Mater. Sci. Eng. C* **2016**, *66*, 25–32. [CrossRef] [PubMed]
61. Mehrasa, M.; Anarkoli, A.O.; Rafienia, M.; Ghasemi, N.; Davary, N.; Bonakdar, S.; Naeimi, M.; Agheb, M.; Salamat, M.R. Incorporation of zeolite and silica nanoparticles into electrospun pva/collagen nanofibrous scaffolds: The influence on the physical, chemical properties and cell behavior. *Int. J. Polym. Mater. Polym. Biomater.* **2016**, *65*, 457–465. [CrossRef]
62. Gessner, A.; Lieske, A.; Paulke, B.R.; Müller, R.H. Influence of surface charge density on protein adsorption on polymeric nanoparticles: Analysis by two-dimensional electrophoresis. *Eur. J. Pharm. Biopharm.* **2002**, *54*, 165–170. [CrossRef]
63. Aggarwal, P.; Hall, J.B.; McLeland, C.B.; Dobrovolskaia, M.A.; McNeil, S.E. Nanoparticle interaction with plasma proteins as it relates to particle biodistribution, biocompatibility and therapeutic efficacy. *Adv. Drug Deliv. Rev.* **2009**, *61*, 428–437. [CrossRef] [PubMed]

© 2018 by the authors. Licensee MDPI, Basel, Switzerland. This article is an open access article distributed under the terms and conditions of the Creative Commons Attribution (CC BY) license (http://creativecommons.org/licenses/by/4.0/).

Article

Elastomeric Polyurethane Foams Incorporated with Nanosized Hydroxyapatite Fillers for Plastic Reconstruction

Lili Lin, Jingqi Ma, Quanjing Mei, Bin Cai, Jie Chen, Yi Zuo *, Qin Zou, Jidong Li and Yubao Li *

Research Center for Nano-Biomaterials, Analytical and Testing Center, Sichuan University, Chengdu 610064, China; linll900730@163.com (L.L.); jingqima2018@gmail.com (J.M.); scumqj@foxmail.com (Q.M.); haisengyan@sina.com (B.C.); ChenJie005@outlook.com (J.C.); zouqin80913@126.com (Q.Z.); nic1979@scu.edu.cn (J.L.)

* Correspondence: zoae@scu.edu.cn (Y.Z.); nic7504@scu.edu.cn (Y.L.); Tel.: +86-136-8342-0973 (Y.Z.); +86-139-0804-2111 (Y.L.)

Received: 30 October 2018; Accepted: 21 November 2018; Published: 25 November 2018

Abstract: Plastic surgeons have long searched for the ideal materials to use in craniomaxillofacial reconstruction. The aim of this study was to obtain a novel porous elastomer based on designed aliphatic polyurethane (PU) and nanosized hydroxyapatite (n-HA) fillers for plastic reconstruction. The physicochemical properties of the prepared composite elastomer were characterized by infrared spectroscopy (IR), X-ray diffraction (XRD), scanning electron microscopy with energy dispersive X-ray spectroscopy (SEM-EDX), transmission electron microscopy (TEM), thermal analysis, mechanical tests, and X-ray photoelectron spectroscopy (XPS). The results assessed by the dynamic mechanical analysis (DMA) demonstrated that the n-HA/PU compounded foams had a good elasticity, flexibility, and supporting strength. The homogenous dispersion of the n-HA fillers could be observed throughout the cross-linked PU matrix. The porous elastomer also showed a uniform pore structure and a resilience to hold against general press and tensile stress. In addition, the elastomeric foams showed no evidence of cytotoxicity and exhibited the ability to enhance cell proliferation and attachment when evaluated using rat-bone-marrow-derived mesenchymal stem cells (BMSCs). The animal experiments indicated that the porous elastomers could form a good integration with bone tissue. The presence of n-HA fillers promoted cell infiltration and tissue regeneration. The elastomeric and bioactive n-HA/PU composite foam could be a good candidate for future plastic reconstruction.

Keywords: elastomeric foam; plastic reconstruction; hydroxyapatite/polyurethane; nanosized dispersion; viscoelasticity; biocompatibility

---

1. Introduction

Plastic materials have a huge clinical market for patients of plastic, reconstructive, and aesthetic surgery [1]. The development of orthopedic materials plays a crucial role in plastic surgery, which benefits from the spring-up of new materials [2]. Maxillofacial, oral, and plastic surgeons keep on looking for ideal materials, which should have osteointegration potential, biological safety, easy molding and good plasticity, appropriate mechanical support, and should be cost-effective for skeleton reconstruction and shape restoration [3,4]. In the late of 1960s, W. L. Gore invented expanded polytetrafluoroethylene (e-PTFE), which has become a popular material in the field of facial plastic and reconstructive surgery [5]. The e-PTFE sheets have been used in the temporal region and have achieved good results, of which the physic trait of the e-PTFE sheet is similar to human soft tissue [6]. Moreover, hyaluronic acid is another useful material for tissue augmentation, because of its unique and favorable biophysical properties [7]. However, the use of hyaluronic acid is very limited because of its very short lifetime, and because of its rapid elimination from the injection site in less than a week [8]. It is regretful that both the e-PTFE and the hyaluronic acid are not bioactive and cannot form a bond with bone tissue that easily introduces the implant migration after implantation.

Calcium phosphate mimics the osteoid compound of mature bone, which is a foundation of the bioactive techniques for alloplastic cranioplasty [9]. The excellent biocompatibility and bone-bonding bioactivity of synthetic hydroxyapatite (HA) have made it attractive in plastic surgery. Yet, the pristine ceramic based on inorganic HA is too brittle, and it is difficult to mold or tailor in the operation [10]. In recent years, hybrid composites composed of polymers with nanosized hydroxyapatite (n-HA) may create a bio-mimetic interface to host tissue, and can be considered as one of the most promising biomaterials. Different n-HA/polymer composites, such as n-HA added poly(lactic-co-glycolic acid) (PLGA) or polycaprolactone (PCL), have been developed and broadly investigated as scaffolds for bone repair [11–13]. But few of them have been used for plastic and reconstructive surgery, because the specific biomaterials need match to the comfortable and aesthetic requirements of tissue augmentation, such as resilience, adjusting physical softness and hardness, contact pressure, uniform block, and easy shaping [14,15]. It is well known that there is a class of polymer materials that can be synthesized as "soft or hard scaffolds" by different segmented structures [16]. The synthetic polymer called polyurethane (PU) has been applied in tissue engineering with the advent of a wide range of mechanical and physical properties by changing the ratio of soft and hard segments and the composition [17,18]. PU exhibit diverse properties with a self-foaming porous and unique segmented structure, excellent and varying mechanical properties, and good biocompatibility, which could be considerable interest to specific tissue [19,20]. In our previous studies, several kinds of polyurethane (PU) rigid foams compounded with n-HA nanoparticles have been developed, chiefly for bone repair and regeneration. As a result, the crosslinked structure based on castor oil as a soft segment, has displayed a high modulus for the PU scaffolds for bone regeneration [21,22]. But for plastic surgery, the moderate flexibility of the PU scaffolds is the potential and preferable characteristic for biomedical applications.

In this study, we designed and prepared a foam based on the elastomeric PU matrix and n-HA crystals, to match with the multifunctional requirements for plastic reconstruction. The elastic foam was fabricated in a non-toxic manner, and the obtained product was expected to be easily tailored and to exhibit both enough mechanical support and elastic adaptability, as well as to provide a bioactive microenvironment for cell growth and tissue regeneration. The elasticity of the scaffolds, including resilience and stretchability, has been deeply studied, for the deformation ability is one of the major factors to consider for plastic surgeons. The elastic n-HA/PU foams have been postulated to have a promising prospect for future plastic surgery or reconstruction fields of specific types of tissue.

2. Materials and Methods

2.1. Materials

Raw reagents that polytetramethylene ether glycol (PTMEG) as a soft segment, isophorone diisocyanate (IPDI) as hard segment, and 1,4-butanediol (BDO) and dibutyltin dilaurate (DBTL) as a chain extender and catalyst, respectively, were purchased from Aladdin Industrial Corporation (Shanghai, China). The n-HA particles were prepared through a chemical precipitation method, as reported in our previous study [23]. All of the reagents were of analytical reagent grade.

2.2. Preparation of Porous Elastomer Foams

The following three types of polymeric foams were prepared: (1) pure PU, (2) PU with 10 wt% n-HA (10HA/PU), and (3) PU with 20 wt% n-HA (20HA/PU). Considering that a higher content of inorganic HA will produce a stiffer scaffold, which is not proper for the plastic elastomer, the maximum content of HA has been set to 20 wt%. The porous elastomers were prepared in a three-necked flask under a nitrogen atmosphere. Briefly, PTMEG and IPDI were first mixed thoroughly at 70 °C. The stoichiometric ratio of [NCO]/[OH], called the isocyanate index, was set at 1.5. Then, certain amounts of BDO, DBTL, and distilled water were added successively as chain extender, catalyst, and foaming agent, respectively [24], followed by polymerization at 70 °C for 2 h under stirring. For the composite foam, the n-HA powder was filled in the mixture during the polymerization process. After polymerization, the mixture was finally formed and dried in an oven at 90 °C for 24 h, in order to obtain the porous PU scaffold or the n-HA/PU scaffolds. The scaffolds were cut into discs (φ 14 mm × 1.5 mm) for in vitro cell culture and (φ 6 mm × 1.5 mm) for in vivo studies.

2.3. Characterization Analyses

Scanning electron microscope (SEM, JEOL 6500LV, JEOL, Tokyo, Japan) with an energy dispersive spectrometer (EDS, INCA ENERGY 250, Oxford, UK) was used to observe the structure and the Ca and P element mapping of the samples. The test samples were lamellar squares (10 mm × 10 mm × 1 mm) and were sputter-coated with gold before the examination. Transmission electron microscopy (TEM, Tecnai, G2 F20, Hillsboro, OR, USA) was used to observe the morphology of the n-HA crystals. To clearly observe the hard and soft segmented structure of the PU matrix, cryoultramicrotomy with a diamond knife was used to obtain a 50 nm thickness ultrathin section. The microphase image of the PU scaffolds was labeled using a staining agent of ruthenium tetroxide (RuO_4), for which the staining and exposure time was 15 min [25].

The chemical groups of the porous materials were recorded using Fourier transform infrared spectroscopy (FTIR, Nicolet 6700, Nicolet Perkin Elmer Co., Waltham, MA, USA), at a wavenumber range of 650–4000 cm^{-1}. The crystallographic structures of the PU composites were tested using X-ray diffraction (XRD) (PANalytical B.V, EMPYREAN, Almelo, Netherlands) with a Cu Kα radiation. The conditions were 40 kV and 25 mA, and the 2θ scan range was from 10° to 60° at a step size of 0.03°. The element spectra of the PU scaffolds were tested using X-ray photoelectron spectroscopy (XPS, AXIS Ultra DLD, Manchester, UK.) at a step of 1000 m eV with a mono Al X-ray source (150 W). All of the above test samples were lamelliform at a size of 10 mm × 10 mm × 1 mm. The thermal stability of the PU composites was measured using a thermogravimetric analysis (TGA) simultaneous thermal analyzer (STA 449 F3 Jupiter®, NETZSCH, Bavaria, Germany), under nitrogen flow rate at 20 mL/min. The TGA temperature range was set from 50 to 600 °C at 10 C/min, after all of the PU scaffolds were dried at 60 °C for 48 h in a vacuum oven so as to eliminate the humidity, and then samples of about 4–5 mg were placed in the aluminum crucible for testing.

The three-dimensional (3D) structure of the scaffold block (φ 6 mm × 12 mm) was reconstructed using a vivaCT80 micro-CT imaging system (SCANCO Medical AG, Brüttisellen, Switzerland). Three groups of foams were scanned, of which a slice increment of 20 μm and a total of 600 microtomographic slices were taken for each sample. The scanning system was set at 45 kV and 175 μA. In addition to the visual assessment of the structural images, the morphometric parameters were directly measured using 3D morphometry from the micro-tomographic datasets, including the density and the fraction of the bone volume/total volume (BV/TV), which means that the materials volume/total volume, and the porosity of the elastomers could be calculated according to the following formula:

$$Porosity_{CT} = (1 - (BV/TV)) \times 100\% \tag{1}$$

The opening pore porosity of the foams was also tested using a liquid displacement method. The scaffold samples had a volume (V) of 10 mm × 10 mm × 20 mm, and their weight (M) was recorded. After the samples were immersed in distilled water, a series of evacuation–repressurization cycles were conducted in order to exhaust the gas and force the liquid into the pores of the scaffolds, until the water was squeezed into all of the opening pores of the porous foams. Then, the weight of the scaffolds was recorded as M_0. The porosity was calculated according to the following formula:

$$Porosity = (M_0 - M)/V \times 100\% \tag{2}$$

The apparent density (ϱ) was determined from the mass and volume data of the samples, and was calculated by the following equation:

$$\varrho = M/V \tag{3}$$

Then, the weight and volume of the scaffolds were recorded as M and V, respectively. Five samples were tested in each group.

To study the resilience of the PU foams, the samples were characterized using dynamic mechanical thermal analysis (DMTA) on DMA Q800 (NETZSCH, Bavaria, Germany). The specimens with a dimension of 40 mm × 10 mm × 1 mm were uniaxially deformed in tension mode at a 1 Hz oscillating frequency. The temperature range was −50 to 130 °C at a rate of 3 °C/min. Three samples were tested in each group.

To study the stretchability of the PU elastomers, the tensile strength and Young's modulus of the elastomeric PU scaffolds were tested using a universal testing machine (AG-IC 50KN, SHIMADZU, Kyoto, Japan). The standard of ASTM D695-96 was adopted, and the dimension of the samples was 40 mm × 10 mm × 1 mm. All of the tests were conducted at the speed of 15 mm min^{-1}, with a load capacity of 250 N at room temperature and the typical humidity of the laboratory air was about 80%. Five independent samples were tested for each group.

2.4. Cell Culture

Cell culture and proliferation: Bone marrow mesenchymal stem cells (BMSCs) isolated from Sprague Dawley rats (one-month-old, male), with a weight of approximately 100 g, were provided by the West China Animal Center of Sichuan University, and the third passage of the BMSCs were utilized in the experiments [26,27]. After ultrasonic rinsing (SB3200D, Ultrasonic cleanser, Ningbo Xinyi, Ningbo, China) in distilled water and autoclave sterilization (GI54DWS, Autoclave, Schneider Electric, Rueil, France), the PU porous specimens were co-cultured with BMSCs (2×10^4 cells/well) in 24-well plates equilibrated in α type minimum Eagle's medium (α-MEM) medium (Gibco, 1 mL/well) in a humidified incubator (37 °C, 5% CO_2). The tissue culture plastic was set as the blank control group for comparison. After seeding, the media was replaced every 48 h for the next 20 days. After 1, 4, 7, and 14 days, the cell proliferation was evaluated using a Cell Counting Kit-8 (CCK-8) kit (Sigma-Aldrich Co., St. Louis, MO, USA). The absorbance values of the water-soluble tetrazolium salt were detected at 450 nm using a multi-label counter (Wallac Victor3 1420, PerkinElmer Co., Waltham, MA, USA). Five samples were tested in each group.

Cell morphology: The extracted solution of the PU and HA/PU samples was used to culture the BMSCs for cell morphology observation, according to the leachate method specified in GB/T 16886.5-2003. The morphology and spreading of the BMSCs growing in the extracted solution of the samples were imaged using a laser scanning confocal microscopy (LSCM; Nikon, A1R, MP+, Toyko, Japan). The BMSCs cultured with the extracted solution at 37 °C for 96 h in a humidified atmosphere of 5% CO2 and 95% air, in between, changed the extracted solution in two days. After incubation, the cells were fixed in a 3.7% (w/v) formaldehyde solution; permeabilized with 0.2% (v/v) Triton X-100/phosphate buffered saline (PBS); and stained successively with 0.25 µM MitoTracker® Green FM probes for chondriosome, 0.15 µM Alexa Fluor® 532 phalloidin for red F-actin, and 5 µg/mL blue fluorescent Hoechst 33342 in 1% (w/v) bovine serum albumin (BSA)/PBS for the nucleus, according to the manufacturer's instructions of Thermo Fisher Scientific Co (Waltham, MA, USA). The cells were then washed with PBS to remove the excess stains. The labeled cells were imaged using LSCM.

Cell differentiation: A quantitative analysis of the ALP and OCN activity for the BMSCs cells cultured on the three scaffolds using enzyme-linked immunosorbent assay (ELISA) has been employed to assess the osteogenic differentiation study [21]. Each sample was seeded with 1 mL of the cell suspension (2×10^4 cells/per scaffold) in 24-well plate on the first day, then the media was replaced every 48 h over the next 20 days. The culture media contained an osteogenic medium (50 µg/mL L-ascorbic acid,

10 nM dexamethasone, and 10 mM β-glycerophosphate) used to promote the osteoblastic differentiation of the BMSCs. The activity of the alkaline phosphatase (ALP) and the content of osteocalcin (OCN) in the BMSCs were analyzed on day 7, 14, and 21. The samples were firstly washed with PBS three times after the culture media were removed, and afterward, the cells were lysed with 0.2% (v/v) Triton X-100/phosphate buffered saline (PBS) on the sample, overnight at 4 °C. The ALP and OCN were quantified separately using ALP activity kits and OCN content assay kits (Shanghai MLBIO Biotechnology CO., Shanghai, China). A series of dilutions were prepared in order to obtain the standard curve, and the absorbance at 450 nm was read using a microplate reader [28–30].

2.5. Osteogenic Capacity

Implantation in mandible defects: Twelve rabbits (New Zealand white, male or female was random, about 2.5 kg) were employed in the animal experiments. The implantation periods were set for 4 and 12 weeks. At each time point, two rabbits were in each group, and three groups, were operated on, according to the control, PU, and 20HA/PU groups. The animal experiments on the PU scaffolds were approved by the Animal Ethics Committee of Sichuan University, according to all of the regulations. After the PU and 20HA/PU foams were cut into discs with a diameter of 6 mm and a height of 1.5 mm, the samples were sterilized by autoclave 15 min, three times before use. A defect similar in size to the sample (φ 6 mm × 1.5 mm) was created on both sides of the mandible of each rabbit, and the foam sample was implanted into the defect of each experimental group. Then, the muscle and skin incisions were sutured layer-by-layer. An intramuscular injection of 1×10^5 units of penicillin was given to each rabbit during the initial three days after the implant operation. At the time points of 4 and 12 weeks, the rabbits were sacrificed via CO_2 asphyxiation, and the samples with the surrounding tissue were harvested.

Histological observation: The collected implantation samples were fixed with 4% buffered paraformaldehyde, decalcified, dehydrated by gradient ethanol, cleaned in xylene, and embedded with paraffin wax. The samples were stained with a hematoxylin-eosin (HE) staining agent (Thermo Fisher, Waltham, MA, USA) after being cut into thin sections (5 µm) along the sagittal plane, and were observed under optical microscopy. A schematic diagram showing the material preparation and animal experiment is given in Scheme 1.

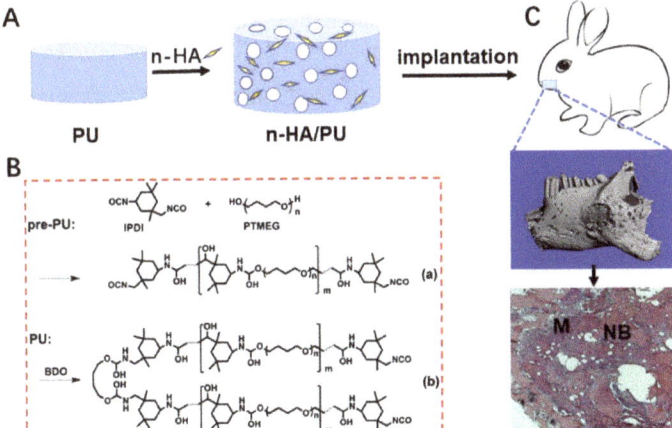

Scheme 1. Schematic diagram showing the material preparation and animal experiment. (**A**) The addition of n-hydroxyapatite (HA) fillers in a polyurethane (PU) matrix to form a porous composite scaffold; (**B**) the chemical molecules and reaction process (step I: prepolymerization process (**a**); step II: the chain extension process (**b**)); (**C**) implantation in the rabbit mandible defect and histological section. M—material; NB—new bone.

2.6. Statistical Analysis

The test results were expressed as the mean ± standard deviation (SD), calculated using Microsoft Excel (Microsoft, Redmond, WA, USA) software. The statistical significance was determined using the statistical analyses that were performed using Origin 2016 (OriginLab, Northampton, MA, USA). Here, $p < 0.05$ (*) indicates statistically significant, $p < 0.01$ (**) indicates very significant, and $p < 0.001$ (***) indicates extremely significant values.

3. Results and Discussion

The prepared foams of PU, 10HA/PU, and 20HA/PU have an interconnected porous structure shown by the micro-CT images in Figure 1A–F. The porosity decreases with the increase of the n-HA content. In Table 1, all of the measured porosities of the PU, 10HA/PU, and 20HA/PU scaffolds calculated using micro-CT are higher than that calculated by the liquid displacement method (LDM). The data difference between the micro-CT and LDM should be caused by the closed pores, in that the latter does not calculate the closed pores. Whether calculated by micro-CT or LDM, the porosity of the foams decreased with the increasing of the n-HA content, and the density value was the opposite (Table 1). The pore size ranged from 100 to 500 μm, as shown in the SEM photo in Figure 1G, which is suitable for new bone formation [29]. SEM–EDX gives the element mapping of calcium (Figure 1H) and phosphorus (Figure 1I), which indicates a uniform distribution of n-HA particles in the scaffold matrix. The TEM micrograph (Figure 1J) and the HRTEM image (Figure 1K) of the 10HA/PU composite further exhibit the presence of the rod-like n-HA particles, at a nanoscale of width 10–20 nm and length 50–100 nm, dispersed in the PU matrix homogenously, and no interface gap or phase separation is present. The nano-domain of the PU soft segments (in black) and hard segments (in white) were also displayed by the HRTEM image (Figure 1L), after the PU matrix was stained using ruthenium tetroxide. The high interpenetration and bridging of the soft and hard segments ensures the good quality and property of the PU matrix.

Table 1. The porosity and density of the foams measured using the micro-CT and liquid displacement method (LDM). PU—polyurethane; HA—hydroxyapatite.

Sample	Porosity by Micro-CT (%)	Porosity by LDM (%)	Density by Micro-CT (g/cm^3)	Density by LDM (g/cm^3)
PU	80.57	78 ± 5	0.16	0.17 ± 0.02
10HA/PU	75.26	71 ± 2	0.24	0.24 ± 0.02
20HA/PU	65.95	62 ± 4	0.34	0.35 ± 0.01

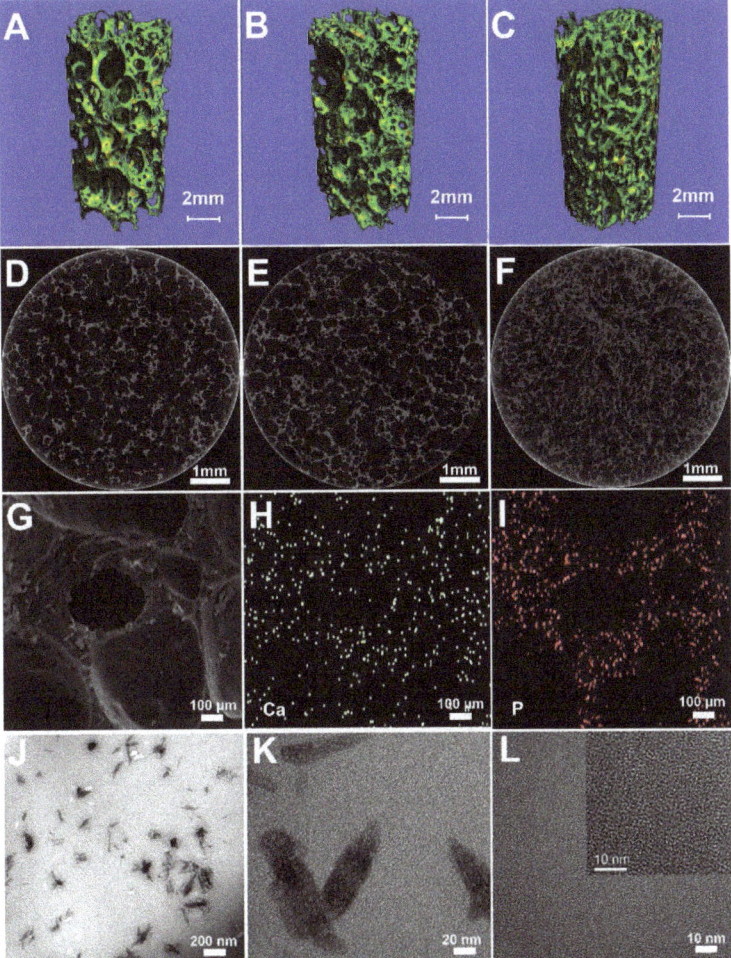

Figure 1. The micro-CT images of PU (**A**, **D**), 10HA/PU (**B**, **E**), and 20HA/PU (**C**, **F**); the SEM photo of 10HA/PU (**G**) and corresponding EDX element mapping of calcium (**H**) and phosphorus (**I**); the TEM micrograph of the 10HA/PU composite (**J**) and HRTEM image (**K**), showing a close interface between the HA nanoparticles and PU matrix; the ruthenium tetroxide stained PU matrix (**L**) showing a homogeneous structure of hydrophilic and hydrophobic phase caused by hard (in white) and soft (in black) segments.

The XRD patterns, FTIR spectra, and XPS spectra were further tested to reveal the phase composition of the foams, and the chemical groups or chemical bonding of the n-HA and PU matrix. As we know, amorphous substances exhibit diffuses and diffraction patterns, whereas crystalline materials show strong diffraction peaks in the XRD spectrogram. In Figure 2A, the pure PU matrix shows only an envelope spectrum with a broad peak centered at 18°, whereas the n-HA particles exhibit a crystallized pattern with characteristic peaks at approximately 25.8°, 32°, 34°, 35.5°, and 40°, and so on. The XRD patterns of 10 HA/PU and 20 HA/PU still exhibit their individual peaks, except for a relative decrease in the peak intensity. The results make it clear that the composite is composed of HA and PU, and their initial composition and chemical structure are maintained in the composite foams [30–32].

The FTIR spectrum can be used to confirm the chemical groups or linkages within a polymeric structure, along with the extent of hydrogen bonding, conformation, and accessibility and interaction between the hard and soft segments in polyurethane [33]. The FTIR spectra obtained from the attenuated total reflection (ATR) methods in Figure 2B confirms the formation of urethane linkages by observing the respective peak positions. The peak at 3570 cm^{-1} represents the –OH groups of HA, and the peak around 3250 cm^{-1} represents the –NH groups of PU. The combined band at 3200–3500 cm^{-1} in the 10 HA/PU and 20 HA/PU composites is because of the overlapping of the –NH and –OH stretching vibrations. The band at 1070–1098 cm^{-1} is the overlapping of the PO_4^{3-} groups of HA and the –O–C=O stretching vibrations of the PU polymer, and the band at 1020–1040 cm^{-1} also belongs to the PO_4^{3-} groups. The peaks at 2945–2900 cm^{-1} and 2830–2800 cm^{-1} are ascribed to eh symmetric and asymmetric vibration of the CH_2 groups of PU [34], which mostly represent the hard and soft segments, respectively. The bands at 1630–1660 cm^{-1} and 1680–1749 cm^{-1} represent the carbonyl C=O stretching of the allophanate or urethane (CONH) groups, and the bands at 1530–1550 cm^{-1} are from the –CN stretching/–NH bending [21]. The presence of the C=O, –CN/–NH, and –O–C=O characteristic bands could confirm the formation of urethane linkages in the n-HA/PU composite scaffolds. The weakening of the –OH peak at 3570 cm^{-1} and the free carbonyl peak at 1747 cm^{-1} in the composites indicates the formation of more hydrogen bonding [35], which could lead to strong interface and intermolecular interactions.

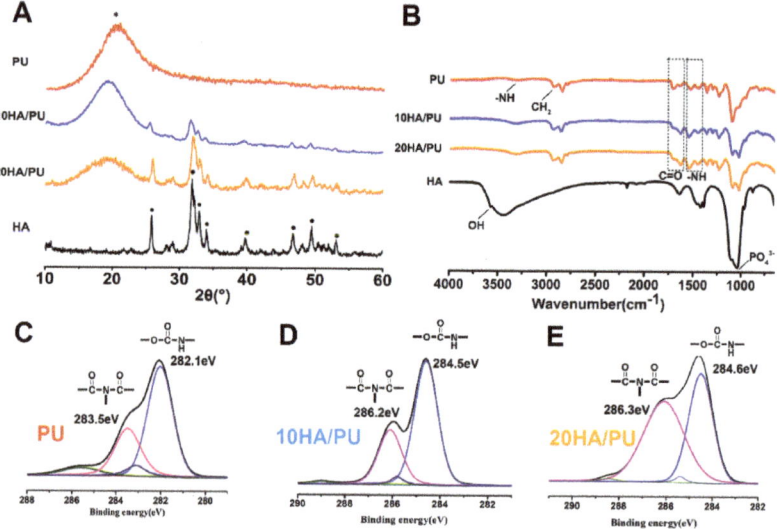

Figure 2. The XRD patterns (**A**), FTIR spectra (**B**), and XPS spectra of the C1s peak (**C**) and the peak fitting of PU (**D**), 10HA/PU (**E**), and 20HA/PU (**F**).

The XPS spectra of PU, 10 HA/PU, and 20 HA/PU are shown in Figure 2C–F. The chemical shifts of the binding energies of the pure PU and n-HA/PU demonstrated that the molecules of n-HA may have chemically bonded with the PU segment groups, as shown in Figure 2D–F. The fitted C1s peaks show that the peak intensity (or concentration) of biuret (CO–N–CO) is obviously increased after the addition of n-HA particles, and the binding energy increases from 283.5 eV of PU to 286.2 eV of 10 HA/PU and 286.3 eV of 20 HA/PU. It suggests that hydrogen bonds have been formed between the n-HA filler and PU matrix. The main peak of HN–CO–O and the small peak of CO–N–CO, in Figure 2D, at 282.1 eV and 283.5 eV, respectively, originate most likely from the allophanate or biuret, created by the reaction of the residual IPDI component with water (foaming agent) or urethane groups in PU. It is worth noting that the binding energy of the polar HN–CO–O and CO–N–CO groups of

10 HA/PU and 20 HA/PU shifts a lot compared with the pure PU, indicating that the addition of n-HA does affect the binding state or linkage of the PU polymer chains.

The dynamic mechanical analysis (DMA) is a useful technique for the evaluation of the viscoelastic property of polymers, by which a sinusoidal force (stress σ) is applied to the polymer, and the resulting displacement (strain) is measured [36]. When polymers composed of long molecular chains have unique viscoelastic properties, the characteristics of the elastic solids and Newtonian fluids combine. The temperature dependency of the storage modulus (E'), loss modulus (E"), and loss tangent (tan δ, from E"/E') of the PU, 10 HA/PU, and 20 HA/PU are shown in Figure 3A–C, respectively. The storage modulus and loss modulus data are originated from the sample nature (elastic PU and rigid n-HA nanocrystals). The storage modulus and the loss modulus of the three foams drop sharply in the temperature range from −50 to 20 °C, in Figure 3A,B. The tan δ curves and data (<1) in Figure 3C indicate that the three foams are viscoelastic solid materials in the temperature range from 0 to 120 °C. The 20 HA/PU holds the highest storage modulus and loss modulus, and the lowest loss tangent, representing that it has a better viscoelasticity and is more rigid than pure PU and 10 HA/PU foams. The glass transition temperature (T_g) values, in Figure 3C, of the 10 HA/PU and 20 HA/PU foams (~90 °C) are higher than that of the pure PU foam (~70 °C), suggesting an enhanced cross-linking in the HA/PU composites. Moreover, the 20 HA/PU sample shows the highest storage modulus resulting from its ability to resist intermolecular slippage, meaning more steric hindrance because of the stronger intermolecular interactions between the n-HA fillers and PU polymer chain. The broad tan δ peaks of 10 HA/PU and 20 HA/PU are indicative of a wider range of polymer chain branching within the PU matrix. The three PU foams also give a wide rubbery plateau, which is dependent on the molecular entanglements or cross-links, indicating that they have a good viscoelasticity at an applicable temperature [37,38].

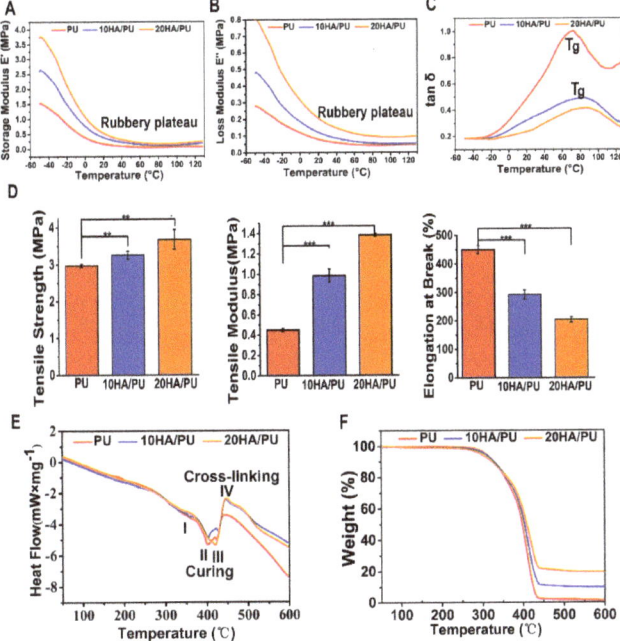

Figure 3. Dynamic mechanical analysis of storage modulus (**A**), loss modulus (**B**), and loss tangent (**C**) against temperature, and the tensile properties (**D**), differential scanning calorimetry (DSC) curves (**E**), and thermogravimetric (TG) curves (**F**) of PU, 10 HA/PU, and 20 HA/PU. Significantly different at * $p < 0.05$, ** $p < 0.01$, *** $p < 0.001$.

Figure 3D shows the mechanical properties of PU, 10 HA/PU, and 20 HA/PU foams. It shows that the pure PU foam has a better elastomeric behavior because of its higher value of strain at break compared with the HA/PU foams; the strain at break for PU is approximately 1.5 times higher than that of 10 HA/PU, and 2.25 times higher than that of 20 HA/PU. The good elastomeric behavior of the PU foam results from the polyol as soft segments and the chain extender [39]. However, with the addition of n-HA particles, the tensile strength and modulus are apparently increased. The tensile strength of PU, 10 HA/PU, and 20 HA/PU is 2.97 MPa, 3.26 MPa, and 3.54 MPa, respectively, and the tensile modulus of the 10HA/PU and 20HA/PU foams are approximately 2.2 times and 3.4 times higher than that of the pure PU foam ($p < 0.05$). The reason for the increase in tensile strength and tensile modulus should be caused by the homodisperse or improved stress dispersity of the n-HA crystals in the PU matrix, and the tight interface bonding between them. It also results from the increased hydrogen bonds between the OH groups of the n-HA and the NCO groups of the IPDI in the n-HA containing polyurethanes [19]. Thus, the tight bonding interface becomes a dominating factor in the mechanical improvement of the HA/PU composites, as shown in the HRTEM image in Figure 1K, as well as in the analysis from XPS spectra in Figure 2. The additional amount of n-HA and the interfacial interaction between n-HA and PU are the major factors that determine the ultimate properties of the composite foams.

The differential scanning calorimetry (DSC) of the thermal analysis is a vital analytical method for understanding the structure–property relationships of different polymeric materials, and an effective method to evaluate materials' thermal stability [40]. The DSC curves of pure PU and HA/PU composites are shown in Figure 3E. The thermal behavior for the three foams can be described in three endothermic and one exothermic process, as follows: I—the endothermic peak at 350 °C, in this process, the chemical bonds between the hard segment and soft segment break down; II—the endothermic peak around 400 °C indicates the breaking process of the hard segment, and the C–O bond of the carbamate groups in the main chain of PU matrix is destroyed, making the PU dissociate first into isocyanates and polyols; III—the endothermic peak from 420 to 430 °C represents the process that the long chain of soft segments break down and the isocyanates break down into carbon dioxide gas evolution; IV—the exothermic peak from 450 to 460 °C, during this period, the residual polyols break down into small molecules and release heat. All of the endothermic peaks and the exothermic peak are affected to some extent by the filling of the n-HA crystals when compared to the DSC curve of pure PU; in particular, the exothermic IV peak is largely strengthened for the two HA/PU composites, as a result of the increased hydrogen bonding between the –OH groups of HA and the HN–CO–O groups of the PU matrix, which enhanced the thermal stability of the composites.

The TG curve in Figure 3F presents a minor mass loss of approximately 4% over the temperature range of 50 to 200 °C, because of the loss of adsorbed water or carbon dioxide. The following thermal degradation of all of the samples is a two-step process, corresponding to the decomposition of the hard segments and soft segments in the PU matrix. In the first stage of decomposition (270–350 °C), the C–O bond of the carbamate groups in the main chain of the PU matrix is destroyed, making the PU dissociate first into isocyanates and polyols, then break down into a carbon dioxide gas evolution. The second stage of mass loss involves the further decomposition of the residual polyol of the PU matrix during 350–450 °C, which results in the 10 HA/PU composite retains nearly 10% of its weight, and the 20 HA/PU composite retains nearly 20% of its weight. The composition of n-HA is stable in such a temperature range. The result indicates that the addition of n-HA has a positive effect on the thermal stability of the PU matrix.

The LSCM images show the cell morphology of the BMSCs after four days of culture with the extraction solution of the three samples (A—blank control; B—PU; C—10HA/PU; D—20HA/PU). The cell morphology, including the red F-actin, blue nucleus, and green cytoplasm, can be clearly observed, as shown in Figure 4A–D. The normal cell spindle shape and well cell spreading with longer filopodia demonstrate that the composition of the n-HA and PU has no negative effect on the growth of the BMSCs.

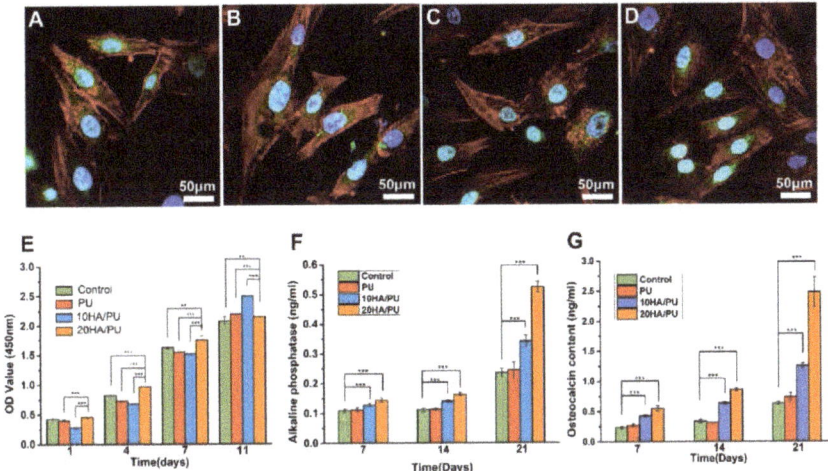

Figure 4. The laser scanning confocal microscopy (LSCM) images showing the cell morphology of the bone marrow mesenchymal stem cells (BMSCs) after four days of culture with an extraction solution of the three samples (**A**—blank control; **B**—PU; **C**—10HA/PU; **D**—20HA/PU). F-actin stained in red, mitochondria stained in green, and the nuclei stained in blue. The cell proliferation of the BMSCs cultured on the PU, 10HA/PU, and 20HA/PU samples via the CCK-8 assay (**E**), and the expression of alkaline phosphatase (**F**) and osteocalcin (**G**) of the BMSCs cultured with the three samples for 7, 14, and 21 days. (* $p < 0.05$, ** $p < 0.01$, *** $p < 0.001$).

The proliferation of BMSCs cultured on the PU, 10HA/PU, and 20HA/PU samples was determined using a CCK-8 assay. As shown in Figure 4E, the cell proliferation of all of the experimental groups steadily increases with culture time from day 1 to day 11, similar to the normal growth trend of the control. The cells cultured on 20 HA/PU show a significant increase compared with the other groups from day 1 to day 7. On day 11, the 10 HA/PU sample holds the highest proliferation, and the 20 HA/PU sample is still higher than the control. This indicates that both the 10 HA/PU and the 20 HA/PU foams display a good cytocompatibility, and can be used for the in vivo investigation.

The ALP activity and OCN content were also measured for the osteoblastic differentiation of the BMSCs cultured on different foams. Both the ALP activity (Figure 4F) and the OCN content (Figure 4G) indicate the successful osteoblastic transformation and good osteogenic activity of the BMSCs on the PU foams during the culture time, for up to 21 days. However, there are differences between the four groups, in which the HA/PU foams exhibit much higher ALP and OCN activities than the pure PU foam and the control, and the 20 HA/PU foam shows the highest ALP and OCN expression. The ALP value of the 20 HA/PU group is approximately 1.5 times higher than that of the 10 HA/PU group and two times higher than that of pure PU and the control groups on day 21 ($p < 0.001$). The OCN value of the 20 HA/PU group is also much higher than that of the 10 HA/PU group, and the pure PU and control groups after cultured for seven days ($p < 0.001$). On day 21, the OCN value of the 20 HA/PU group is about two times higher than that of the 10 HA/PU group, and three times higher than that of the pure PU and the control groups ($p < 0.001$). The results indicate that the terminal marker OCN of osteoblastic differentiation can be up-regulated notably in the HA/PU composites compared with the pure PU, revealing that the presence of n-HA crystals in the matrix does promote the osteoblastic differentiation of BMSCs. Furthermore, each group was also observed with Trichrome dyeing after incubation for four days.

To compare the osteogenic capacity of the different foams, the PU and 20HA/PU foams were implanted into the mandible defects of New Zealand rabbits for 4 and 12 weeks. The histological sections of the harvested samples are shown in Figure 5. All of the three groups show an increasing growth of new bone during this time. However, the 20 HA/PU sample shows a better material–bone bonding interface. After four weeks of implantation, no inflammatory response could be observed in all of the groups, and the new bone tissue appeared in the defects of the control (Figure 5A), PU (Figure 5B), and 20HA/PU (Figure 5C). After four weeks, the new bone continuously grew with the implantation time (Figure 5D–F). Direct contact with the new bone can be observed for the 20 HA/PU sample (Figure 5F), while there is a gap present between the material and new bone for the PU sample (Figure 5D), suggesting that the 20 HA/PU sample has a better osteogenic capability because of the addition of n-HA crystals. The results of the physical and chemical analyses, in vitro cell culture, and the in vivo animal experiment indicate that the prepared HA/PU composite has a good flexibility, cytocompatibility, and bone-bonding bioactivity, which may be suitable for future plastic surgery and tissue regeneration.

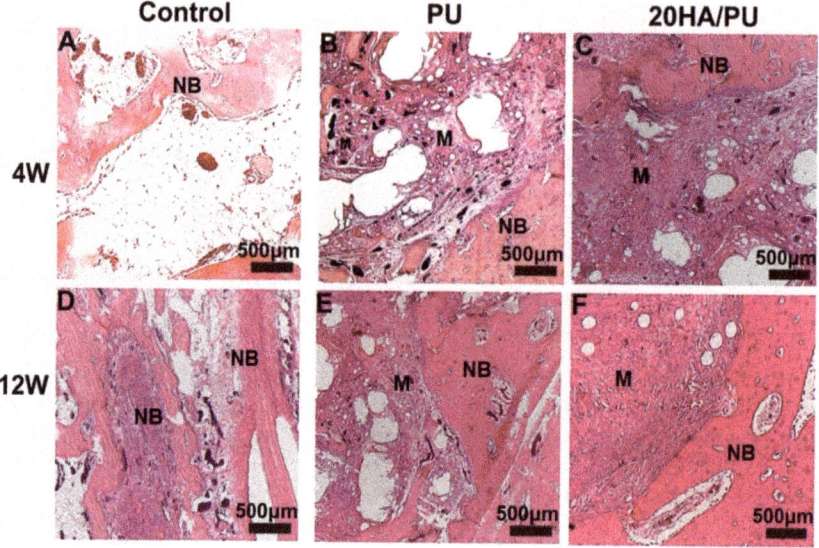

Figure 5. The hematoxylin-eosin (HE) stained histological section images of the control group, and the PU and 20HA/PU foams after implantation in mandible defects for 4 (A–C) and 12 (D–F) weeks. M—material; NB—new bone.

4. Conclusions

We designed and fabricated a porous system based on elastomer polyurethane, for potential use in plastic surgery. The n-HA/PU porous composite has good elasticity, flexibility, and supporting mechanical strength. The uniform distribution of the n-HA crystals in the PU matrix, while a homogeneous structure of hydrophilic and hydrophobic phase, was caused by hard and soft segments of PU. Furthermore, the tightly bonding interface on the nano particles and polymeric matrix is a dominating factor in the mechanical improvement of the HA/PU composites. The pore size (100–500 µm) and porosity (70%) are efficient for cell growth and for the regeneration of bone tissues, and the addition of n-HA fillers promoted cell infiltration and tissue regeneration as designed. The composite elastomers show good resilience to hold against the general press and tensile stress, as well as having the desired clinical manipulation. Further specific molding, such as the three-dimensional printing technique based on the n-HA/PU elastomer, will be a good candidate for plastic reconstruction.

Author Contributions: Conceptualization, Y.Z.; methodology, L.L.; software, J.M.; validation, Q.Z.; formal analysis, Q.M.; investigation, J.C.; data curation, B.C.; writing (original draft preparation), L.L.; writing (review and editing), Y.L. and Y.Z.; visualization, L.L.; supervision, J.L.; project administration, Y.L.; funding acquisition, Y.L.

Funding: This work is supported by the China NSFC fund (no. 31670965) and a project from the National Key Research and Development Program of China (no. 2017YFC1104300/2017YFC1104303).

Conflicts of Interest: The authors declare no conflict of interest.

References

1. Glass, G.E.; Mosahebi, A.; Shakib, K. Cross-specialty developments: A summary of the mutually relevant recent literature from the journal of plastic, reconstructive and aesthetic surgery. *Br. J. Oral Maxillofac. Surg.* **2016**, *54*, 13–21. [CrossRef] [PubMed]
2. Carle, M.V.; Roe, R.; Novack, R.; Boyer, D.S. Cosmetic facial fillers and severe vision loss. *JAMA Ophthalmol.* **2014**, *132*, 637–639. [CrossRef] [PubMed]
3. Harris, D.A.; Fong, A.J.; Buchanan, E.P.; Monson, L.; Khechoyan, D.; Lam, S. History of synthetic materials in alloplastic cranioplasty. *Neurosurg. Focus* **2014**, *36*, E20. [CrossRef] [PubMed]
4. Lee, S.K.; Kim, H.S. Recent trend in the choice of fillers and injection techniques in Asia: A questionnaire study based on expert opinion. *J. Drugs Dermatol. JDD* **2014**, *13*, 24–31. [PubMed]
5. Ye, J.; Meng, Q.; Yang, Y.; Co, Z.J. Application of Expanded Polytetrafluoroethylene(e-PTFE) in the Field of Plastic Surgery and Cosmetic Domestic. *Chem. Prod. Technol.* **2016**, *23*, 24–27.
6. Yoo, Y.C.; Chung, S.I.; Yang, W.Y.; Ko, B.M. Experience with use of Expanded Polytetrafluoroethylene(Gore-tex(R)) in Cosmetic Facial Surgery. *J. Korean Soc. Plast. Reconstr. Surg.* **2003**, *30*, 7–14.
7. Woo, H.D.; Lee, H.J.; Lee, J.W.; Son, T.I. Injectable photoreactive azidophenyl hyaluronic acid hydrogels for tissue augmentation. *Macromol. Res.* **2014**, *22*, 494–499. [CrossRef]
8. Jones, D. Volumizing the face with soft tissue fillers. *Clin. Plast. Surg.* **2011**, *38*, 379–390. [CrossRef] [PubMed]
9. Vallet-Regí, M.; González-Calbet, J.M. Calcium phosphates as substitution of bone tissues. *Prog. Solid State Chem.* **2004**, *32*, 1–31. [CrossRef]
10. Le Nihouannen, D.; Goyenvalle, E.; Aguado, E.; Pilet, P.; Bilban, M.; Daculsi, G.; Layrolle, P. Hybrid composites of calcium phosphate granules, fibrin glue, and bone marrow for skeletal repair. *J. Biomed. Mater. Res. A* **2007**, *81*, 399–408. [CrossRef] [PubMed]
11. Encarnacao, I.C.; Xavier, C.C.F.; Bobinski, F.; dos Santos, A.R.S.; Correa, M.; de Freitas, S.F.T.; Aragonez, A.; Goldfeder, E.M.; Cordeiro, M.M.R. Analysis of Bone Repair and Inflammatory Process Caused by Simvastatin Combined with PLGA plus HA plus beta TCP Scaffold. *Implant Dent.* **2016**, *25*, 140–148. [CrossRef] [PubMed]
12. Chuenjitkuntaworn, B.; Osathanon, T.; Nowwarote, N.; Supaphol, P.; Pavasant, P. The efficacy of polycaprolactone/hydroxyapatite scaffold in combination with mesenchymal stem cells for bone tissue engineering. *J. Biomed. Mater. Res. A* **2016**, *104*, 264–271. [CrossRef] [PubMed]
13. Do, A.V.; Khorsand, B.; Geary, S.M.; Salem, A.K. 3D Printing of Scaffolds for Tissue Regeneration Applications. *Adv. Healthc. Mater.* **2015**, *4*, 1742–1762. [CrossRef] [PubMed]
14. Kelts, G. Review of facial plastic and reconstructive surgery: A clinical reference. *JAMA Facial Plast. Surg.* **2017**, *19*, 544. [CrossRef] [PubMed]
15. Danilova, D.A.; Gorbunova, L.I.; Tsybusov, S.N.; Uspensky, I.V.; Kravets, L.Y. Materials for Plastic Surgery of the Dura Mater: History and Current State of the Problem (Review). *Sovrem. Teh. V Med.* **2018**, *10*, 194. [CrossRef]
16. Janik, H.; Marzec, M. A review: Fabrication of porous polyurethane scaffolds. *Mater. Sci. Eng. C* **2015**, *48*, 586–591. [CrossRef] [PubMed]
17. McBane, J.E.; Sharifpoor, S.; Cai, K.; Labow, R.S.; Santerre, J.P. Biodegradation and in vivo biocompatibility of a degradable, polar/hydrophobic/ionic polyurethane for tissue engineering applications. *Biomaterials* **2011**, *32*, 6034–6044. [CrossRef] [PubMed]
18. Yang, B.; Zou, Q.; Lin, L.; Li, L.; Zuo, Y.; Li, Y. Synthesis and characterization of fluorescein-grafted polyurethane for potential application in biomedical tracing. *J. Biomater. Appl.* **2016**, *31*, 901–910. [CrossRef] [PubMed]
19. Chen, Q.; Liang, S.; Thouas, G.A. Elastomeric biomaterials for tissue engineering. *Prog. Polym. Sci.* **2013**, *38*, 584–671. [CrossRef]
20. Da, L.C.; Gong, M.; Chen, A.J.; Zhang, Y.; Huang, Y.Z.; Guo, Z.J.; Li, S.F.; Li-Ling, J.; Zhang, L.; Xie, H.Q. Composite elastomeric polyurethane scaffolds incorporating small intestinal submucosa for soft tissue engineering. *Acta Biomater.* **2017**, *59*, 45–57. [CrossRef] [PubMed]

21. Li, L.M.; Zuo, Y.; Zou, Q.; Yang, B.Y.; Lin, L.L.; Li, J.D.; Li, Y.B. Hierarchical Structure and Mechanical Improvement of an n-HA/GCO-PU Composite Scaffold for Bone Regeneration. *ACS Appl. Mater. Interfaces* **2015**, *7*, 22618–22629. [CrossRef] [PubMed]
22. Yan, Y.; Zhang, Y.; Zuo, Y.; Zou, Q.; Li, J.D.; Li, Y.B. Development of Fe3O4-HA/PU superparamagnetic composite porous scaffolds for bone repair application. *Mater. Lett.* **2018**, *212*, 303–306. [CrossRef]
23. Yubao, L.; Wijn, J.D.; Klein, C.P.A.T.; Meer, S.V.D.; Groot, K.D. Preparation and characterization of nanograde osteoapatite-like rod crystals. *J. Mater. Sci. Mater. Med.* **1994**, *5*, 252–255. [CrossRef]
24. Du, J.J.; Zou, Q.; Zuo, Y.; Li, Y.B. Cytocompatibility and osteogenesis evaluation of HA/GCPU composite as scaffolds for bone tissue engineering. *Int. J. Surg.* **2014**, *12*, 404–407. [CrossRef] [PubMed]
25. Princi, E.; Vicini, S.; Stagnaro, P.; Conzatti, L. The nanostructured morphology of linear polyurethanes observed by transmission electron microscopy. *Micron* **2011**, *42*, 3–7. [CrossRef] [PubMed]
26. Chen, W.C.; Liu, J.; Manuchehrabadi, N.; Weir, M.D.; Zhu, Z.M.; Xu, H.H.K. Umbilical cord and bone marrow mesenchymal stem cell seeding on macroporous calcium phosphate for bone regeneration in rat cranial defects. *Biomaterials* **2013**, *34*, 9917–9925. [CrossRef] [PubMed]
27. Hu, Q.H.; Zhu, H.T.; Liu, Z.W.; Chen, K.L.; Tang, K.F.; Qiu, C. Isolation and culture of mesenchymal stem cells from mouse bone marrow. *J. Gastroenterol. Hepatol.* **2013**, *28*, 621.
28. Zhao, G.; Raines, A.L.; Wieland, M.; Schwartz, Z.; Boyan, B.D. Requirement for both micron- and submicron scale structure for synergistic responses of osteoblasts to substrate surface energy and topography. *Biomaterials* **2007**, *28*, 2821–2829. [CrossRef] [PubMed]
29. Sidney, L.E.; Kirkham, G.R.; Buttery, L.D. Comparison of Osteogenic Differentiation of Embryonic Stem Cells and Primary Osteoblasts Revealed by Responses to IL-1 beta, TNF-alpha, and IFN-gamma. *Stem Cells Dev.* **2014**, *23*, 605–617. [CrossRef] [PubMed]
30. Li, L.M.; Zhao, M.H.; Li, J.D.; Zuo, Y.; Zou, Q.; Li, Y.B. Preparation and cell infiltration of lotus-type porous nano-hydroxyapatite/polyurethane scaffold for bone tissue regeneration. *Mater. Lett.* **2015**, *149*, 25–28. [CrossRef]
31. Qidwai, M.; Sheraz, M.A.; Ahmed, S.; Alkhuraif, A.A.; ur Rehman, I. Preparation and characterization of bioactive composites and fibers for dental applications. *Dent. Mater.* **2014**, *30*, 253–263. [CrossRef] [PubMed]
32. Li, L.M.; Zuo, Y.; Du, J.J.; Li, J.D.; Sun, B.; Li, Y.B. Structural and Mechanical Properties of Composite Scaffolds Based on Nano-hydroxyapatite and Polyurethane of Alcoholized Castor Oil. *J. Inorg. Mater.* **2013**, *28*, 811–817. [CrossRef]
33. Boissard, C.I.R.; Bourban, P.E.; Tami, A.E.; Alini, M.; Eglin, D. Nanohydroxyapatite/poly(ester urethane) scaffold for bone tissue engineering. *Acta Biomater.* **2009**, *5*, 3316–3327. [CrossRef] [PubMed]
34. Das, B.; Konwar, U.; Mandal, M.; Karak, N. Sunflower oil based biodegradable hyperbranched polyurethane as a thin film material. *Ind. Crops Prod.* **2013**, *44*, 396–404. [CrossRef]
35. Gang, H.; Lee, D.; Choi, K.-Y.; Kim, H.-N.; Ryu, H.; Lee, D.-S.; Kim, B.-G. Development of High Performance Polyurethane Elastomers Using Vanillin-Based Green Polyol Chain Extender Originating from Lignocellulosic Biomass. *ACS Sustain. Chem. Eng.* **2017**, *5*, 4582–4588. [CrossRef]
36. Rice, M.A. Dynamic mechanical response of plasticizer-laden acoustic polyurethanes extrapolated to higher frequencies using time-temperature superposition technique. *J. Therm. Anal. Calorim.* **2014**, *118*, 377–396. [CrossRef]
37. Saucedo-Rivalcoba, V.; Martínez-Hernández, A.L.; Martínez-Barrera, G.; Velasco-Santos, C.; Castaño, V.M. (Chicken feathers keratin)/polyurethane membranes. *Appl. Phys. A* **2010**, *104*, 219–228. [CrossRef]
38. Fernández, M.; Landa, M.; Muñoz, M.E.; Santamaría, A. Thermal and Viscoelastic Features of New Nanocomposites Based on a Hot-Melt Adhesive Polyurethane and Multi-Walled Carbon Nanotubes. *Macromol. Mater. Eng.* **2010**, *295*, 1031–1041. [CrossRef]
39. Roy, P.; Sailaja, R.R.N. Mechanical, thermal and bio-compatibility studies of PAEK-hydroxyapatite nanocomposites. *J. Mech. Behav. Biomed.* **2015**, *49*, 1–11. [CrossRef] [PubMed]
40. Herrera, M.L.; Anon, M.C. Crystalline fractionation of hydrogenated sunflower seed oil. II. Differential scanning calorimetry (DSC). *J. Am. Oil Chem. Soc.* **1991**, *68*, 799–803. [CrossRef]

© 2018 by the authors. Licensee MDPI, Basel, Switzerland. This article is an open access article distributed under the terms and conditions of the Creative Commons Attribution (CC BY) license (http://creativecommons.org/licenses/by/4.0/).

MDPI
St. Alban-Anlage 66
4052 Basel
Switzerland
Tel. +41 61 683 77 34
Fax +41 61 302 89 18
www.mdpi.com

Nanomaterials Editorial Office
E-mail: nanomaterials@mdpi.com
www.mdpi.com/journal/nanomaterials

www.ingramcontent.com/pod-product-compliance
Lightning Source LLC
LaVergne TN
LVHW070742100526
838202LV00013B/1285